Is the universe around us a figment of our imagination? Or are our minds figments of reality?

In this refreshing new look at the evolution of mind and culture, bestselling authors Ian Stewart and Jack Cohen eloquently argue that our minds necessarily evolved in an inextricable link with culture and language. They go beyond conventional reductionist ideas to look at how the mind is the response of an evolving brain trying to grapple with a complex environment. Along the way they develop new and intriguing insights into the nature of evolution, science and humanity.

FIGMENTS OF REALITY

Ian Stewart and Jack Cohen

Figments of reality

The evolution of the curious mind

PUBLISHED BY THE PRESS SYNDICATE OF THE UNIVERSITY OF CAMBRIDGE
The Pitt Building, Trumpington Street, Cambridge, CB2 1RP, United Kingdom

CAMBRIDGE UNIVERSITY PRESS
The Edinburgh Building, Cambridge, CB2 2RU, United Kingdom
40 West 20th Street, New York, NY 10011-4211, USA
10 Stamford Road, Oakleigh, Melbourne 3166, Australia

First published 1997

Printed in the United States of America

Typeset in Swift 9/14pt

A catalogue record for this book is available from the British Library

Library of Congress Cataloguing in Publication data

Stewart, Ian, 1945–
Figments of reality : the evolution of the curious mind / Ian Stewart and Jack Cohen.
p. cm.
Includes bibliographical references.
ISBN 0 521 57155 3 (hardcover)
1. Cognition and culture. 2. Consciousness. 3. Intellect. 4. Genetic psychology. 5. Pluralism (Social sciences) I. Cohen, Jack. II. Title.
BF311.S679 1997
153.4′2–dc21 96–49357 CIP

ISBN 0 521 57155 3 hardback

Contents

ZARATHUSTRAN THEORY OF EVERYTHING:

$$E = 8$$

Preface

Isn't it strange that the animal we used to be developed into the creature that we now are? How – and why – did human intelligence and culture evolve? How did we evolve minds, philosophies and technologies? And now that we have them, where are they taking us?

The orthodox answer to these questions looks inside our brains to see what they are made of and how the various components operate. This leads to a story based upon DNA biochemistry, the evolution of nerve cells as pathways for sensory information, and their organisation into complex networks – brains – that can manipulate neural models of natural objects and processes. Mind is seen as a property of an unusual brain – complex enough to develop culture – but here the 'reductionist' story starts to lose its thread. Many people see mind as something that transcends ordinary matter altogether. Philosophers worry that the universe around us may be a figment of our own imagination.

In *Figments of Reality* we explore a very different, but complementary, theory: that minds and culture co-evolved within a wider context. Every step of our development is affected by our surroundings. Our minds are rooted in ordinary matter; they are complex processes – or complexes of processes – that happen in material brains. Our brains are linked to reality by their molecules; but they are also linked to reality on another level, their ability to model reality within themselves.

Those links have had important effects on the evolution of the brain and the mind. For example, even our sense organs are not totally pre-programmed: far from it. Instead, as we grow, our senses are 'tuned' to detect particular features of our surroundings. Mind is not immaterial transcendence: it is the response of an evolving brain to the need to survive in a complex environment. And with the evolution of culture, that environment has become self-modifying and self-referential, and human minds have done the same.

Evolution and tuneable senses have produced minds that can grapple with reality by operating upon *features* – high-level structures/processes in the brain that correspond to large-scale regularities in the surrounding world. For example, a goat eats leaves because they *look* like leaves, not because its nerve cells have a chemical affinity for chlorophyll. If plants had evolved differently, using a purple chemical for photosynthesis, then goats would be looking for purple leaves instead; but otherwise they would be much the same as present-day goats. We shall investigate how the mind explores its own mental landscape and works with the features that it finds there. This leads to a new theory of the relation of individual minds to the human culture in which they reside.

This is a different view from that of current physics, which, for instance, sees a table as 'mostly empty space' because of atomic theory, and thereby directs our attention *away* from important human-scale features such as 'wooden', 'solid', 'brown'. Such 'commonsense' features were important for evolution, and remain important for understanding many areas of science. For example, the evolution of the goat as a successful herbivore depended upon its ability to perceive leaves, not upon its understanding of biochemistry.

How can a conscious, intelligent mind evolve? Instead of giving a reductionist answer based upon internal fine structure we take an external, contextual view. We see the accumulating knowledge of generations of intelligent beings as a thing or process with its own characteristic structure and behaviour: *extelligence*. Extelligence constantly modifies and organises itself through continuing interactions with innumerable individuals. As a result, extelligence has become greater, more permanent, and far more capable than any individual intelligence. However, extelligence makes no sense without intelligences to interact with it: the two are 'complicit'. The developing mind of each child interacts with extelligence by way of language, and the two-way flow between individuals and their surrounding culture changes both. Intelligence is fostered in the child, and extelligence is fostered in the culture. Thus the evolution and structure of the brain cannot be divorced from the evolution and structure of human society and its environment, the universe.

Our minds co-evolve with everything that influences them. Minds *are* figments of reality, processes going on inside structures made from ordinary matter whose behaviour evolved *in order to* mimic, model, and manipulate natural processes. This explains why they are 'unreasonably effective' ♪[1] at perceiving and reorganising their environment. The human condition is a complicit interaction between culture and individual minds, each shaping the other.

Culture depends upon communication, which we achieve with language. Language, the first step towards extelligence, co-evolved with brains and made minds, complicit with hands and technology, and the discovery of patterns

1 The symbol ♪ ('note') indicates that there is a note at the back of the book which takes the appropriate topic further, or provides a reference. Various technicalities are relegated to the Notes, as are discussions of interesting distractions.

and laws. Mind can only think about mind once language equips it with a recursive (that is, self-referential) feature-detection system. Once it has this, *self*-awareness is an immediate, essentially trivial property, because 'self' is a feature too. The existence of features makes it possible to employ a mental map instead of the real territory.

The greatest single step in organic evolution was the aggregation of different bacteria to make the nucleated cell. Similarly, the greatest step so far in our cultural evolution has been the aggregation of different cultures to make multicultures. There are many kinds of multiculture, ranging from multinational corporations to major cities like New York. But the self-complication of human culture will not stop there, because it is a self-propelled process. Today's multicultures are like the creatures of a colony, coexisting as more or less isolated 'ghettos'. Tomorrow's multicultures will be more like genuine multicellular organisms, in which extelligence is specialised like the different tissues of a complex animal. Our new communication technologies are beginning to knit all of the different multicultures into a new entity, a superculture: Humanity.

This will be our story.

And here is the place to thank everybody who has contributed to it. JC is grateful for the hospitality of the University of Warwick, which provided him with a room and a phone. IS had a room and a phone too, but then, he works there. The manuscript of *Figments of Reality* was subjected to critical reading by a variety of people. We are grateful to them for their sterling efforts, which improved the book considerably. Naturally we take full responsibility for any remaining errors or infelicities (unless they were the other author's fault). Our editors at Cambridge University Press, Tim Benton and Barnaby Willitts, deserve special mention since they were exposed to more than one version of the manuscript. In alphabetical order, the other readers were: Daniel Goldenberg, Steve Gould, Mac Hanson, Rabbi Dr Margaret Jacobi, Mike Leci, Mal Leicester, Christine McNulty, Alan Moore, Alan Newell, David Poyser, Terry Pratchett, Helen and Gareth Rees, Lena Sarah, Paulo Sousa, Heather Spears, Colin Touchin, and Elizabeth Viau.

A word about the brief stories that head each chapter. They are there because they illuminate, perhaps indirectly, the main themes of the chapter concerned. All of them are, to the best of our knowledge, true. (Except one, which we invented, for very good reasons – only to find that it contained more truth than we had thought. We confess the falsehood early in the appropriate chapter, and explain the unexpected germ of truth shortly afterwards.) Some of our readers complained that one or two of these stories were not 'politically correct'.

However, we left them unchanged, because we feel that political correctness should be secondary to truth. We emphasise that the stories are not intended to be derogatory or offensive, and if you think any of them is, then you're reading things into it that we didn't intend. For example, we mention a woman scientist who becomes emotional. We cast her as a woman because, as it happened, she was. The emotion, to us, is a positive and necessary feature of the story, and if the same thing had happened to a male scientist, he'd have been just as emotional, and we would have told it that way. Several stories rest upon human frailties or idiosyncracies, but we are not holding anyone up to ridicule. The aim of those stories is to illustrate what a strange but wonderful animal we humans are.

Figments was written in a variety of places – lots of trains between Coventry and London Euston, benches in St. James's Park in London, benches on the Thames Embankment – even sometimes at a desk. Many of the places were aeroplanes – Ryanair from Birmingham to Dublin, American Airlines from Chicago to San Francisco, Delta Airlines from Salt Lake City to Cincinnati ... One of the ironies of the complex world of the late twentieth century is that one of the best places to find the solitude to write is at 35,000 feet travelling at 550 mph.

Not for long, we fear: already telephones are proliferating across the backs of aircraft seats.

A tropical island would be more comfortable, but comfort is not conducive to wordage. Stuck on a tropical island, it is too easy to consume coconuts and rum punch all day long without ever having the stimulus to put pen to paper. Stuck in a Boeing 767 on an overnight flight, with eight hours to pass and about ten cubic feet of space to pass it in, one's only companion a heap of gin miniatures and a can of tonic water, the attraction of taking refuge in the world of the imagination is far stronger. All you need is a legal pad and a pen, or for technofreaks, a laptop.

We used both. We're not fussy and we type fast.

IS & JC
Coventry, September 28, 1996

Figure Acknowledgements

Figure 3
From *Wonderful Life: The Burgess Shale and the Nature of History*, by Stephen Jay Gould. Copyright © 1989 by Stephen Jay Gould. Reprinted by permission of W. W. Norton & Company, Inc.

Figure 21
Jack Cohen, *Reproduction*, Butterworths, London 1977, p.179, Figure 10.1.

Figure 22
Jack Cohen, *Reproduction*, Butterworths, London 1977, p.186, Figure 10.7.

Figure 24
Reprinted by kind permission of Alice O'Toole. From *Scientific American* April 1996, p.24.

Prologue

Fifteen thousand million years ago the universe was no bigger than the dot at the end of this sentence.

A tiny, tiny, *tiny* fraction of a second before that – but there was no fraction of a second before that. There was no time before the universe began, and without time, there can be no 'before'. (As well to ask what lies north of the North Pole.) There was no space, no time, and no matter. But when the space that was coextensive with the universe had grown to the size of a dot, time had already begun to tick. The temperature within the dot was far too high for matter to exist, but there was plenty of what was required to create matter: radiation. The primal dot seethed with radiant energy.

During time's first duodecillionth (10^{-39}) of a second of existence, the universe was a 'false vacuum', a state of negative pressure in which every fragment of space repelled every other fragment. Space exploded exponentially, and in that near-infinitesimal instant the universe inflated from a tiny dot to a ball many light-years across as its negative pressure literally blew it apart. As the temperature dropped the false vacuum gave way to a true vacuum, a state of zero pressure, and the era of inflation ceased. The universe, now large enough to be interesting, continued to expand under its own momentum – but more sedately, at a rate of a few thousand kilometres per second.

When time was one ten thousandth of a second old, the temperature of the universe dropped to a trillion degrees. Pairs of particles, one of matter, one of antimatter, were winking into existence and out again, born in and dying as fluctuations of radiant energy. Matter and radiation were in perfect balance. However, the balance between matter and antimatter was imperfect. For every 999,999,999 antiprotons there were 1,000,000,000 protons. From that imbalance came everything that we know.

When time attained the grand old age of one second, the temperature of the nascent universe had fallen to a mere ten billion degrees. Electrons and antielectrons, colliding in pairs, filled the universe with bursts of neutrinos and antineutrinos. Neutrons, no longer stable, decayed into protons and electrons.

Two minutes after time began (some say one and a half minutes, others three) the universe had cooled to one billion degrees, and matter as we know it began to assemble. Neutrons paired incestuously with their proton offspring to form creation's first atoms – heavy hydrogen, otherwise known as deuterium. Deuterium fused into helium and matter began to diversify.

After half an hour the universe changed: now it was three quarters hydrogen, one quarter helium. The pace of change slowed. It took seven hundred thousand years before the universe cooled enough to become transparent to light. By then, matter had formed itself into almost a hundred different elements. It took a hundred million years for that matter to clump itself into galaxies, and for the first stars to shine.

Ripples in the early fabric of spacetime, amplified by the inexorable tug of gravity, folded in on themselves, collapsing under their own mass, leaving huge voids hundreds of millions of light-years across, bubbles of emptiness filling the universe like foam. On the surfaces of the bubbles, matter condensed into vast sheets and filaments. One such structure – let us call it the Distant Superattractor – made itself felt a billion light-years away, as its gravitational attraction sucked matter inwards towards its centre. There was nothing that greatly distinguished it from trillions of equally enormous clusters of matter.

Smaller – but still many thousands of light-years across – was a clump of matter known as the Great Attractor. Like all of the matter in a region of space billions of light-years across, it streamed towards the Distant Superattractor. Within and around the Great Attractor, matter arrayed itself into a hierarchy of ever-smaller clumps, which were sucked towards the Great Attractor even as it made its way towards the Distant Superattractor. One such was the Local Supercluster, a group of tens of thousands of galaxies that surged collectively towards the constellation Virgo at 700 kilometres per hour. The Local Supercluster was composed of more than a hundred galactic clusters, none differing significantly from the rest – the M101 cluster, the M81 cluster, the Virgo cluster, the Local Group ... A typical cluster is several million light-years across, and is composed of hundreds of galaxies; an individual galaxy is some hundred thousand light-years in diameter, a vast swirl of matter that rotates once every quarter of a billion years.

In the Local Group were two dozen galaxies: Andromeda, M33, and one – not greatly different from any other – known simply as 'The Galaxy'. Like most galaxies it was spiral in form, although unusually it had two smaller close companions known as the Greater and Lesser Magellanic Clouds. Its spiral arms – like those of most other galaxies – were density waves, places where its component atoms piled up against each other. Along the crests of those waves the pressure became so intense that it sparked nuclear reactions, and stars came into being.

In this one galaxy there were more than a hundred billion stars.

One such star – not especially different from its companions – had spectral class G2, meaning that its surface temperature was about average (six thousand degrees) and the light that it emitted (at a level of brightness that was also close to the average) showed a prominent trace of calcium. Like many stars, it was enclosed in a cloud of cosmic debris – stardust blown across the intergalactic space in shockwaves generated by explosions in the galactic core. All of the different chemical elements born in stars' nuclear furnaces were present – some in abundance, others the merest traces. Among them, fused into existence by a coincidental resonance of nuclear vibrations, was the element carbon.

As the universe grew older, and colder, and larger, this particular cloud of stardust – like many others – began to condense, the grains sticking to each other, to form irregular lumps of methane ice, dense clouds of gas, fragments of rock. As it condensed, it also collapsed into a flattish disc, spinning on its axis, a swirl of cooling matter that collided, bounced, broke, stuck, aggregated. As time passed, a mere instant on cosmic scales, the clumps became fewer, but bigger. Crushed under their own gravity, they formed flattened spheres – planets. The G2 star acquired a solar system.

None of this was especially unusual.

Each planet, forming in its own particular place, found itself in possession of the features that its mode of formation would naturally create – a rocky core, a methane-hydrogen atmosphere, a surface flowing with molten metal or dotted with lakes of acid, encircling companions ... Each planet acquired its own identity. This in particular was true of the third planet, counting outwards from the central star. Much of its surface was covered by a thin layer of water. It had an atmosphere, mostly nitrogen. And its surface temperature was within the range at which water remained liquid. Although no other planet in this particular solar system resembled the third in these respects, it was probably much the same as many other planets around many other stars in many other galaxies.

Everywhere, even in the depths of intergalactic space, atoms bumped against each other and stuck to form molecules. On the third planet this happened more often than in the vacuum between the stars, because there were more atoms to bump into. The individual features of the third planet constrained the kind of molecule that occurred, producing structures that would not have occurred on a methane world or an ice giant. One day there arose a collection of molecules that could make copies of itself – a replicating system. Perhaps it came together accidentally in the primal soup of the oceans, perhaps it was given a helping hand by the receptive surfaces of rocks or clays. However it happened, the replicator did what replicators do – it replicated. Over and over again. After a fairly short time the planet became distinctly unusual, its chemistry subverted and reorganised by the voracious replicator. The replicator made the occasional mistake, but some mistakes could also replicate, and soon a kind of long-term War of the Replicators was under way, as ever more sophisticated molecular collectives did battle for the right to continue replicating.

It all got rather complicated.

For instance: one group of replicators acquired the knack of converting starlight into food.

For instance: an early success, the bacterium, attained such numbers that one of its metabolic by-products, the corrosive gas oxygen, came to occupy a substantial portion of the planet's atmosphere.

For instance: other groups of replicators evolved the ability to leave the solid ground and soar upon the gases of the atmosphere.

For instance: sixty-five million years ago an especially successful type of replicator was exterminated, planetwide, by the impact of a large rock. Other tiny hairy warm-blooded replicators suddenly found that their main competition had vanished from the face of the third planet, and their rapidly diversifying successors exploded across continents and oceans.

For instance: today, two of the descendants of those tiny creatures are busy delineating their own limited version of the entire story in strange, angular geometric symbols, impressed in contrasting pigment upon sheets of compressed white vegetable matter, in the hope that other creatures of similar kind will scan the sheets with their light-detecting sensors – and in some inexplicable manner imbibe meaning and significance and make them part of themselves. Typically for these replicators we find a tiny portion of the ungraspable universe making a glorious, hopeless attempt to encapsulate that awe-inspiring whole inside its own tiny form, improbably employing weak electrical impulses that scuttle along a network of a trillion tiny fibres – vibrant, alive, and even more ungraspable than the universe that it is attempting to grasp.

A circle closes.

A mystery opens.

1 The Origins of Life

A woman scientist had been working for some time with a chimpanzee, teaching it to carry out various tasks such as opening a box and rewarding it with fruit. One day, after a session with the chimpanzee, she came into the coffee room half laughing and half crying, obviously very emotional. Her colleagues, a little alarmed, finally managed to get out of her what had happened. She had decided to leave the laboratory area temporarily, and had undone the bolt on the door – whereupon the chimpanzee had solemnly handed her a stick of celery.

Our prologue is one way to tell the story of who we are and how we got here. Such a story has several virtues: it demonstrates how utterly incomprehensible the universe in its entirety is, and how difficult it is for a newly intelligent upright ape to close the conceptual circle by encapsulating the sheer vastness of that universe inside its tiny brain case. It encourages humility. It is the cosmological story as we currently conceive it, the best guess that today's science can make about a past that we cannot revisit and distances too enormous for us to cross. It is a story so strange that we may be tempted to dismiss it as wild speculation, but that will not make the strangeness go away, because if that story is false then the true story must be even stranger.

Assuming there is such a thing as *the* true story of the origins of the universe, which is debatable.

From our own point of view, however – we mean the human race, not JC & IS – this story is impersonal and back to front. It starts with nothing, and ends with each one of us as some kind of accidental by-product of forces beyond our wildest imagination. It describes a universe that is largely alien to the one that we inhabit, which is a private universe filled with very different, human-scale things – friends, spouses, children, pets, plants, bricks and mortar. Each of us inhabits a personal universe; in a sense each of us *is* a personal universe – for if we are destroyed then our personal universe vanishes with us. The universe of cosmology is made of fundamental particles, such as electrons, and radiation, such as light; but our personal universes are made of very different kinds of things. We don't mean that our own universes aren't made from ordinary matter – we mean that this matter is organised in a different manner. Most of the interesting features of our personal universes are people and their activities – friends and lovers, enemies and acquaintances from our work or our play. Because most of us live in cities the typical personal universe is urban, composed of buildings, rooms, out-of-town shopping centres ... What occupies most of our daily thoughts is *people* – their influence upon us, and ours upon them. There are babysitters to arrange, theatre tickets to book, bosses to placate, bank managers to be persuaded that a loan would be a sound business proposition ...

Sometimes the external 'non-people' world intrudes, but even then it normally does so by way of a human-made artefact: the car needs new tyres, the lawn needs mowing, a sudden attack of 'flu needs medication. Changes arising outside our own small circle affect our lives in ways we do not anticipate and of which we may not approve – new machinery makes our job unnecessary, anti-pollution laws add to the cost of doing business, a new disease infests our food supply, vandals cut our telephone wires, or people from a country thousands of miles away, which we have never visited, start dropping bombs on us. When the outside world intrudes upon our personal universe we become conscious that the outside exists, but most of the time we still interpret the intrusion in personal terms. We look for a new job that suits our abilities, we hire a lawyer to help us avoid our expensive new legal obligations, we temporarily stop eating burgers, we call the telephone repair man, we build bomb shelters and sit in them cursing the enemy while the bombs fall.

But we do more than that. Many other creatures look up into the night-time sky and see the stars, but we stare at them, wonder how many there are, wonder how far away they are, wonder how they got there, wonder what they are made of, wonder – indeed – why they are there at all. We link them into simple patterns and weave stories around them to help us to rationalise their existence and to remember which pattern is which – the Hunter, the Hero, the Princess, the Bear, the Swan. Although we cannot get inside other animals' heads, we see no evidence that any other creature looks outside its personal universe in this manner. Maybe chimpanzees and dolphins do; maybe the whale's enigmatic and interminable song is an exercise in submarine philosophy – but maybe it's just the whale's way of saying 'Hi, anybody out there? This is me.' Chimps and dolphins and whales don't build astronomical observatories, they don't make calendars to predict the seasons, they don't carve symbolic versions of their thoughts on rocks. Maybe they're wiser than we are, having fun instead of agonising about their place in the vast uncaring universe; but wiser or not, even the bright ones behave differently from us.

When we look outside our personal universe, we find that the external world is organised in its own characteristic way. It has gravity, ecology, dinosaurs, $E = mc^2$, angles of a triangle adding up to 180°, and so on. It is impersonal: while it is perfectly reasonable to argue with your bank manager that she should increase your overdraft above £180, it is fruitless to argue with a triangle in the hope of increasing the sum of its angles above 180°. On the other hand, the external universe links into our personal world in many ways: calories in food, digital music on CDs, passenger jets, television. All these technologies depend on science, and science is our most successful way to dig into the structure of that external, impersonal universe. Television strengthens the connection between the personal and impersonal worlds by providing science programmes on how the world began or how it will end, and natural history programmes – like our pets and aquariums, house plants and gardens – provide tenuous links with the rest of living nature. All this notwithstanding, we are much more concerned about how we fit into our personal circle of friends than about how we all fit into the complex ecology of our own planet.

Those of us who are scientists behave in exactly the same way, but we tend to be more bothered by it, because we have real trouble understanding why we're doing it. Our scientific instincts tell us that the real universe out there is actually far more important, on any serious scale of events, than whether Mary told her mother she was dieting ... but somehow questions on the level of Mary's diet take up much more of the scientist's time than the whys and wherefores of galactic superclusters – even when the scientist is a cosmologist.

We lead a dual existence – *in* nature but not *of* it, perpetually reacting to our estimate of what the world will be rather than what it is right now. We mirror the world outside us with another in our heads: our perceptions of that world. It's a distorting mirror, an imperfect representation, but to us it seems *real*. In a funny self-centred way we see ourselves as existing slightly to one side of the rest of the universe. We are in control of our world, we can make choices, we have *minds* that we can make up or change. Everything else is just following the inexorable impulses of nature. When we think of an amoeba, a fox, an oak tree, or a dinosaur, we think of them as a part of nature. The amoeba fiddles about putting out pseudopods and ingesting food particles, and that's about it. The fox runs through the bushes chasing a rabbit for dinner, and when it encounters the occasional bunch of subhumans on horseback it's too busy running from the dogs to debate the morality of blood sports. The oak tree is just sitting there synthesising, drawing in water from its roots and carbon dioxide from the air, and if it's worrying about anything it's about the impending winter and dropping its leaves – not whether the neighbouring oak tree thinks it's a cad for fertilising too many of its acorns. We see dinosaurs as eating, breathing, multiplying, and dying out against the great backdrop of natural forces, like the K/T meteorite that hit the Earth 65 million years ago and caused mayhem all over the planet. Gary Larson's 'Far Side' cartoons often work by imputing human-type motivation to animals, and they are funny because we know that most animals *don't* worry about their circle of friends.

All very well. But how much of our belief that we are special is grounded in fact, and how much is just a comfortable illusion of superiority? The belief that we are superior to other animals is a human value judgement, and as such is likely to be biased in our own favour, but there can be little doubt that we are *different* – in important ways – from the other animals on our planet. These differences must be explained. Their explanation is made more difficult, but also much more interesting, by the fact that human beings have not always been as they are now. Few of us doubt that we evolved from creatures that, like most animals, related directly to the natural world and thereby avoided all of the social problems that occupy our every waking minute and even assail us in our dreams.

How did that happen?

This question is the central issue that will shape our narrative. What was it about this particular lump of rock, in this particular spiral arm of this not terribly special galaxy, that made us the way we are? How is it *possible* for inanimate matter to turn into complex creatures like us with their own inner worlds of mind and imagination? Given that it is possible, why did it happen? Why us?

Some will ascribe it to God and be satisfied: we have nothing to say to them.

Some will ascribe it to inexorable consequences of the fundamental laws of physics, and be satisfied: we have nothing to say to them either.

We *do* have something to say, however, to those who find either answer incomplete, people who think that our presence on this planet and our curious mental abilities deserve to be explained rather than explained away. In *Figments of Reality* (henceforth abbreviated to *Figments*) we attempt to explain the evolution of human beings from a new point of view – one that differs considerably from the usual scientific story, although it retains many points of contact with it. More accurately, we shall look at the questions of mind and culture from *two* disparate viewpoints, which complement rather than contradict each other. One is the conventional scientific viewpoint: take the system to bits – in a conceptual sense – and see how those bits fit together. The other, less conventional but in our opinion equally important, is to look at *context*, and see how the system is shaped by what lies around it.

Along the way we shall be forced to reassess the orthodox scientific stories about how things work, many of which are little better than myths. We don't think that such reassessment makes the orthodox stories any less 'true' (we'll air some of our prejudices about truth later), and we certainly don't think that it makes them any less 'scientific'. The point is that if you approach the questions from different directions you may find yourself wanting different kinds of answer, just as 'God' may satisfy a priest in search of virtuous living but not a programmer in search of virtual reality. We think that such changes of perspective help to make many problems of human evolution and cultural development seem less puzzling. In particular they will help us to tell the story of human mind and culture in a more accessible way – one that explains, rather than just asserts, the scientific bases of our world and of ourselves.

We'll give you the bare bones of the story now, to act as a 'road map' for the rest of the book. First, we look at the origins of life and its evolution – both on Earth, the story of how we came into being, and elsewhere, the story of what might have happened instead and what might be happening right now on a planet of some distant sun. We describe the evolution of senses – in particular sight, hearing, and smell – showing how they have influenced the evolution of networks of nerve cells, leading to that most flexible and enigmatic of all organs, the brain. We demonstrate that, far from being mere passive observers of reality, our senses are fine-tuned during development to emphasise those features in which our brains have an especial interest. By manipulating these mental features we construct 'conceptual maps' of the reality around us, which enable us to make up our minds (take decisions) and change our minds (modify our choices in response to the consequences of those decisions). We do not so much *observe* reality as put together our personal representation of it and drape that back on to our perceptions of the external world. This facility is moderated by intelligence – the ability to reason, to solve problems – which is not merely a structural feature of large brains with intricate networks of nerves. Intelligence arose in intimate association with a marvellous non-genetic trick used by parents to provide their offspring with a head start in life, a trick that we call 'privilege'. Privilege begins with yolk and nests, and culminates – so far – in culture. We further claim that it is not intelligence alone, or culture alone, that leads to mind, but both – interacting 'complicitly'.

A feature of our minds that is often singled out as *the* thing that makes us uniquely human is language. Some scientists think that language is a necessary prerequisite for intelligence, and others that intelligence is a necessary prerequisite for language. We think that both are right – and so both are wrong, for each thinks the other mistaken and both are mistaken about 'prerequisite'. Language and intelligence evolved *together*, both being inextricably linked to culture.

Finally, we tell of the rise of human culture, the techniques that cultures employ to survive in a changing world, and the effect of cultural differences on displaced ethnic groups, leading to multicultural societies in which individuals grapple with changes in their cultural identity. We tell of the growth of global communications that lock the multiculture in place, so that we cannot go back even if we wish to. We take a brief look at the future of human multiculture. And we wrap the entire package up and tie it with a neat bow, by means of a unifying concept – extelligence – that is the contextual and cultural analogue of internal, personal intelligence.

To kick the whole story off, we now ask a 'warm-up' question: how did inanimate matter give rise to life? In the Prologue we described the current view of the origins of the universe, the 'Big Bang' theory as it is called. Space, time, and matter arose from nothing; then the simple kinds of primal matter that existed at the prevailing high temperatures began to combine to make all of the different chemical elements – hydrogen, helium, lithium, beryllium, boron, carbon, nitrogen, oxygen ... These different atoms then combined to form chemical molecules – two hydrogens plus an oxygen to make water, one carbon and two oxygens to make carbon dioxide. The bodies of living creatures are made from millions of different molecules, all of which trace back to the nuclear reactions in the cores of stars. Literally, 'we are stardust', as Joni Mitchell sang about Woodstock.♪

Particles building into atoms, and atoms into molecules – these we can comprehend, they're just like bricks building into a house. But houses don't develop a will of their own, get up, and walk away. Living creatures did, and that's a real puzzle. How did inanimate, inorganic chemistry somehow generate the rich flexibility of life? Not all at once, that's for sure. There was no wondrous, special moment, pregnant with significance, at which life suddenly appeared on the planet. Instead, life emerged gradually from non-life. In this respect the origins of life are a bit like the origins of a person's life. There was a time when Maureen didn't exist. At what time did the egg, embryo, fetus, child, become Maureen? At what time did it become human? Surely there was not a specific *moment* of becoming Maureen – though people who don't know about it do talk of 'the moment of fertilisation' – except in a legal sense, at her naming ceremony. A person is like a painting or a novel: it progressively comes into being. Maureen started as not-Maureen and gradually became Maureen. So it was with the origins of life.

We can't go back and see what actually happened, but we can infer the kind of molecular game that must have been played out upon the primal Earth. In particular, we can understand that life could reasonably have come into being gradually and spontaneously as a consequence of perfectly reasonable chemistry. Four billion years ago, the Earth was a very different place. Its surface was barren rock, sandy desert, bubbling tar-pit, smoking sulphur-hole. Its oceans were a watery layer of chemicals dissolved out of the rocks and injected into the ocean depths by underwater volcanoes. All of the diversity of chemical elements that we find today was already present then – for apart from a continual infall of meteoric dust and a slow leakage of the lighter gases, the atoms that make up today's world are the same ones that were present four billennia ago. The difference between that ancient Earth and the one we inhabit today lies not in its atoms, but in its molecules. They are much more diverse now, and – absolutely crucially – they are organised in much more complicated ways.

Textbooks tell you that a molecule is a system of atoms connected together by interatomic forces – 'bonds'. This is true – as much as any human statement about nature is true – but it is not the whole truth. Another part of the story is that unlike atoms, molecules can become more complex. Atoms, left to themselves, do not produce types of atom that have never existed before – although some atoms can change by way of nuclear reactions, with uranium turning into lead, for example. But atoms can rather easily produce entirely new types of molecule by combining in new ways, and those molecules can also go on to produce new molecules – a process that continues to this day. If the only thing you knew about the Earth was a catalogue of its molecules, you would be able to see a distinct difference between today's catalogue and that of four billion years ago. Today's catalogue would include many enormous molecules, such as proteins and DNA, that would be missing from the early version.

So over the billennia, molecules have become more complex. However, that is by no means the whole story, because there is much more going on than mere complexity. That four billion year-old catalogue of molecules would include some amazingly complicated ones too, for instance innumerable weird conglomerations formed in the tar-pits. Similarly today's catalogue would be littered with molecules like toffee, a disordered mass of one-off constructions whose greatest similarity to each other is that every single one of them is totally boring. No, the molecules that are of greatest interest are not just *complicated* –

they are organised. They are, in fact, machines – the first machines that appeared on Earth. To be sure, they don't look much like the machines with which we are familiar – lawnmowers, cars, aeroplanes – but they have a basic property in common with these human-made devices. They can perform functions, a fancy way to say that they do things. A function is an operation which, when presented with certain inputs, produces various outputs in a reliable manner. The most obvious function of a lawnmower, for example, is to mow a lawn: here the input is a lot of straggly grass and the output is a neat, tidy swathe of green. A lawnmower can perform other functions too: propping open the door of the garden shed or holding down a pile of plastic sacks when a breeze is blowing.

Molecules, too, can perform functions, because they interact with other molecules. And because molecules have definite shapes, these interactions are different for different molecules. For example, molecule A may have some kind of dent in its surface, just the right shape to fit a bump in molecule B. If so – and if the interatomic forces are suitable – then you would expect to find many molecules that are made from A and B fitted together. This sort of 'plug and play' construction of molecules is going on all the time. It is to some extent counterbalanced by the tendency of molecules to fall apart for various reasons, so we don't just get the whole of terrestrial existence locked together in a single supermolecule.

Molecules can also have moving parts. The bonds that join their atoms together can bend and twist, to a limited extent, and sometimes atoms can even revolve on their bonds like propellors on a spindle. This flexibility provides a lot of scope for making chemical machines with interesting functions. Some molecules can make other molecules fit together, or pull them apart. After performing their function they remain unchanged, and are ready to carry it out again and again. Such molecules are called 'catalysts'. Catalytic molecules act like a production line: provided they are supplied with the right 'raw materials' they can go on turning out copy after copy of their favoured molecule, indefinitely.

Carrying out a function is quite different from having a purpose. Molecular machines do not carry out functions because they want to do so: they carry them out because this is how they are made. Indeed it it impossible for them *not* to carry out their functions. In the same way, a rock carries out the function of rolling down a hill because it is suitably rounded and has significant enough mass for gravity to latch on to. But it does not have that rounded shape for the *purpose* of rolling down a hill. We mention this because human beings seem to have an innate tendency to confuse functions with purposes – so that, for example, 'the sun keeps us warm' becomes 'the sun was placed in the sky *in order to* keep us warm'. This kind of purpose-centred thinking can easily lead to people worshipping the sun-god, not realising that the sun can perform the function of keeping them warm without either wishing to do so, or requiring worship to continue doing it.

At any rate, four billion years ago there were pretty much the same atoms around as there are now, but not in the same combinations, and not organised like they are now. The complex molecules that occur in living organisms and in pseudo-living entities such as viruses are known as 'organic' molecules. The atom that makes all organic molecules possible is carbon: carbon atoms have the ability to stick together and form huge, stable skeletons, to which other atoms can attach. Even carbon can perform this task only within a narrow range of temperatures, and other atoms can't do it at all, with the possible exception of silicon. This is not to say that carbon is essential for life; just that it is essential for *our* kind of life, which is the only kind we know about, and it generally looks like rather good stuff to make life from. However, the kind of organisation that we call 'life' might in principle arise in other ways – silicon-based molecules, interacting trains of electrons in metallic crystals, colliding plasma vortices in the corona of a star ... The possibility of complex molecules is important because *some* complex molecules can perform more sophisticated tasks than simple ones. Upon these more sophisticated tasks does the peculiar form of matter that we call 'life' depend. Living organisms are much more than just formless bowls of molecular soup: the manner in which their molecules are arranged is at least as important as what those molecules are. But without the potential complexity that carbon provides, molecules complicated enough to get themselves organised into organisms like us would not exist.

Life seems very different from inorganic matter – it can move of its own volition, reproduce itself, consume other substances, respond to its environment. It is therefore hardly surprising that some people think that living material is simply a different *kind* of stuff from non-living matter. This belief is known as vitalism. Its greatest defect is that there is no evidence in its favour: none of this different kind of stuff has ever been isolated. If you take a living organism to bits, right down to the molecular level, all you find is ordinary matter. We humans are made from the same atoms as the rocks, water, and air around us. The inevitable conclusion is that it is not the ingredients that differ: it is how they are organised. A living creature can be killed by bashing its head with a rock: it is hard to see how such a crass act can devitalise its esoteric immaterial substance, but easy to see how it can wreck its organisation.

In the same manner a car is made from the same atoms as the sheets of metal, sacks of aluminium powder, and cans of polymer from which it is assembled. Its ability to move does not arise because it is made from a different kind of matter: it is merely a consequence of how that matter acts when it is put together in a particular manner. An automotive engineer would be able to explain, in more than enough detail to send any partygoer in search of the drinks tray, what is involved in this organisation. But nobody ever made a car by going out and looking for a new kind of matter that has the ability to move when petrol is poured into it.

There is a danger with the 'car' analogy if it is pushed too far. To some people, organisation implies the existence of an organiser, as the existence of a watch implies that of a watchmaker. This is a seductive line of argument, but there is no compelling reason to accept it. One of the most remarkable features of organic matter – and, we now realise, inorganic matter too under suitable circumstances – is its ability to organise *itself*. So in some ways a better analogy than a car would be a whirlpool, a tornado, or a flame: an organised structure that comes into being without conscious intervention. Our intuition is upset by self-organisation, probably because we seldom experience such behaviour directly: in our everyday world the only way to produce organisation is to work pretty damned hard to make it come about. Nevertheless, we are surrounded by and made from matter that is highly organised, and it must have got that way by some route. Either it has been organised by an organism-maker, or it has organised itself.

The problems with the 'organism-maker' hypothesis have been rehearsed by philosophers and theologians for as long as anyone cares to remember. Its obvious advantages (it 'solves' the problem to many people's satisfaction) are countered by its equally obvious defects. For instance, who or what organised the organiser? And where *is* the organiser? The 'self-organisation' hypothesis has far more to offer to those who share the scientist's wish to understand nature and not just postulate it. It is a daring hypothesis, which does not solve the problem unless we can explain *how* and *why* living matter self-organises. It is becoming clear that there is nothing inherently self-contradictory in the idea that organisation sometimes comes 'for free', and it is also becoming clear that limited laboratory-scale systems and computer simulations indulge in self-organised behaviour far more often than we might have anticipated. *Why*, we are still unsure, but we know that it is so. Perhaps our universe is special in being like that; perhaps all universes must be. Which, we don't know.

The self-organising ability of life becomes clear only over long timescales: compare an organism such as a mouse, today, to a lump of rock four billion years ago. One of the most obvious 'unusual' features of life, however, can be seen on far shorter timescales: its ability to reproduce. Life makes new life – and pretty much the *same* life. People make new people, cats make new cats, nematode worms make new nematode worms, and amoebas make new amoebas. This is an amazing ability, and it certainly looks very different from ordinary chemistry.

However, we tend to underestimate what 'unaided' chemistry is capable of, and that distorts our assessment of how amazing or unlikely life is. Thirty years ago, biology was thought to be very complex and chemistry relatively simple. The chemical story of the origins of life seemed to require the construction of a conceptual pyramid of ever-complicating processes, rising from the lowly plains of test-tube chemistry to the lofty heights of biology. Nowadays we understand that this picture is wrong. 'Unaided' chemistry – chemistry that does not require a living organism to make it happen – goes all the way up. Even simple unaided chemistry is a lot more complicated than the textbooks would have us believe. For example, if a mixture of two parts hydrogen to one part oxygen is ignited, then it explodes, giving water. The old textbooks see this as a single chemical reaction: $2H_2 + O_2 \rightarrow 2H_2O$. (We don't write this in the apparently simpler form $H_2 + O \rightarrow H_2O$, by the way, because reactions are about *molecules*, and a molecule of oxygen is O_2, not O.) Newer textbooks will tell you that there are at least ten other molecules involved as intermediaries, and the more closely you look, the more of them you will find. The old textbooks tell you what to start with and what it ends up as, but not what happens in between. When reactions as basic as this turn out to be so complex, it is not surprising that more sophisticated kinds of chemistry are *far* more complex. Moreover, as our understanding of the complexity of chemistry grew, we also came to recognise that biochemistry is a lot closer to 'unaided' chemistry than we used to think. In fact modern industrial processes, which make extensive use of catalysts, sit right at the junction of 'unaided' chemistry and very similar biochemistry.

Another reason why we are so puzzled by life arising from 'mere' chemistry is that it is very difficult to find, on the Earth, now, the kind of chemistry that long ago gave rise to life. This is because life has invaded all of the possible habitats for such chemistry, from the deep oceans, tens of miles deep in granite cracks, to high in the atmosphere – so their chemistry has been changed out of all recognition. Rusting would be a good example, except that on Earth it is nearly always 'assisted' by bacteria, who take a tithe of the energy. So let's imagine iron rusting on the surface of a lifeless planet. Recall the concept of catalysis: a molecule is a catalyst if it assists in the production of another molecule, or molecules, without itself being used up in the process. Sterile rusting proceeds by autocatalysis – given a bit of rust on iron it catalyses more of *itself*. Such a process is recursive, it pulls itself up by its own bootstraps, so you need a bit of the product to get it started. (Stop worrying: we never said that that initial bit of product was produced by the *same* recursive process. See later.)

Many recursive systems are known in real chemistry and technology, but they are largely missing from school or college chemistry because they don't fit the simplified theories being taught there. The catalytic convertor in a car oxidises pollutants using just such a system. The catalytic surface does its work in a series of expanding rings, just like the very best example of this kind of chemistry, the Belousov-Zhabotinskii (BZ) reaction of figure 1. This is an extremely photogenic instance of recursive chemistry, with expanding rings of blue in a rusty red solution. For forty years after such systems were first described, most chemists did not believe they could work: they seemed to be contrary to that most famous – and misunderstood – of scientific laws, the Second Law of Thermodynamics.

They are not, and neither is life.

Thanks to the epic researches of Maurice Wilkes, Rosalind Franklin, Francis Crick, and James Watson in the 1950s, we know that one remarkable molecule – more properly, a family of very similar molecules – underlies almost all terrestrial life. That molecule is DNA, whose initials stand for 'deoxyribose

Figure 1 *Typical 'target' patterns in the Belousov-Zhabotinskii reaction. As time passes, the rings expand.*

nucleic acid' (or 'deoxyribonucleic acid' according to taste). DNA forms the genetic material of almost all organisms. A few viruses use RNA, 'ribose nucleic acid' or 'ribonucleic acid', but DNA and RNA come from the same molecular stable. DNA has a simple but clever molecular structure in which twin strands spiral like a staircase. The treads are made from four types of molecule called 'bases', held together by a framework of sugars and phosphates. This structure allows DNA to do two important things: encode information, and replicate. The information is represented by the sequence of bases, and includes such things as the structure of key proteins without which organisms cannot be built, and sequences that determine *when* they will be built. DNA replicates by separating the two strands, in which the bases are complementary to each other, and re-creating a matching strand for each, thereby producing two copies of the genetic information from one original. (This description, though standard, is an oversimplification, but it is sufficiently accurate for our present purposes.) Throughout *Figments* we shall distinguish replication, the creation of exact or nominally exact copies, from reproduction, the creation of *similar* copies – in particular, similar enough that they too can reproduce. Normally DNA replicates, but when the occasional inevitable copying error – the technical term is *mutation* – creeps in, then the molecule is better thought of as reproducing.

Although it is often described as such, DNA is not a *self*-replicating molecule: leave a mass of DNA in a beaker and you won't get more of it. It replicates only with the aid of many other molecules, known by names like transfer RNA, messenger RNA, and enzymes. We mention these merely to drive home that DNA needs an entire 'support team' in order to replicate: it no more makes copies of itself than a document in a photocopier makes copies of itself. Moreover, the fact that DNA contains 'information' is far less important than the physical (that is, chemical!) form that the information takes. All molecules 'contain' information – the positions of their atoms, for example, are a kind of information, as you will quickly discover if you build molecular models. The information in DNA is useful *not* because it is information, but because it is information stored in a form that other chemical machines can manipulate. As an analogy, the positions of the wood fibres that make up this page encode a huge amount of information, but when you read the page the only *useful* information – for you – comes from the letters printed on it.

The process that allows DNA to replicate is another autocatalytic recursive cycle, only here it is a *collection* of molecules that catalyses itself. The DNA contains the defining information for the molecules in the support team. The support team helps DNA to replicate, and the DNA helps to replicate its own support team. Recursion often feels disturbing, but how *else* could a replicative process work? What makes recursive processes disturbing is the feeling that they can never get started – the 'chicken and egg' problem. Actually that's not a serious problem at all, just a case of sloppy thinking caused by incorrectly extrapolating the process backwards. It's relatively easy to get a replicative process *started*. What you can't do – without destroying the process – is *stop* it. The way to start a chicken-and-egg process is to create a suitable start-up configuration, one that is part of the process only the first time round. For example a non-chicken might be persuaded to lay an egg that grows into a chicken, whose eggs also grow into chickens, and so on forever. Clearly you can't play this trick if you start with a perfectly replicating non-chicken and absolutely nothing untoward happens to its egg; but if it is a reproducing non-chicken, subject to variations that do not affect the reproductive abilities of its offspring, there's no conceptual problem at all – just a technical one of actually making the trick work. The answer to the hoary philosophical teaser then becomes no more than a question of definition. Is a chicken egg one that was laid by a chicken, or is it an egg that grows into a chicken? In the former case, the chicken came first (from a non-chicken egg); in the latter case, the egg came first (laid by a non-chicken).

There are other ways to get a replicative or reproductive system started. One is for it to 'piggyback' on a pre-existing replicative or reproductive system. This is how documents replicate: they piggyback on photocopiers, which are replicated by humans working in factories. The photocopiers in turn piggyback on human reproduction. Of course it's not possible for every replicative/reproductive process to piggyback on a previous one, or else there is a genuine chicken-and-egg problem, so at least one process has to get started some other way (and act as a start-up configuration for everything that subsequently piggybacks on it). That other way is best described as 'scaffolding': *before* the replicative loop closes up, the process is assisted by something else, which drops out of the loop permanently *after* it is closed. Once a system acquires the ability to replicate, it spreads rapidly and takes over any disorganised substrate.

Although the loop formed by DNA and its support team is in principle replicative, in practice it is 'only' reproductive. The procedure is so complex that it seldom takes place without errors. Moreover, in sexually reproducing organisms, the reproductive procedure introduces 'mix-and-match' modifications. This should not be thought of as a defect. Reproductive systems are much more interesting than mere replicative ones, precisely because they can change. Replication is just the same thing repeated forever. Reproduction has room for flexibility – it can produce a chicken from a non-chicken's egg.

That possibility leads to evolution, which in various ways forms the subject of the next three chapters. Before tackling such a subtle subject we shall deal with a more down-to-earth question: how did DNA replication get started? The process looks too complex to have arisen from raw scaffolding: most probably it piggybacked. There are hints of possible precursors in the DNA replication process itself. Over the years, many different proposals have been made, and we mention them here to show that there are *several* plausible solutions to the problem of how life got started on its reproductive path.

One is the 'RNA world'; a second, due to Graham Cairns-Smith, is clay; and a third is Stuart Kauffman's concept of an autocatalytic network of molecules. The RNA world is a hypothetical period of evolution when DNA did not yet play a role in the replication of proto-living forms: instead, the simpler molecule RNA held centre stage and reproduced without help from DNA's band of molecular assistants. Back in the 1950s Stanley Miller, a student of Harold Urey, performed experiments showing how amino acids – the building blocks for proteins – arose spontaneously in a simulation of the Earth's primal chemistry. Variations on this system have provided all the raw materials for life, either DNA-based or RNA-based. The possibility of an RNA world, predating today's DNA/RNA combination, first became apparent in the 1980s when Tom Cech and Sydney Altman discovered special RNA molecules now called ribozymes. These acted as a catalyst in a reaction that snipped out parts of themselves – one element of the recursive process needed for replication. Jack Szostak then employed a laboratory version of molecular evolution to produce more efficient ribozymes which could copy long RNA sequences. In 1996 David Bartel found some that are as effective as some modern protein enzymes. RNA 'self'-replication – employing molecular assistants, but not DNA – has not yet been achieved, but it looks far more plausible.

In May 1996 the chemist Jim Ferris discovered a way in which long RNA strands (10–15 bases in length) might have formed in the primal environment. If he added montmorillonite – a kind of clay – to the chemical mix, then long RNA chains formed on the surface of the clay. This was especially interesting in view of Cairns-Smith's earlier speculations that clay might provide a replicative structure upon which RNA could piggyback, and we will briefly describe what he had in mind. Clay is a complex combination of aluminium, silicon, oxygen, magnesium, calcium, iron, and many other elements. Clays can dissolve in water and precipitate out again. Their crystalline forms employ rarer elements to structure themselves into exotic shapes: scrolls, curlicues, spirals. Like most crystals, these shapes can act as templates to produce more shapes of the same kind, building up on top. When some external event causes the crystal to break, each piece can act as a template for further growth, so these clay forms can replicate – indeed, reproduce (figure 2). They can even compete with each other, because some shapes are better at extracting particular substances from solution. Clays are probably the nearest thing on Earth to a silicon-based life-form, a replicating system upon which others can piggyback. As Cairns-Smith realised, carbon compounds naturally stick to the surfaces of clay crystals, and they catalyse organic reactions. In particular they catalyse processes of polymerisation, in which molecules of the same kind are added to each other, forming long chains and other structures. By this process amino acids could become proteins, and simple bases could link up to form RNA, DNA, or – mostly – other nucleic acids. As Ferris showed, Cairns-Smith's chemical intuition was justified, which adds weight to his view that the origin of our kind of life was subsequent to a much more primitive kind of clay life, a story that he calls 'genetic takeover'. It is a story of a smooth transition from inorganic chemistry to our kind of life, eventually resulting in creatures about as organised as a bacterium, without a nucleus – what biologists call a prokaryote. We can even have our cake and eat it too: perhaps DNA piggybacked on RNA and RNA on clay.

The autocatalytic network idea is rather different: it presents a set of circumstances in which 'scaffolding' is almost inevitable, rather than being just a convenient coincidence. A replicating molecule would be one that catalyses itself, but that's just a bit *too* convenient and seems not to happen naturally. (Rust doesn't really count: it needs iron, water, and oxygen too, not to mention bacteria.) However, it's *much* easier to come up with a 'support team' of molecules in

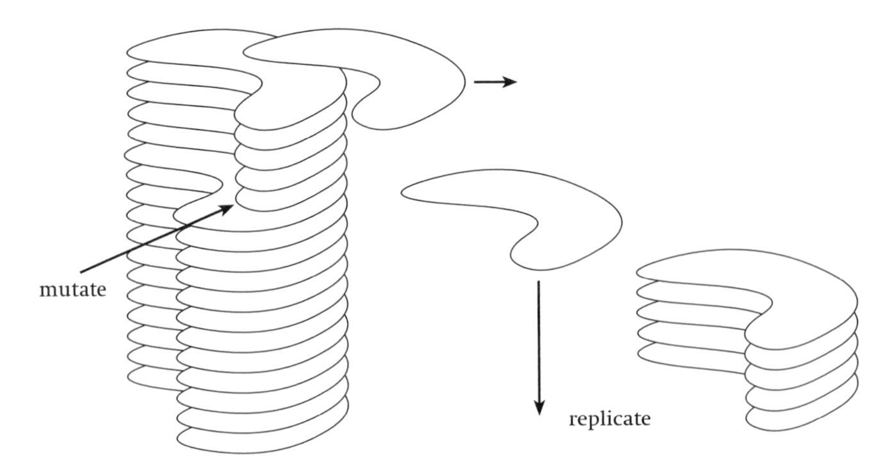

Figure 2 *Mutation and replication in stacks of clay platelets.*

which each member catalyses other members. Such a team 'closes up' into an autocatalytic network if *every* member of the team is catalysed by some other combination of members. Then the entire team acquires the ability to replicate.

Autocatalytic networks, in particular, illustrate just how close 'unaided' chemistry can get to genuine life. Now: suppose that we could provide such a network with one more feature, its own 'identity', so that it could exist – and replicate – as a well-defined entity, instead of just dispersing into the vast chemical ocean. Then we would have a rudimentary chemical 'life-form'. Here's a possible scenario, proposed some years ago by Alexander Oparin: actual events were probably more complex and possibly quite different. There are molecules known as lipids – fats – which look rather like tadpoles. Their heads are attracted to water molecules but their tails are repelled. Moreover, lipids quite like to stick together. So lipids at the surface of a watery environment – such as the primal ocean – naturally align themselves in sheets, with all the heads on one side and all the tails on the other. These sheets are biological membranes, and their key property is to separate regions of space from each other. In our previous terminology, they are naturally occurring chemical 'machines' whose function is to separate the watery ocean into distinct regions. (In real biological membranes there is a *double* wall, its molecules aligned tail-to-tail with heads on both 'external' surfaces, but the general function is the same.) Doron Lancet suggested that such membranes might also close up into tiny capsules, and other molecules could diffuse in or out. Now, said Lancet: suppose that, by chance, the molecules trapped inside such a capsule when it first forms happen to make up an autocatalytic network. Then – fuelled by raw materials that diffuse in from the diverse but disorganised primal ocean – the network will replicate. Suppose further that the lipid whose molecules create the capsule is also part of that autocatalytic network. Then the capsule will swell up as its contents replicate, and eventually it will grow to such a size that it becomes unstable, in the sense that a large capsule will tend to break up into smaller ones – each containing the chemical support team needed to make the process continue. There is no obvious end to the process: what we have is in effect a prototype cell, which sucks in nutrients, grows, and divides into cells of the same type. So, given a fair sea and a following wind, autocatalytic networks can in effect organise their own spatial geometry to form replicating organisms – or, at least, proto-organisms. More complex replicating molecular systems now have something to piggyback *on*.

All the above is speculative: its purpose is to show you how *easily* life might get its act together as a result of natural combinations of ordinary physical and chemical features of the inorganic world. What actually happened? In Chapter 4 we shall take a look at how and why this kind of piggybacking eventually led to life as we know it – with the driving process, of course, being evolution. For the moment we shall set the scene by leaving out the evolutionary element, describing what seems to have happened without asking why.

Lipid capsules filled with autocatalytic networks of chemicals are close enough to genuine organisms of a bacterial grade of complexity – prokaryotes – that it is easy to imagine how prokaryotes might have come about, although the most plausible theories of their origins are less simple – as we see shortly. At any rate, whichever route it was that produced prokaryotes, we know that they appeared, multiplied, evolved, and took over the surface of the Earth soon after there was a liquid sea. We know that it happened very quickly, though we can't be sure whether it happened as soon as there was liquid water, or whether it required a few million years after that. On geological timescales the difference is immaterial; the point is that it happened so fast that the process must have been chemically and physically 'easy' – for an entire planet, cooling down from a bombardment of meteorites. In fact some kinds of meteorite contain organic compounds, so, for all we know, Molecules from Outer Space may have been the initial scaffolding, as Fred Hoyle and Chandra Wickramasinghe suggested many years ago.

Before more complex forms evolved, prokaryote life dominated the seas for three billion years. During that time many different kinds invented photosynthesis, a way to power their recursive chemistry by extracting energy from sunlight. In so doing they excreted a highly toxic waste product – oxygen. At that time few, if any, organisms made use of oxygen, a highly reactive chemical: it still causes problems today, because it lets things catch fire. The build-up of oxygen changed our atmosphere completely, to the extent that it is a long way from chemical equilibrium – that is, without life the level of oxygen would decrease considerably as it reacted with minerals, oxidising them. Life doesn't stop those reactions happening, but it puts the oxygen in faster than unaided chemistry can take it out again.

About 1.5 billion years ago new forms of life, with much more complicated recursive chemistry, arose to exploit the new reactions made possible by oxygen. These were the eukaryotes, and their most important feature was the possession of a nucleus. People often talk of bacteria as being 'unicellular' and creatures like us as being 'multicellular', as if you can evolve a human being by sticking a lot of bacteria together, but this is quite the wrong image. Bacteria are *not* single-celled organisms, because they are not cells. They have a few features in common with cells, and they seem to have evolved into cells, but even a single cell is considerably more sophisticated than a bacterium.

Eukaryotes can be single-celled – a well-known example is *Amoeba* – but they can also be many-celled. The eukaryote cell differs in significant ways from a bacterium. It is larger – typically about 10,000 times as large by volume. Even in a single-celled eukaryote, the cell possesses a range of 'organelles', component sub-units with some special function, such as the nucleus (which contains most of the cell's DNA) and mitochondria (which protect the cell against oxygen and provide much of its energy). The currently accepted theory,[♪] which goes back at least a century and was revived in 1967 by Lynn Margulis, is that the cell arose from independent bacteria of various kinds by a process of symbiosis, which may have started out as parasitism. A simple, but misleading, way to say this is that various bacteria 'got together' to produce a cell. A more accurate way to say it is that cells *emerged* from the coevolution of bacteria. We don't just mean 'appeared'; we are using the word in the sense of 'emergent phenomenon'. This term comes from philosophy, and is used when the behaviour of a system appears to transcend anything that can be found in its components – where the whole seems 'greater than the sum of its parts'. Here the point is that if you put a lot of bacteria together and wait long enough, then the overall system will home in on the cell as a viable way to organise its business.

We shall have plenty more to say later about emergence.

In a similar manner, multicellular creatures mostly arose *not* by combining separate cells together into a colony, but by starting with a single cell and letting it *divide* repeatedly – 'multiplication by division'. In this manner a single large cell became an aggregate of sub-cells *with the same DNA*, a useful degree of genetic coherence that made it possible for the entire system to co-evolve simply and naturally. But now each sub-unit was free to specialise, if the result helped to keep the organisms' reproductive cycle going, so eukaryotes evolved different types of cells, with different capabilities. Just as molecules added entirely new dimensions of complexity to atoms, eukaryotes added entirely new dimensions of complexity to organisms. The new atmosphere opened the way to oxygen-breathing organisms with a faster lifestyle; life embarked upon a wild romp of self-complication.

Sometimes an apparently minor change had major implications on a global scale. At some point some varieties of marine organism stopped excreting their wastes in liquid or semi-liquid form, and instead produced them in solid form. Probably the first such animal was a swimming worm, but it might conceivably have been a trilobite – the timing is right, but the evidence is slim.

This minor change in water content made a huge difference, because the solid wastes sank to the bottom of the shallow seas, forming an anaerobic layer for soft-bodied organisms to graze. One animal's waste became another's resource – just as had happened earlier for the toxic oxygen wastes of the bacteria.

In 1909 evidence for one of the more curious stages in the evolutionary process came to light in Yoho National Park, high in the Canadian Rocky Mountains. Charles Walcott, secretary of the Smithsonian Institute and America's leading palaeontologist, discovered a large number of unusual fossils in a rock formation known as the Burgess Shale. The story has been grippingly told by Stephen Jay Gould in *Wonderful Life*. The fossils were unusual because they were formed from soft-bodied creatures. Normally conditions are unsuitable for soft parts to fossilise, but in this case something like a mud-slide had overwhelmed the pool in which they were living. Walcott took a cursory look, assigned them to various known groups of organisms, filed them away in drawers and forgot about them. In 1971 Harry Whittington of Cambridge University recognised that the Burgess Shale organisms are far more interesting than Walcott had supposed. They represent an early explosion of multicellular life: the anatomical diversity in that one small pool was much greater than that over the entire global ecosystem today. Not in terms of the number of species, but the number of phyla. A phylum is one of the largest units into which organisms are classified. For example today's many-jointed 'arthropods' – members of the three phyla of crustaceans (shrimps and the like), chelicerates (spiders, scorpions and their kin), and uniramians (insects and more) – *all* evolved from three groups present in the Burgess Shale. However, more than twenty other radically different arthropod designs are found in the Burgess Shale creatures too, only one of which – the now-extinct trilobites – went on to establish itself as a major player. In recent years several palaeontologists have suggested that the diversity of the Burgess Shale organisms is not *quite* as great as was first supposed, but they are without doubt a weird and varied bunch: for instance figure 3 illustrates *Opabinia*, which has a nozzle at the front, a claw at the back, five eyes, gills on top of its body, and a three-segment tail.

Just *one* of the Burgess Shale creatures was part of the evolutionary lineage that led to humanity. Since most of the Burgess Shale creatures quickly died out, for no obvious structural reasons, Gould deduces that it was largely a matter of luck which of them survived – and he inferred that our presence on this planet owes much to Dame Fortune and not much to Good Design. But is this really so? We give our answer in Chapter 5. But whatever the interpretations placed on it, the Burgess Shale fossils show that around 570 million years ago, at the start of the Cambrian period, life suddenly became *enormously* more diverse. This geologically rapid change is known as the 'Cambrian explosion'.

The Burgess Shale is a dramatic demonstration that evolution can produce rapid and far-reaching changes. But what actually 'drives' evolution? Human beings, conscious of their personal mortality, are somewhat obsessed with the Grim Reaper, if only because they dimly see Him coming and they don't like it. In consequence the Grim Reaper plays a central role in humanity's usual story of evolution: 'nature red in tooth and claw', where creatures strive to out-compete each other in a desperate no-holds-barred battle for survival. Only the winners of these battles, it is said, get to perpetuate their kind: the losers just die, and in this way organisms with 'good genes' proliferate at the expense of all the rest. It's a simple, compelling picture, which seems to explain the general increase in the complexity of life-forms. However, we will see shortly that this picture of evolution is misleading, even for simple organisms.

In fact, in the evolution both of complex organisms and of mind, the central role is played not by the Grim Reaper, but by the Grim Sower, who starts things up by their billions so that nearly all of them have no option but to die before they have reached maturity. The popular view of 'natural' animal lives has been romanticised to such an extent that they are universally seen as idyllic, whereas actually the reverse is the case. *Nearly all wild creatures die without breeding.* For example, from the 10,000 eggs that a female frog lays during her lifetime, on average 9,998 die for each pair that survives to replace the parents and breed. A more extreme case still is the cod: a single female lays forty million eggs, of

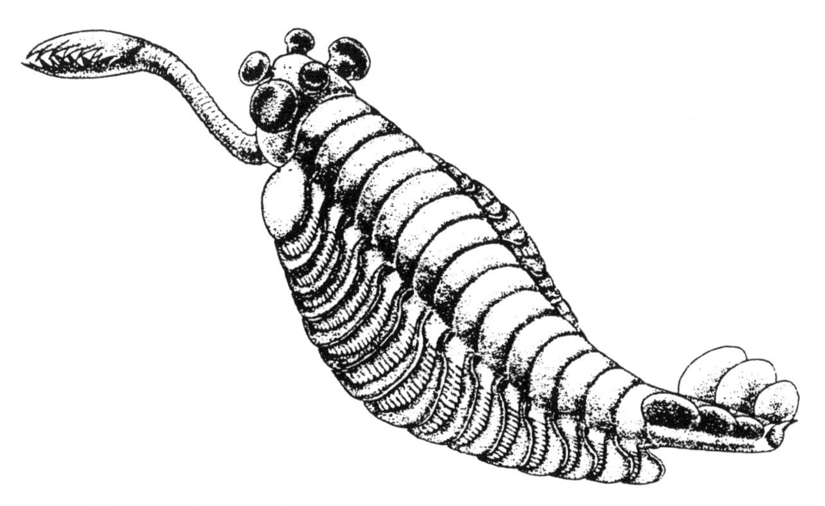

Figure 3 *The Burgess Shale organism* Opabinia *drawn by Marianne Collins.*

which 39,999,998 die for each pair that survive to breed. This is what food chains are all about, and it's the system that started with eukaryotes, who made death a necessary part of the system of life. The Grim Sower provides life with a slightly counter-intuitive route to self-complication. Stop thinking about the 'waste' involved in making 39,999,998 surplus cod eggs. Biology is a reproductive system, and it can make that number as easily as it can make two. It isn't a nineteenth century factory owner who loses money for every reject. However it does have one feature in common with the factory owner: if the cost of maintenance is high, or if raw materials are difficult to come by, then it can no longer afford to be so sloppy. So to begin with we'll consider creatures that invest very little in either the production or the care of their individual offspring. Here the Grim Sower has a field day, because there is a huge advantage to making vast numbers of potential offspring and throwing most away. The advantage is that you can be selective about which you keep. So instead of investing in a high-quality production line that turns out perfect, polished items, you can make large amounts of trash, cheaply, and sift through them for the occasional accidental good one. Indeed you can produce a few high-quality items even if the 'technology' needed to make thousands of them reliably does not exist at all.

As simple eukaryote organisms began to live their lives more and more in the fast lane, it became advantageous to have sensory and computational systems, and nerve cells evolved. If you are a predator, and your prey is busy evolving a nervous system, then there's not much future for you unless you keep up. The same holds true with the roles interchanged, so an evolutionary 'arms race' soon developed, driving all of the organisms towards ever more sophisticated neural machinery. This arms race has so far got to the stage of producing brains – who knows what it will lead to next? – although some organisms, such as echinoderms (starfish and their kin) are very effective with no brains at all.

The brainiest animals on the planet – dolphins, gorillas, chimpanzees, humans – are all mammals. Mammals have had a chequered history, which may have encouraged them to develop very effective brains. The Cambrian explosion, preserved for posterity in the Burgess Shale, is not the only rapid burst of diversity in the fossil record. Another, pivotal in the evolution of mammalian brains, happened shortly after the dinosaurs died out – killed, the evidence now suggests, by a modest-sized meteorite which hit the Earth just off the Yucatan coast of Mexico. The dinosaurs were too specialised to cope with the climatic changes induced by this collision, which had the energy of several hundred Hiroshima-sized atomic bombs. Their departure, it is said, released numerous tiny furry creatures – the mammals – from their dinosaurian thrall. The mammals had existed alongside the dinosaurs for tens of millions of years, but they could occupy only niches suitable for tiny animals. One theory, due to Harry Jerison[♪], is that at first most mammals lived their lives in daylight hours, relying on sight as their main sense; but when the dinosaurs really got going, mammals were forced to become nocturnal, falling back on their less developed sense of hearing, and refining it. With the dinosaurs out of the way, the mammalian population found itself able to employ both senses extremely effectively, and that gave them such an evolutionary advantage that they 'radiated' – underwent a rapid split into innumerable new species. The mammals interacted with each other and with everything else, and the whole system thereby drove itself into ever more complex forms, until eventually minds appeared.

Many animals have brains, but few appear to have minds. A mind is a higher level of organisation altogether, bearing much the same relation to a brain as an antelope does to an amoeba. The main question that we try to answer in *Figments* is: where did mind come from? We contend that an important part of the answer is 'the Grim Sower'. Not only is He responsible for the transition from simple organisms to complex ones: He is also responsible for the transition from brains to minds. Here is an outline of the argument: we will pick it up again in Chapter 6.

The transition from brains to minds can be traced back to the time when animals came up with non-genetic routes to protect their offspring. This is a useful trick for more complex animals because they are rather more like the nineteenth century factory owner, and cannot 'afford' the attentions of the Grim Sower to such an extreme extent as cod. Some animals, mammals being the best example for our purposes, evolved the trick of protecting their offspring, and so could produce fewer of them. Among the inventions that made this trick possible were the uterus, a controlled environment for the growing embryo, and milk, a way to provide the young animal with instant food. The young of such creatures were 'privileged' – given special care at the parents' expense. Because the young developed under the care of the mother, it was possible for more complex developmental routes to remain reliable – mother in effect providing the control devices needed to keep baby's development on track. Mother was able to do this reliably because *her* mother had done it for her, and so on. (If you're wondering how it all got started, you haven't absorbed the lesson of chicken and egg: the idea you're missing is 'scaffolding'.)

By a whole series of gradual becomings, which are now moderately well understood, parental privilege led to a whole new kind of intelligence, involving the new tricks of *learning* and *teaching*. Parents became part of the behavioural context of their offspring. Other intelligent animals do not do this; for example octopuses and mantis shrimps have also developed very versatile brains – intelligence – but only by virtue of internal circuitry that learns from their environments. In contrast, mammals and some birds make the parents part of the context for intelligence. Rats and wolves, cats and dolphins are born into a parental environment in which they pick up new tricks 'in the nest' – new alarm cries, new ways of catching fish, new ways to cadge food from people.

And that is where mind comes into the picture. It is from this kind of *cultural* transmission of special forms of behaviour, we shall argue, that the specifically human mind has evolved. Mind is not just a matter of sophisticated brain structure; it is something that arose through the cultural trick of passing on behaviour through teaching and learning. The *contextual* element is crucial: mind cannot arise in isolation.

Just as life did not appear fully formed, but gradually *became* from non-life, so our minds arose gradually among organisms that came to transmit behaviours more and more via learning rather than in the form of instantly accessible hereditary programmes. The main thrust of *Figments* is to show that the apparently unique features of human imagination, human creativity, and human morality are 'emergent' features of complex processes such as evolution and culture: they have *not* been developed gradually from small beginnings and many important features cannot sensibly be traced back to precursors. The crucial trick has been our development of specific cultural 'Make-a-Human Kits', which arrange that our juveniles pass through a succession of experiences which result in the kind of adult that supports the society that produces that kind of human being.

It is, of course, another recursive process.

There will be more to discuss towards the end of *Figments*, in particular the future of our increasingly multicultural world. But we can't give away the *entire* plot right at the start. We hope that the aims of this introductory chapter have now been attained – to give you some 'deep background', to raise the main themes that we shall develop in the rest of the book, and to provide you with a 'road-map' of where we are going. However, we have no intention of revealing right now everything that we shall encounter along the way.

Only one task remains for this first chapter, and that is to 'bring on the aliens'. In *The Collapse of Chaos* (henceforth abbreviated to *Collapse)* we found that our imaginations – and, we hope, yours – could be liberated by having a standard mechanism for indulging in wild, unbridled speculation without having constantly to remind you that 'this bit isn't scientifically factual, it's here to provoke imaginative lateral thinking'. We did this by the literary device of dramatic interludes, written in script form, whose central characters were weird alien beings from the planet Zarathustra, resembling fluffy yellow ostriches but with much stranger habits. You should be warned that the Zarathustrans have a cultural obsession with the number eight, as betrayed by their conversation, which is liberally spattered with the syllable 'oct'. For example, they are perennial octimists – except for a few septimist heretics. They also like/dislike (delete whichever is inapplicable) multiple choice questions and responses. And they are generally found in association with 'The Regulations', a large purple sluglike composite hive-bureaucrat.

The aliens are back.

The cruise-vessel *Watcher-of-Moons*, crewed by a standard octuplet of Zarathustrans, is passing through an obscure but ordinary spiral arm of their obscure but ordinary galaxy. The journey has hitherto been uneventful – or as uneventful as any journey made by an octuplet of Zarathustrans can be – but now the vessel's sensors alert the crew to an unexpected discovery.

For simplicity we shall refer to the aliens by their *roles* in the octuplet, and not by their names. This is probably a good thing since their names are typically something like 'Nifflepuffltrentleboffle-Pufflex Third Remove'. We abbreviate the roles once they have been introduced. The octuplet's Destroyer-of-facts reports the surprising discovery to the Master-of-rings. 'Destroyer-of-facts' is Zarathustran for 'scientist': the phrase carries no negative overtones.

Destroyer-of-facts [*Politely.*] We have located an oxycarbon planet, Ringmaster. [*This is the disjunctive/informal mode for 'master-of-rings', used among members of the same octuplet when extraoctuplectual beings are not explicitly present.*]

Ringmaster Wonderful. I have not tasted fresh bacterial soup for *ages*! Instruct our Hewer-of-wood to send out a scoop–

DoF The ecology has advanced a little beyond the primal soup stage, Ringmaster. It has reached at least the herbivore-carnivore weapons-race. I have observed mobile plant-predators in large numbers. There are also carnivorous metapredators preying on the herbivores, which hunt either singly or in small packs.

R All of them octually reproducing, of course.

DoF No, they are binary.

R Binary? How dazzlingly unoctimistic! Surely Aristoctle demonstrated the logical superiority of octual reproduction long before the– oh, never mind, the ecology will be contaminated beyond repair anyway. I shall have to wait until our period of public service is complete, and then go on a bacteria-binge on one of the isolation farms–

Liar-to-children [*Rushing in excitedly. 'Liar-to-children', a highly respected role, is analogous to our 'teacher'.*] Ringmaster, there are radio signals emanating from the planet!

R But any civilisation advanced enough to produce radio waves would have learned to contain them for security reasons.

DoF No doubt, Ringmaster – except that this one has not.

R That is remarkable / ridiculous (delete whichever is inapplicable). I hope that this will not prove to be a wild groose chase. [*The groose chase is an old Zarathustran ritual in which children learn self-control by stalking – very slowly and quietly – the utterly inoffensive groose. The exercise is deemed a failure if the groose notices that it is being stalked. In order to be successful, a groose chase must not be carried out wildly.*]

DoF Look at this reconstruction. [*A holovisual image projector displays a short segment of badly distorted, flickering forms. A curious, two-limbed creature with loose, flappy multicoloured skin and large clumpy hooves adorned with knotted tendrils stands in front of a flat screen upon which various symbols are placed. The ship's translator – an electronic device found on all Zarathustran vessels – has by now learned the language. It announces that the creatures call themselves 'humans' and their planet is 'Earth'.*]

DoF [*Hesitates.*] I am not quite sure how to say this, but from the linguistic structure it would appear that these 'humans' do not share their minds octually.

R What? Dualminds? But that is almost unheard of.

DoF No, singleminds.

R [*Shocked.*] That *is* unheard of.

Translator 'And that's the weather for today, 24th September 2359. Rain with intermittent showers in the North, showers with intermittent rain in the South-East. Heavy thunderstorms coming in from the West. This is Trent Severn on the Sunshine Channel: to all you lovely people out there – good night!' [*The screen changes to an incomprehensible series of images which Destroyer-of-facts tentatively identifies as some kind of statement about the virtues of an aromatic compound for suppressing pheromones, called 'Banish-for-Men'. The screen changes to a series of more recognisable images, culminating in a crude but vivid representation of the planet itself against a background of fake stars, while a string of angular alien symbols scrolls across it:* FIGMENTS OF REALITY, PART 17 OF 932.]

R [*Turns to their Liar-to-children.*] What is this?

LtC A continuing educational narrative of some kind, Ringmaster. Based upon a revered / reviled (delete whichever is inapplicable) ancient text. [*Watches the screen and interprets the tale that unfolds – a long and dramatic story of an exploding universe, elements born in stars, complex carbon-based molecular machines, a doubly-helical genetic molecule, the origins of life, evolution, sense organs, brains, minds, and intelligence.*]

R What a fascinating narrative.

LtC And such a convincing story.

DoF Such vigour and power! Such unified scientific insight!

R Not a word out of place, no loose ends – amazing!

[*They all look at each other, then simultaneously say:*]

ALL [*In unison.*] Must be wrong, then.

R [*Interpreting this joint statement so that his fellows can understand it.*] Any true story has loose ends all over the place.

DoF Not so much loose *ends*, Ringmaster. Loose *surroundings*. The contextual element is missing. There is no examination of alternative scenarios, merely a bare linear story-line. Each step makes logical sense, but in no case are the surrounding possibilities explored. And did you notice the biggest flaw, right at the end?

R I did. Of course. Er– explain it for the benefit of the others.

DoF It was a wonderful / woefully deficient (delete whichever is inapplicable) explanation of the origins of intelligence.

R It certainly was ... [*His voice tails off.*]

DoF *Human* intelligence, of course. [*Affects a significant posture.*] These aliens acknowledge no other. [*Gestures to emphasise the point.*] Intelligence among these *binary, singleminded* humans.

R [*As ever, slow on the uptake, but finally gets there.*] Ah, I see. Always did, of course, now that you have explained it to me so graphically. I shall inform Liar-to-children, so that he will have understood it too. [*Turns.*] Liar: the linear logic of the aliens leads to an explanation of the origins of intelligence in a two-sex species, each member having its own isolated mind.

LtC No mention of octuality?

R None.

LtC No group intelligence?

R None.

LtC Then the humans' story is incomplete. It cannot explain *us*.

R Correct. These humans seem not to be aware of the immense space-of-the-possible that surrounds every instance-of-the-actual. They are singleminded not only in mentality, but in attention.

LtC How can that be? Since their every act is *surrounded* by a space-of-the-possible, why can they not sense its presence?

DoF I believe their situation to be analogous to that of a gravid wompus bug that is so strongly focused upon its discovery of the perfect spawning-root that it fails to observe the covert approach of a hairy snuffler–

LtC And gets snuffled instead of propagated?

DoF Precisely. Ringmaster? I defer to your wisdom for a summation, to be graffited on to The Regulations.

R [*Thinks for a moment, and then complies.*] Humans do not detect that they are surrounded by a space-of-the-possible because they are not looking in that direction.

2 The Reductionist Nightmare

According to the opening paragraph of Stephen Hawking's *A Brief History of Time*, a famous scientist – possibly Bertrand Russell – was giving a public lecture on astronomy. He described the structure of the solar system and its place in the galaxy. At the end of the talk, a little old lady at the back stood up and complained that the lecture was utter rubbish. The world, she pointed out, was a flat disc riding on the back of four elephants, which in turn rode on the back of a turtle.

'But what supports the turtle?' the scientist objected, with a superior smile.

'You're very clever young man,' said the woman, 'but you can't fool me. It's turtles all the way down!'

(Actually Hawking tells the story with 'tortoise' where we have put 'turtle', and unaccountably omits the elephants. We have rewritten the story slightly in order to pay proper deference to Great A'Tuin – whom, of course, you recognise as the turtle who supports Discworld in the fantasy series by Terry Pratchett.)

To many people, science is seen as a source of certainty, a box full of answers that can be trotted out when dealing with life's many questions. Most working scientists, however, see their subject in a very different light: as a method for navigating effectively in an uncertain world. Whatever science may be, it is not just a matter of assembling 'the facts'. The outside world seldom presents us with unequivocal facts; instead it provides a variety of indicators whose interpretation is usually open to debate. Is the world warming up as the result of human activity? Is the human sperm count in decline? Do emissions from cars cause asthma attacks? Is mad cow disease transmissible to humans? In a world of facts, all of these questions would have simple yes/no answers. In the real world, in which facts are replaced by interpretations of evidence, they are sources of genuine controversy, in which conflicting points of view can all point to reasonable observational support.

What science really offers is not facts, but understanding; not answers, but contingency plans. Everything we think that we know about the world is based upon assumptions – that our senses do not deceive us, that equipment works the same way wherever it is placed, that patterns of behaviour established in a limited number of experiments hold good in general, and so on. Science studies the consequences of those assumptions, tests them, and discards any that are found wanting. Its aim is to devise coherent points of view that can be used to understand the workings of the world around us.

In this chapter we take a look at scientific inference, explanation, and understanding. We discuss the nature of scientific theories, and set up some mental images and concepts that we need in order to get *Figments* off the ground. In particular we will distinguish two ways to think about natural phenomena. One, known as reductionism, explains *how* a system functions by finding out what its components are and how those components fit together. The alternative, which for want of a better word we shall call 'contextualism', explains *why* a system functions in terms of the circumstances in which it operates or has come into being. Reductionism looks at the 'insides' of things, contextualism looks at the 'outsides' – in a conceptual, not a literal, sense.

In discussing such questions we shall be happy to reinvent philosophical wheels, in the hope that they have not been punctured irrevocably long ago. That is, we shall not give chapter and verse for who first came up with what idea – or something vaguely like that idea that might be seen as containing its germs – and we have not rummaged through the literature to look for devastating counter-arguments. Instead, we shall 'wing it', on the grounds that the lines of thought are in any case more interesting than any particular conclusions. To continue with the 'wheels' metaphor, however, we would argue that wheels often have to be reinvented. Michelin and Dunlop make much of their money that way. You see, the old wheels were cart- and carriage-wheels, inappropriate to today's empirical roads: we need new wheels, suitable for modern technology and science.

The idea of embedding a system in its surroundings – thereby studying not just what it does, but what it might have done in slightly different circumstances – opens up completely new ways to think about the universe. It can be formalised in terms of what mathematicians call 'phase spaces' – in *Collapse* we called them 'spaces of the possible', but here we shall revert to the more standard term. We will make considerable use of phase space imagery throughout *Figments*, and here we introduce it in terms of strategies for games – a key image for much of what comes later, especially evolution.

We start with some thoughts about the nature of scientific explanations. The conventional structure attributed to science is that of a body of knowledge which evolves by a process of theory and experiment. Experiments test potential weakness in theories, and theories suggest experiments to be performed. A theory that does not fit observations must be rejected, or at least modified – unless it can be shown that the observations are incorrect or have been misinterpreted, in which case an improved experiment is in order. In this view, theory and experiment seem to be on much the same footing; indeed in universities throughout the world you will find departments of theoretical physics alongside departments of experimental physics, and the two are treated by the authorities in much the same way. However, this symmetry is illusory. The role of experiments is clear and simple: it is to ensure that people do not believe things to be true merely because they want them to be. Experiments provide a 'reality check' that helps us avoid one of the great pitfalls of the human psyche: the belief that we know the answer before we start working on the problem. The role of theory is more subtle, and it is here that the conventional view of science omits a crucial feature. Everyone accepts that theories provide explanations of observed 'facts' by fitting them into a coherent conceptual framework. The true role of theories, however, lies rather deeper: without a theoretical framework, the meaning of observations may not be clear. For example, suppose you look outside your window and notice that the branches of the trees are swaying. Without a conceptual framework, the motion of the branches tells you very little; but if you know about the existence of wind, and you know that it imparts a force to any object that it encounters, and if you know that tree branches sway when they are pushed by a force ... well, then your observation acquires meaning. You could use it to estimate the speed of the wind or the elasticity of the branch. Not only that: your observation tells you something about the *air*, not just the tree – so a conceptual framework can make a visual observation indicate properties of an invisible substance.

More strongly, it is often impossible to *make* observations without relying upon a conceptual framework. Until the advent of modern electronics, the way to measure the strength of an electric current was to employ an instrument containing a magnet. The electric current produced a force which acted on the magnet, giving it a push: this motion was transmitted to a pointer on a dial. The stronger the current, the bigger the push, and the further round the dial the pointer moved. But without a theory linking electricity to magnetism, this observation would be solely about pointers on dials: it could not bear any relationship to an electric current. (*With* modern electronics the measurement is now made in other ways – but the role of theory is if anything more crucial, since without it we would have no idea of what the electronic circuits were doing, or how to build them in the first place.) Scientific 'facts' are thus context-dependent: many of them are inferred from an accepted scientific viewpoint, or 'paradigm', rather than being observed directly.

The realisation that most scientific observations are context-dependent has led some philosophers♪ to argue that science is a social construct which has nothing to do with reality and is solely a matter of human convention. This argument stems from the entirely sensible modern perception that scientific 'truth' is not absolute, but depends upon having some agreed common conceptual framework. However, the belief that science is *solely* a construct, which by implication could be whatever scientists decided to agree upon, is really very silly – however elegantly it may be phrased – because it ignores a very important aspect of these conceptual frameworks. They are not arbitrary: they are the outcome of a previous scientific process. For example, scientists cannot make objects float skywards merely by agreeing among themselves that the force of gravity acts up rather than down.

There has to be a reality check.

Science has more stringent reality checks than any other area of human activity, and applies them more frequently. Religion hinges upon faith, politics hinges upon who can tell the most convincing lies or maybe just shout the loudest, but science hinges upon whether its conclusions resemble what actually happens. Not so long ago we might have said 'whether its conclusions are true', but the idea of science as *absolute* truth has gone the way of the dodo. Because human beings experience reality indirectly, through the medium of their senses, there is room for genuine and reasonable disagreement about the nature of reality.

Even a reality check must have a contextual element.

At any rate, the role of science is not to establish some kind of factual data-bank about nature, but to help us *understand* nature. There are many different degrees of understanding, and many different *kinds* of understanding. Some aspect of nature may be not understood, or slightly understood, or fairly well understood – although it is unlikely that it will be fully understood. More subtly, it may be well understood from one point of view but an impenetrable mystery from another. Thus we may understand that a flower petal is red because it contains a particular pigment, but fail to understand why the flower uses that pigment and not a different one.

Let's sort out a few ground rules about 'understanding'. Understanding something is quite different from describing it in full detail. You *can't* describe nature in full detail anyway, it's too complicated; but in any case the aim of understanding is to capture certain features of the thing being explained and to illuminate *those*. For example in Chapter 1 we reminded you that old textbooks describe the formation of water from oxygen and hydrogen as a single chemical reaction, but newer textbooks admit that at least ten other molecules are involved. Each of these descriptions provides a different level of understanding. The old textbooks write down a chemical equation, $2H_2 + O_2 \rightarrow 2H_2O$. This equation tells a story: two molecules of hydrogen and one of oxygen get together, swap atoms and move bonds around, and produce two molecules of water. This story is not *true*: it is a teaching myth, a lie-to-children in the Zarathustran sense. It teaches us the ingredients and the relevant proportions, but it creates the wrong impression about the process that puts them together. The ten-reaction network of the modern textbooks explains that process much more accurately – but it is *also* a lie-to-children. In fact it makes the same mistake of conflating many different reaction routes to create a single idealised network.

So what 'really' happens? Any description suitable for human minds to grasp must be *some* type of lie-to-children – real reality is always much too complicated for our limited minds. But we can get an idea of how the reality differs from simple equations and reaction networks by means of an extended analogy. Imagine an eighteenth century ballroom filled with dancers. The ladies (hydrogen atoms) have only one free hand (bond): the other holds a fan. The gentlemen (oxygen atoms) have both hands free. Molecules of hydrogen consist of two ladies, dancing hand in hand. Molecules of oxygen similarly consist of two gentlemen with both hands clasped – no doubt whisking each other round energetically in one of those robust, manly, all-good-chums-together country dances that break chairs and scare the ladies, you understand. The simple textbook equation makes the whole reaction appear to be a choreographed ritual. Two pairs of ladies and one pair of gentlemen come together. They circle each other, stop, bow, release hands, and rejoin them differently: each gentleman steps onwards with a lady on each arm. You can imagine long lines of them, in strict formation across the room, all doing the same thing at the same time. The experts' multi-reaction network image is similar, but now much larger choreographed groups come together, and they go through an intricate chain of steps and formations – all the gentlemen in a ring with half the ladies inside and half outside, ladies twirling in pairs along an archway formed by pairs of gentlemen ... and every so often a gentleman leaves the group with a lady on each arm, until everybody has dispersed.

Those are the lies-to-children: choreographed archetypes, one simple, one less so. But the reality isn't choreographed at all. Instead, in various places throughout the packed ballroom, fights break out – between gentlemen, between ladies, between both. In the general pushing and shoving, nearby pairs are disrupted and people grab hold of each other to stay upright. Struggling waves of people surge to and fro across the ballroom, bouncing off the walls. Some ladies drop their fans for a few moments, others pick them up again. In the odd corner there may be a pocket of tranquility, a small bunch of ladies sitting hand in hand – but everywhere else the battle rages. Temporary groups form, break up, reform, collide with each other. Some people's hands get stuck to the wall or the floor. But slowly, stable combinations of one gentleman and two ladies (each with a fan) assemble here and there, and the fighting dies down as more and more such triples form. Even then, the triples continue to bounce off each other (steam is a gas); and as the people run out of energy and the ballroom settles down, triples still break apart and recombine (liquid water has a highly complex fluctuating molecular structure). Nevertheless, the dominant pattern is now one gentleman and two ladies (H_2O). It's not so much a ballroom dance as a bar-room brawl.

Understanding, then, comes at many levels. Our level and type of understanding can be gauged by the extent to which we can answer questions about nature, and the type of question that we can answer. There are (at least) three different types of question, and each requires a different type of answer. The simplest questions are 'how' questions: *how* does the speed of a falling rock vary with height, *how* does haemoglobin capture and release oxygen molecules, *how* do geese string themselves out into loose V-shaped skeins as they wing their way across the sky? The next simplest are 'why' questions: *why* do rocks fall, *why* does the body use haemoglobin, *why* do geese form flocks? The third, and the most difficult, type also comprises 'why' questions, but posed on a different level – the 'philosophical why'. *Why* is our universe so richly structured that it contains falling rocks / reacting molecules / geese?

The first type of question is usually answered by a description of the internal workings of the system under consideration, an approach known as reductionism. 'The law of falling bodies is ...', 'Here is a three-dimensional animation of a haemoglobin molecule capturing an oxygen molecule ...', 'Here is a computer simulation which shows that if geese obey *these* simple rules then ...' We may ask supplementary questions requesting further details, but the answer itself is concrete, testable, and self-contained.

Answers to 'why' questions are much more open-ended, and one thing they almost always have in common is context. By assuming a particular context, such answers describe some chain of logic that explains why this particular structure or process is to be expected. 'Rocks fall because they are attracted by the Earth's gravitational field ...', 'Animals use haemoglobin because they need to carry oxygen round their bodies to various organs, notably the brain ...', 'Geese form flocks because natural selection has favoured that strategy as a protection against predators ...' Because a given system can have many contexts, the answers to 'why' questions are non-unique and open-ended. The flocking of geese might be explained by social behaviour instead of evolution, by features of their visual system, by learning processes in their brains, or by the laws of aerodynamics. Such explanations complement each other: there is no reason to suppose that only one of them can be 'true'.

What about the third type of question? These are questions along the lines 'why do we live in a universe so rich that ...?', and we shall call them 'Deep Thought' questions, in honour of Douglas Adams' fictional supercomputer in *The Hitch Hiker's Guide to the Galaxy*. Deep Thought questions are pretty much unanswerable. Indeed it is difficult to imagine what form an answer can take, other than 'it's just like that'. Many people favour the answer 'God', which in Deep Thought questions is just a culturally-loaded way of saying 'it's just like that', and in 'how' and 'why' questions is simply an unsatisfactory cop-out. An alternative is the 'anthropic principle' – 'If it wasn't that sort of universe then it wouldn't contain creatures intelligent enough to ask that kind of question'. Some people seem to think this is a Deep Answer to a Deep Thought, but we think it is just *another* elaborate way to say 'it's just like that' – only this time accompanied by supporting evidence that it *is* like that. The anthropic principle still begs the deeper question 'Yeah, I agree the universe must be rich enough to contain wonderful things like me, but *how come* it's rich enough?' And here the anthropic principle offers no answers, just as religions stand mute when confronted by '*Why* did God create it that way?' We (JC & IS) can do no better. We will do our best to answer pertinent 'how' and 'why' questions, but we can answer Deep Thought questions only by our own version of 'it's just like that'.

There really seems to be no other answer.

There is actually a fourth class of question, which at first sight looks like Deep Thought but on closer examination involves neither depth nor thinking. An example is 'What happened before time began in the Big Bang?' If time has not begun then there can be no sensible meaning for 'before', so this question is merely nonsense. Its context is not deep, but self-contradictory. Pointing this contradiction out is not a mere verbal quibble (we recall the physicist and science writer Paul Davies responding with some heat to the suggestion that it is). It is a *logical* quibble, and a fatal one. As noted in the Prologue, you might as well ask what lies North of the North Pole – or what part of Britain lies five hundred miles from the nearest coast. What is rather odd about this particular question (the origin of time) is that most people seem perfectly happy with 'it's always been like that', finding no difficulty in conceiving of a universe that goes back forever.

Why shouldn't they?

Well ... We opened this chapter with the story of the old lady who argued that the only way to support the Earth would be 'turtles all the way down'. Nearly everybody finds an infinite pile of turtles highly incongruous – as an explanation it's a non-starter. So why are we so happy with an infinite pile of causality: today's universe riding on the back of yesterday's, which rides in turn on the day before's?

It's universes all the way back!

Mathematicians, whose understanding of the slippery nature of the infinite is more fully developed, feel that a universe that has always existed needs just as much explaining as one that suddenly pops into being from nowhere (and no-time). An entirely reasonable mathematician's question about a system that goes back forever is: 'Yes, but where did it *all* come from?'. Infinity is not an impassable barrier, it's just a name that sums up a way of thinking about recursive processes. At any rate, the Big Bang is no more inexplicable than the 'steady state' ever-present universe; it's just that we tend not to ask awkward questions about things that have always been around. Another question of the same kind is 'what use is half an eye?', usually produced triumphantly as a refutation of evolutionary theory. But evolution never produced half an eye: what it did was produce structures somewhere between a flat light-sensitive surface and a complete eye, whose light-detecting ability was half that of a present-day eye. Such questions are not how, why, or Deep Thought. They are Silly Questions. Their silliness is not always immediately apparent, but once it has been exposed, there is no need to answer them at all.

Another way of asking Silly Questions is not so obvious. It goes 'Ah, yes, you *say* you don't believe in ghosts / UFOs / the Loch Ness Monster ... but what if you *saw* one?' Usually said in a tone of triumph, as if the statement is significant. It's not. The fact that your questioner can imagine something to be possible does not imply that it actually *is* possible, so its hypothetical existence is just that – hypothetical. It is not necessary to answer hypothetical questions, though it might be honest to say 'of course, if I did see one then I'd have to change my mind, but right now I don't believe they exist so I don't think *anybody* will ever see one, let alone me.'

So much for questions: what about answers? The philosophy of reductionism focuses on answering 'how' questions. It finds its ultimate expression in fundamental particle physics, whose practitioners are currently seeking a 'Theory of Everything' that will unite all the phenomena of the physical world into a single deep, fundamental equation that could be worn on a T-shirt. It might, for instance, take the form of the Quantum Wave Function of the Universe – a description in the terms employed by modern physics of everything there is. There are valid intellectual reasons for seeking such a unification – but they are emphatically *not* that it will explain everything, which is what many of these practitioners seem to think, or at least how their rhetoric goes when they are trying to obtain funding. The reason is simple: the chain of logic that leads from the equation on your T-shirt to the T-shirt itself must be so intricate, so incredibly long, that no human mind could encompass it. To work out how your T-shirt hangs on your body, starting from the Theory of Everything, requires an atom-by-atom description of you and your shirt and a massive computation of the interatomic forces before the chain of logic can even get started ... In order to understand where the cotton thread in your T-shirt came from you must work your way up from the Theory of Everything to principles of molecular structure and the entire DNA sequence of cotton, and use these to compute the effect of water molecules and sunlight on the growing shoot ... We recognise that the reductionist approach is more subtle than this, but we suspect that devotees of a Theory of Everything haven't quite noticed the consequences of that subtlety for their programme.

We will find it useful, as we proceed, to contrast two equally reductionist points of view: 'bottom-up' and 'top-down'. The bottom-up approach can be visualised as the 'Tree of Everything', a conceptual map of the world (figure 4) in which arrows of explanation twirl upwards through funnels that represent logical chains of cause and effect. The base of the Tree is the hoped-for Theory of Everything, and the explanatory chain works its way up from that foundation, step by step, until it ultimately gets to DNA, cotton plants, T-shirts, people, planets, galaxies, and the universe as a whole. This viewpoint is more intelligent than naive reductionism: it adds a hierarchical structure, in which each layer depends upon general principles deduced from the layers immediately below.

For example, given the sequence for cotton DNA the user of the Tree of Everything does not go right back down to the subatomic level to apply the T-shirt formula: instead he or she relies on general principles about the development of organisms. Richard Dawkins calls this way of thinking 'hierarchical reductionism', and it captures in a very sensible manner the actual structure of most scientific explanations. (*Except*: there are often huge gaps between neighbouring stages of the hierarchy, which are dealt with by talking very rapidly about something else. For example we have very little knowledge of how an organism's DNA sequence affects its development, beyond a few ideas about how protein structures are specified. How does the protein get to the right place at the right time? We talk a blue streak about 'regulatory genes' but most of it's waffle. We cannot even derive the structure of such a simple form of organised matter as a crystal from the quantum wave functions of its atoms.) Even if these gaps can be filled,

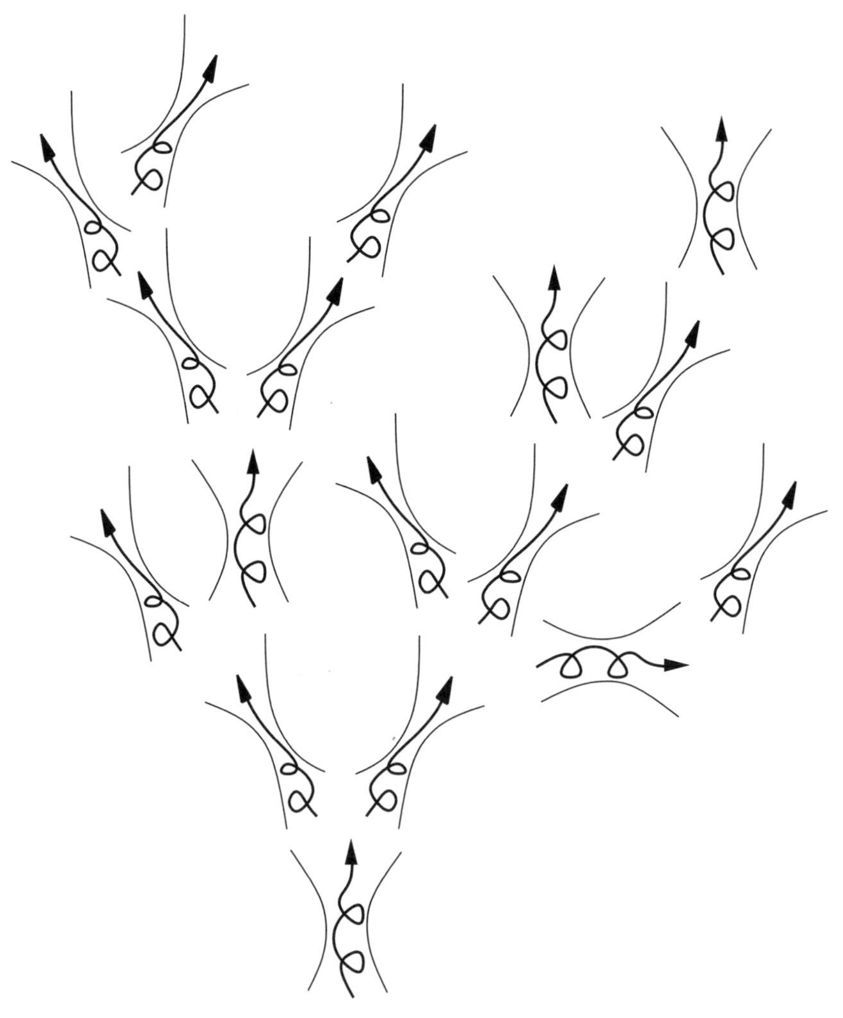

Figure 4 *Is there a single Theory of Everything at the bottom of every explanation?*

hierarchical reductionism differs enormously from the rhetoric of a Theory of Everything – because once you set up such a hierarchy, for most practical purposes you can throw the lower layers away. Given the sequence for cotton DNA, what do you do? As we've just stated, you do *not* apply the Theory of Everything. You apply standard developmental biology.

So why do we need the T-shirt formula?

Yes, it's intellectually satisfying. Yes, it may well give us a much better understanding of subatomic physics. No, it doesn't explain everything. In fact, it hardly *explains* anything. Though it does give us a nice, comfortable feeling that the universe rests on firm philosophical foundations rather than an infinite pile of turtles.

The top-down version of reductionism starts with large-scale phenomena and 'reduces' them to simpler ones. It takes the point of view that the T-shirt is made from cotton fibres on a loom, and the fundamental formula is inked on it using certain chemicals. In order to explain the structure of the shirt by reductionist methods we must analyse cotton – its growth patterns, susceptibility to various diseases and parasites, which strains produce the most suitable cotton ... We must analyse the structure of looms – their mechanical construction, how to operate them, how to deal with broken threads ... We must analyse the chemistry of ink and esoteric techniques for printing on fabric ... The interesting thing is that the industry that manufactures T-shirts, and those industries that serve it, actually do depend upon such knowledge. As they become more technological, they may even penetrate down another layer or two, into the biochemistry of the cotton plant and the proteins and enzymes that it employs to go about its business ... Soon they may wish to use information about the DNA sequence of the plant – indeed genetic engineering companies are no doubt already doing so. Notice that the T-shirt gets manufactured entirely effectively without *using* the formula printed on it: the aim of the game is to stay as near to the *top* of the explanatory tree as possible, delving into the depths only when there is a need to.

However, all too often the pursuit of ever-branching trees of logical implications, if carried out with too much uncritical enthusiasm, leads not to understanding, but to a tangle of ever finer detail – which in *Collapse* we called the 'reductionist nightmare'. Its branches ramify indefinitely (figure 5) as it seeks in vain to connect the everyday world to the Tree of Everything that allegedly underpins all understanding. In principle such a connection may well exist – but there is no hope of ever mapping it out completely. That is certainly an obstacle to human comprehension, but nobody ever said the universe had to be comprehensible to humans.

These two versions of reductionism, bottom-up and top-down, are not competitors. Clearly they complement each other. But to what extent does the reductionist vision of science, in either version, correspond to how the universe actually operates?

To tackle this question, stop reading and listen for a moment. What do you hear? Maybe the wind in the treetops, or birds singing, or vehicles roaring past along a nearby road, or the kids playing pop records. Such a simple action on your part, carried out without conscious thought. You *know* it's wind, or birds, or a truck, or a techno remix of sixties standards, without having to stop and think about it. Now think about how complex it all is when considered on a reductionist level. A great tit pronounces its simple, unimaginative, repetitive song – a series of more or less identical *tweetles* traditionally represented as 'teacher, teacher'. It has taken four thousand million years of evolution to produce that tit, and its song. The muscle movements required to produce that simple note are

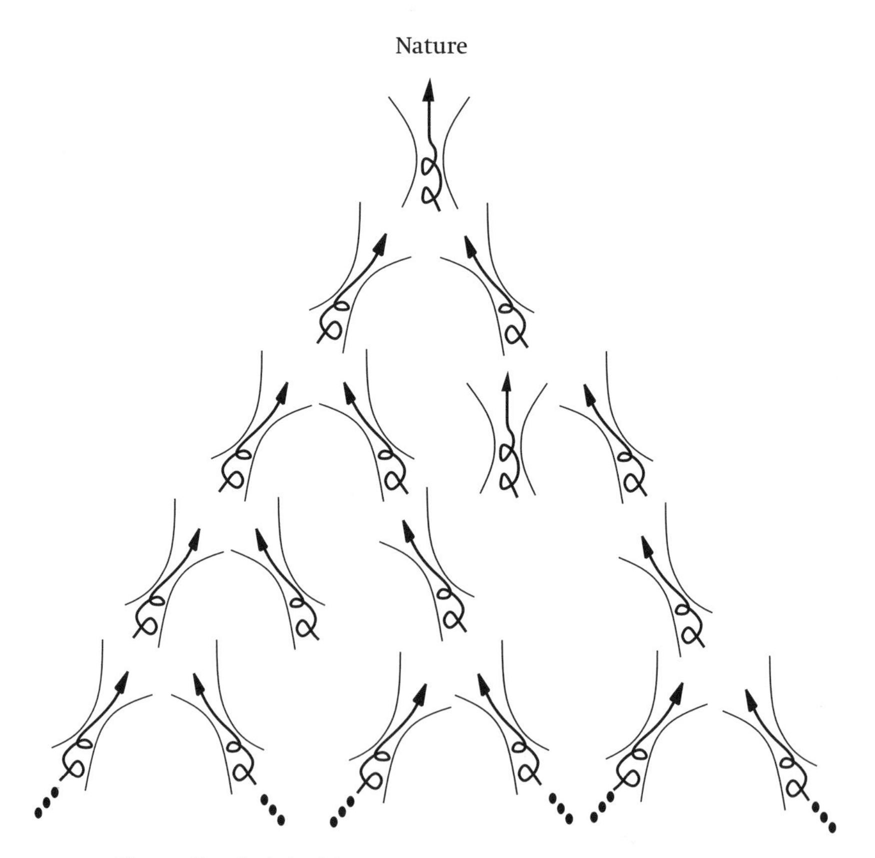

Figure 5 *The reductionist nightmare.*

more intricately choreographed than *Sleeping Beauty* – every tiny muscle fibre must contract at the right moment. A vast, unfathomable sea of molecules sloshes around inside the bird's body in unknown patterns to control those muscles. Electrical pulses flash along its neural pathways. Somewhere in its brain there is circuitry that stores the patterns of movement. Nobody yet knows exactly where, or exactly how – but even those details are starting to fall into place, thanks in particular to new methods for imaging the brain's activity. At any rate, all of that activity reaches its culmination:

Tweetle.

The sound radiates outwards, conveyed by vibrations in the air. The air exists because it condensed, along with nearly everything else on the planet, from primal dustclouds surrounding a nascent star. The mathematical equations for the movement of that dustcloud are simple in principle: a few symbols expressing how one particle of matter attracts any other. However, in order to capture reality *exactly* they have to be applied to ungraspably large numbers of particles, every atom in the solar system for a start (and what about the tiny effect of distant galaxies?).

The sound waves propagate through a medium that consists not of some nice indivisible fluid – as the 'wave equation' commonly used to analyse them assumes – but through an intricate, fizzing madhouse of gas molecules, charging around like children at a party, bouncing off each other, perpetually colliding with their neighbours, everybody pushing everybody else out of the way. Somehow the wave of sound navigates across this molecular bedlam to impinge upon your ear. Now it is funnelled deep into your middle ear, where it excites a tiny system of bones and membranes. Complicated sensory organs in your inner ear, which we describe in Chapter 6, turn the sound waves into electrical impulses that travel into your brain. There a network of nerve cells – or 'neural net' – which has also taken four thousand million years to evolve, analyses the babble of impulses into constituent sounds: the hum of distant traffic, the buzz of a nearby bee, the beating of your own heart. These neural computations exceed the capabilities of any of today's supercomputers, in complexity of circuitry, speed, and memory capacity. The algorithms employed – if in fact they *are* algorithms in any meaningful sense – exceed by an even greater amount the programming abilities of today's top computer scientists. And out of it all, in some manner – we have no clear idea how – our brain isolates one packet of impulses and labels it 'birdsong'.

Tweetle.

Now you hear the bird. Immediately other circuits in the brain flicker into activity, identifying the sound, recalling that after your last country walk you looked it up in the *Field Guide to British Birds*, and that it is the song of the great tit.

There is no serious prospect of filling in *every* detail in a reductionist description of how you, and the great tit, do all of the above. And everywhere you look there are things like molecules, soundwaves, and birds – all much too complicated even to begin to understand. What little we know about them involves sensitive and sophisticated processes that can be followed only in very simplified versions by experts. So instead of seeing simplicities down the reductionist funnels, we are trapped in the reductionist nightmare. This makes us feel that we live in an overwhelmingly complicated world, and the best we can hope to do is to try to understand very tiny bits of it, and to take advice from people who understand other tiny bits. As science progresses, the amount that we have to know in any such tiny field increases steadily, so the area that any individual can comprehend gets smaller. But, scientists are confident, deeper ... of course. It is an effective mechanism for generating scientific publications, but not for generating understanding. To some extent, these problems are unavoidable: like the universe at large, science is sometimes just 'like that'. But are we making rods for our own backs by adopting the methodology of the reductionist nightmare? After all, the great tit manages to produce its song without understanding the nightmare that reductionist analysis reveals (or is it 'creates'?), and we similarly pick out its repetitive note and identify the perpetrator with no more real understanding than the bird has. The world that we inhabit seems, to us, to be filled with simplicities.

We need a different approach, then; something that offers insights when reductionism offers only an impenetrable mess. One of the 'hot' research areas that offers a possible complement to reductionism is known as Complexity Theory, which models 'complex systems' in terms of more or less simple interactions between very large numbers of more or less simple 'agents'. It is the large number of agents that justifies the adjective 'complex'. Many real world systems have that kind of structure – people in a crowd, ants in an ants' nest, traders in the stockmarket, nerve connections in the brain.

What we *really* need, however, is Simplicity Theory, an effective and relatively painless way to extract the big simplicities from the underlying rules. Its beginnings already exist. Over millions of years of evolution, human beings have faced the identical problem of making their way in a highly complex and uncertain world, and they have done so by exploiting the great simplicities of nature, not the deep complexities. The human brain has evolved a non-reductionist approach to the world – 'quick-and-dirty', but very effective nevertheless. For instance, you knew exactly what we meant when we told you to put down the book and listen; and your reaction to what you heard was – to you – equally straightforward. 'There's a bird out there somewhere, singing,' is a simple enough thought. It doesn't take a mass of higher mathematics to hear the bird – just a functioning pair of ears. The *detailed* analysis of what those ears are doing, and what they are responding to, is far beyond the reach of any mathematical equations, but that didn't stop you hearing the bird.

The key here is not to ask what goes on inside the bird, or inside the sound waves, or inside the listener. It is to ask what goes on outside them. In what context are they operating, and what constraints does that context impose upon their behaviour? This line of thought leads to an alternative non-reductionist approach to explanation, which we flagged earlier as 'contextualism'. (There doesn't seem to be a standard name that really captures this style of explanation, so we invented this one. 'Holism' is an obvious possibility, and it comes close, but to our mind it is a little too closely associated with various kinds of wishy-washy thinking, such as 'alternative' medicine that often has no rational basis whatsoever, and uncritical tree-hugging mysticism.) Instead of looking (conceptually) inside things to work out how they tick, and then inside the bits to see how *they* tick, and so on, contextualism looks at their outsides. What external constraints moulded this thing as it developed, or are governing its behaviour now?

From the contextual point of view the explanation of the formula-bedecked T-shirt involves the fact that humans need to keep warm and have devised tricks like cloth to do so; that they have found farming to be an effective way to exploit the planet's biological resources and cotton happens to be a plant suitable for agriculture; that for cultural reasons they like to decorate their clothing with patterns and images; and that for reasons related to commercial advertising in American culture it is customary for T-shirts to carry humorous, cute, or clever messages – or, at least, messages perceived by the wearer to be humorous, cute, or clever. And what could be more clever than the Theory of Everything?

We are not suggesting that contextualism should replace reductionism. Reductionist science has been very successful, in its own sphere of influence. But there are other spheres, and it would be nice to have influence in them too. There is no inconsistency in holding both the reductionist and contextualist views at the same time. The bird *does* contract a vast number of muscles, it *did* take four thousand million years to evolve. But, equally, it is singing *because* that's what great tits do, and you put down the book to listen *because* you wanted to follow what we were saying and it seemed the best way to achieve that.

The most attractive feature of contextualism is that it places the emphasis squarely on *simplicity*, which is what the human mind needs to create understanding. So, whether or not the deep structure of the universe matches the reductionist rhetoric precisely, we are forced to conclude that (apparent) simplicities are commonplace in our complicated world. Can we understand why? It is difficult to steer a sensible path through this philosophical minefield; and all too often any attempt to discuss it degenerates into mysticism or cross-paradigm noise. For example, we could summarise our own position by saying that we don't believe the reductionist rhetoric, that we don't think the universe results from the blind pursuit of low-level laws by countless tiny particles. Indeed we don't think that, but it would be easy to misunderstand what we mean when we say so. 'Do you mean that when the molecules of air interact with the molecules in the bird's larynx, they don't obey the laws of physics? Are you saying that molecules behave in one way in the laboratory and in another way in a bird?' Leave aside the question of how accurate a model of molecular interactions the 'laws of physics' provide: are we attempting to deny that nature obeys laws, or that it obeys them in laboratories but not birds? No. Our argument heads off in a very different direction, as will unfold. It is that the reductionist rhetoric is not necessarily the only, or the best, way to understand the interactive dynamics of birds and soundwaves. While birdsong may indeed 'really' be a logical consequence of the molecular structure of birds and air molecules, nobody in their right mind would actually contemplate thinking about it on that level.

The distinction we are trying to draw here is not about 'what nature really does'. It is the distinction between a theory being true (in the sense that it seems to match nature sufficiently well – we doubt very much that absolute truth makes much sense) and its having explanatory value. Our brains cannot handle complexities *as such*, so in order to comprehend our world we extract simple features from one level of explanation and build the next level on top of them. Even if a valid Theory of Everything exists, science must develop that kind of hierarchically simplified vision of the universe, before anyone will be able to understand what that theory actually tells us.

There is a long way to go. Reductionist thinking has become very sophisticated over the centuries, and it has acquired a high degree of formalism – for example, the underlying 'laws' are typically stated as mathematical equations. At first sight there can be no such formal structure to contextual thinking: how can one formalise something as nebulous as a context? But over the last century science has been moving steadily towards just such a formal structure. The key idea was that the way to understand complicated systems is to embed them in a surrounding 'phase space'. For the next few pages we shall examine where this idea came from and show you what it is good for. It provides one of the central images for the rest of this book.

The phase space concept emerged from dynamics – the mathematical theory of systems that change over time. It was introduced by one of the most original and imaginative mathematicians the world has ever known, Henri Poincaré. In Poincaré's day dynamics was reductionist: if you wanted to understand how a system behaved, then you began by writing down its mathematical equations and solving them. Then you looked at the solutions and worked out what they actually did. Poincaré changed all that. He did so because all too often the programme falls apart: you can write down the equations, but you can't find the solutions. His search for an alternative was motivated by the three-body problem in celestial mechanics: given three masses attracting each other by Newtonian gravity, what do they do? For two bodies the answer is simple: the equations can be solved, and the solutions tell us that the bodies follow elliptical orbits. But for three bodies the equations *couldn't* be solved. Mathematicians grappled with them for well over a century before they realised that *there is no solution*.

How can you work out what the bodies are going to do, then, if you can't solve the equations? That was Poincaré's great insight, and it led to today's geometric approach to dynamical systems. It goes like this. The state of such a system is described by various 'state variables' which change over time. The changes obey a prescribed rule, or *dynamic*: 'if the current state is *this*, then the next state will be *that*.' Now, you can introduce a visual point of view by thinking of the state variables as coordinates for a point that moves around in a fictitious mathematical 'phase space'. As time passes, those coordinates change, so the representative point *moves*, tracing out a path. In fact the entire temporal development of the system boils down to a wiggling wormlike path through phase space. Phase space includes not just the *actual* values of the state variables, but all the *potential* values: it is a formalisation of the notion of context. By studying the geometry of phase space for the three-body problem, Poincaré recognised the existence of what is now called 'chaos' – dynamical behaviour so complex that it that appears to be random, but having entirely deterministic (non-random) causes.

Dynamics is where the phase space image originated, but for our purposes there is a simpler way to understand how it works, why it is needed, and what it can do for you. Our key to phase space will be the theory of games. Games will provide important metaphors for *Figments*, so we will spend the rest of this chapter setting up an appropriate image upon which we can build. The ingredients for a game, in a mathematical sense, are two competing players (or teams), a concept of 'win' or 'lose', and rules of conduct. The theoretical analysis of a game involves writing down the entire 'game tree', a diagrammatic representation in which possible positions in the game are linked by 'branches' corresponding to moves. We'll demonstrate the general principle using the game of 'Yucky Choccy'. Here the players are faced with a chocolate bar, a rectangle divided into an array of smaller squares. Each in turn is allowed to break off a lump of chocolate, which they must eat. The break must be a single straight line cutting right across the rectangle along the lines between the component squares. However, the square in one corner contains some nasty substance, and whoever is forced to eat this square loses the game. The game tree has a reasonably simple form, summarised for small bars of chocolate in figure 6. The black arrows show an actual game, and the grey arrows show all the other moves that could have been made instead.

The visual imagery here is vivid and useful – for example we shall shortly see that the loser made a bad mistake – but what *is* a game tree? It is *the phase space of the game*, the space of all possible states (positions). What is the dynamic on phase space? It is the rules that tell you how to get from one state to the next – and that's just the rules of the game. So we can intepret a game as a kind of dynamical system. However, there are some technical differences, the most important being that the rules of a game often permit several choices of move, only one of which can actually be made – whereas in a Poincaré type of dynamical system all of the 'moves' are completely determined by the initial state. This difference is irrelevant to the uses we shall make of phase space imagery, and will be ignored. In fact, for our purposes, games are a wonderful example of the phase space concept, precisely because they incorporate an element of choice, so that they correspond very closely to the kind of natural system that will interest us, such as an ecosystem or an evolutionary system. We will develop this 'game-theoretic' viewpoint as the book progresses; for now we content ourselves with some introductory examples to lay down the key concepts.

The concept upon which we wish to focus is that of a *winning strategy* – a way to force a win no matter what your opponent does. The idea of a strategy is on a 'higher' level than the rules of the game, and it brings the need for a phase space into sharp focus. Of necessity, the concept of a strategy involves the overall structure not just of one game, but of all possible games. Imagine yourself playing a game – say chess or draughts (checkers in the US). Most of the time you spend planning your next move goes into 'what if' questions. 'If I advance my pawn what could his queen do? If I don't protect my king soon, that rook is going to cause trouble.' Tactics and strategy centre around what moves you or your opponent could make; they do not depend solely upon the moves that they do make. So in playing a game you *must* be aware of more of phase space than just the part that actually gets used; and the same goes if you are watching a game, otherwise the moves don't make any sense.

There is an elegant theory of strategies in games that have a couple of nice mathematical properties – namely that they cannot continue forever, and eventually one player or the other must win. There is an analogous theory if drawn games can occur, as they do for example in chess, but for simplicity we

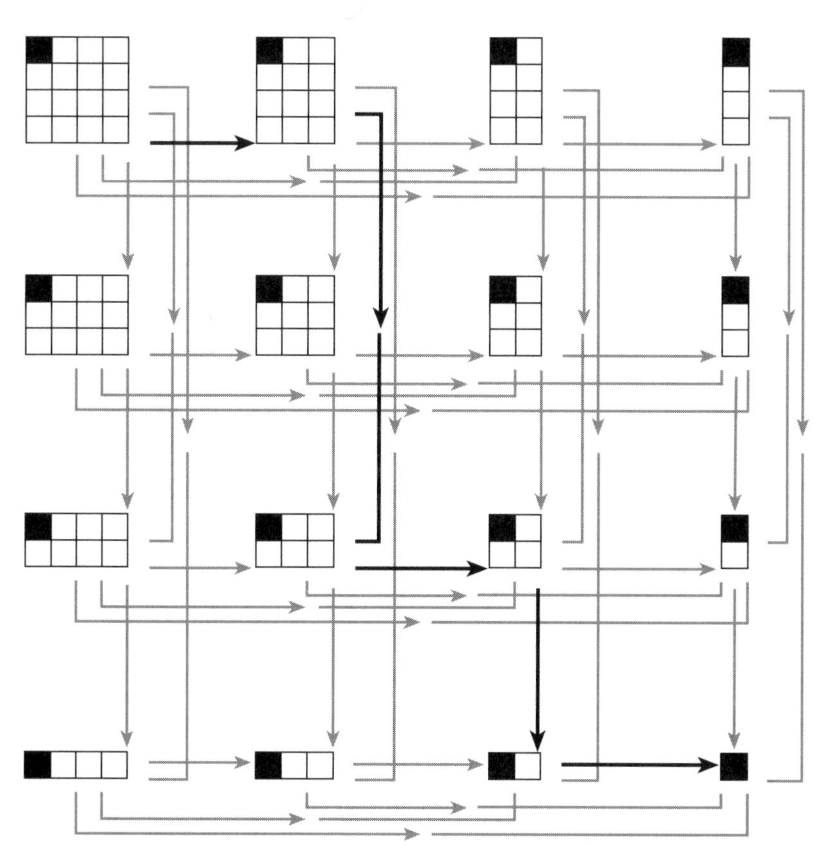

Figure 6 *Game tree for Yucky Choccy. This portion of the full game tree shows all bars up to size 4×4. Arrows indicate legal moves: the piece removed is eaten. The square shown in black is the nasty one. Black arrows indicate an actual game, grey arrows alternative moves that could have been made instead.*

will eliminate draws. (An easy way to do this is to add a new rule: drawn games don't count and you have to start again.) At first sight, this theory seems extremely powerful. For instance, in principle it is straightforward to decide whether a given position is one from which you can force a win (a winning position) or not (a losing position). All you need are two obvious principles:

1. A position is a winning one if you can make *some* move that places your opponent in a losing position.
2. A position is a losing one if *every* move that you can make places your opponent in a winning position.

The logic here may seem circular, for these two sentences merely affirm the same platitude in two equivalent ways. Actually that's not so: the logic is recursive, not circular, and those two sentences between them let you work out whether any given position is a winning one or a losing one. The extra feature that turns circular reasoning into recursive reasoning is that with recursive reasoning you have somewhere to start.

Let's see how it works on Yucky Choccy. We'll try to find a winning strategy by starting from the end and working backwards – 'pruning the game tree', as those in the business say. How? Well, we know that one piece ■ is a losing position. We can mark the chocolate bars in the game tree of figure 6 to show what we know so far, which isn't very much:

*	*	*	*
*	*	*	*
*	*	*	*
*	*	*	L

Here 'L' means 'losing position', * means 'don't know yet', and 'W' will mean 'winning position' once we've found some. That's not hard to do: in fact ■□ ■□□ ■□□□ are winning positions, because you can break off all the white squares in one move to leave your opponent with the losing position. Equivalently, there are arrows in the game tree that lead from those positions directly to ■, and by principle 1 all such positions are winners. For similar reasons the same positions rotated through a right angle, such as ■□, are also winners. So now we know this much about the status of the positions:

* * * W
* * * W
* * * W
W W W L

What about ? Well, the only moves you can make are  or 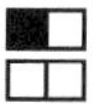 , and when you remove the all-white piece you leave a winning position for your opponent. So principle 2 tells us that is a loser, and we now have

* * * W
* * * W
* * L W
W W W L

This in turn implies that 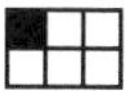 , 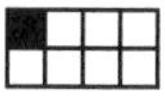 and so on are winners (break off a chunk to leave) and gives us

* * W W
* * W W
W W L W
W W W L

By pruning the game tree in this manner you can work out the win/lose status of any position. So the logic doesn't run in circles, but in interlocking spirals, climbing back along the game tree from leaf to twig, from twig to branch, from branch to limb ...

Starting from the *end*. Which is the snag. As a player, what you would really like to know is the status of your opening position – and what move to make to realise that status. For games with a small phase space this is no problem: for example by continuing the above process you can easily find the status of all positions in figure 6. What you get is

L W W W
W L W W
W W L W
W W W L

If you try larger bars of chocolate, you'll quickly find that the same pattern emerges: the losers live along the diagonal line sloping from left to right; all other bars are winners. Now the bars on that diagonal are the square ones: 1×1, 2×2, 3×3, 4×4 ... So, leaping to conclusions, we seem to have come up with a general strategy: square bars are losers, rectangular ones are winners. The leap is justified: in this case pruning the game tree leads to a pattern, whose beginnings can be seen in the list of winning and losing positions just obtained, and that pattern does indeed provide a complete strategy. Once you've noticed this pattern, you can check that it really works *without* considering the entire game tree: all you have to do is verify the defining properties 1 and 2 of winning and losing positions. That's actually rather easy. Any rectangle (winner) can always be converted to a square (loser) in one move. In contrast, whatever move you make from a square (loser) you leave your opponent a rectangle (winner). finally ■ is square, hence a losing position. All this is consistent with principles 1 and 2, so we deduce (recursively) that *every* square is a loser and *every* rectangle a winner. We now see that the second player's first move in figure 6 was a mistake, and threw away the whole game. That's what phase spaces can do for you.

Now consider how this procedure would apply to any game, not just Yucky Choccy. The opening position in the game corresponds to the 'root' of the tree. At the other extreme are the tips of the branches, the outermost twigs, which terminate at positions where one or other player has won. *We know the win/lose status of these terminal positions.* We can then work *backwards* along the branches of the game tree using the recursive principles 1 and 2 for winning and losing positions to label positions 'win' or 'lose'. The first time round we label all positions that are one move away from the end of the game. The next time, we label all positions that are two moves away from the end of the game, and so on. Eventually we must get to the root of the tree, the opening position. If this is labelled 'win', then the first player has a winning strategy; if not, then their opponent has a winning strategy. We can even say what the winning strategy is. If the opening position is 'win' then you should always move to a position labelled 'lose' – which your *opponent* will then face. Because this is a losing position, every move they make presents you with a 'win' position ... Now you can repeat the strategy until the game ends. Similarly, if the opening position is labelled 'lose', then the second player has a winning strategy. So in finite, drawless games, pruning the game tree decides the status of all positions, including in particular the opening one. The elegant structure of the self-referential win/lose loop carries a general mathematical implication: that in every such game there must exist a winning strategy for one, and only one, of the players. The proof, in

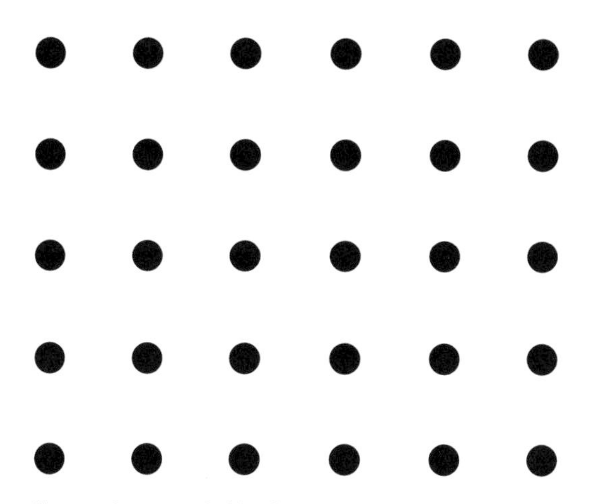

Figure 7 *Opening grid of dots for Boxes.*

essence, is that the only way for there *not* to exist a winning strategy for your opponent is if there exists one for you. The proof does not say who has a winning strategy, or what it is, but it does provide an algorithmic construction, obtained by pruning the game tree using the simple recursive principles 1 and 2.

Yucky Choccy is misleading in one respect, however, precisely because its entire strategy can be summed up in a simple recipe. We now contrast it with a game whose rules are just as simple, but where pruning the game tree does not lead to a compact strategy: the child's game of Boxes, in which players start with a grid of dots like figure 7.

They take turns to draw a line joining two dots that are adjacent either vertically or horizontally. A player who completes a box – a unit square – writes their initial inside it and gets another turn – whether they want it or not. At the end, whoever has completed the most boxes wins. No simple winning strategy is known, even for quite small grids, and most people play the game very poorly. You may not believe that, so let us ask the Zarathustrans to draw back a tiny corner of the veil that conceals the virtually endless possibilities of Boxes.

Destroyer-of-facts [*Who has been watching a game of Boxes on terrestrial children's television.*] Pah! That is a far more subtle game than most humans imagine.

Liar-to-children Yes, but it is being played by tadpoles. You should make allowances, Destroyer. Tadpoles usually play games badly.

DoF I accept that criticism, but to my amazement human adults generally fare no better. Their openings are rituals, based on a complete misconception of the subtleties of the game. It is so singleminded – or, to be precise, two minds with half a thought each. [*He laughs at this novel concept.*] Most players keep making moves that do not offer their opponent an opportunity to make a box, until they are forced to give some boxes away; then they give away as few as possible. As a result both players pursue opening moves that create a kind of snaky maze of almost-completed boxes, split into a number of distinct, disconnected regions.

LtC I do not apprehend your drift.

DoF Why do you not play a game against Creator here? Then I can comment on your stupidity.

Creator-of-creations [*Zarathustran for 'artist'.*] That sounds like fun. Come on, Liar. [*They play for a time, until Creator is faced with* figure 8:]

DoF It is time for me to comment, for both of you have already made a complete mess of your strategy. Observe that two boxes have already been completed – one per player – and the grid has been divided into two rectangular 'mazes'. Creator: you are in deep doo-doo, as those humans put it.

CoC Mmmm. Doo-doodly-*doo*. *Any* move that I make in either one of those mazes will present Liar with the entire maze.

LtC Why?

CoC Because you can just pick off the boxes one at a time in order around the maze. For example if I play *this* (figure 9)
then you can grab all of *these* (figure 10):

LtC That is an excellent / foolish idea (delete whichever is inapplicable). Let me explore the space-of-the-possible. [*Makes these moves.*] Oh.

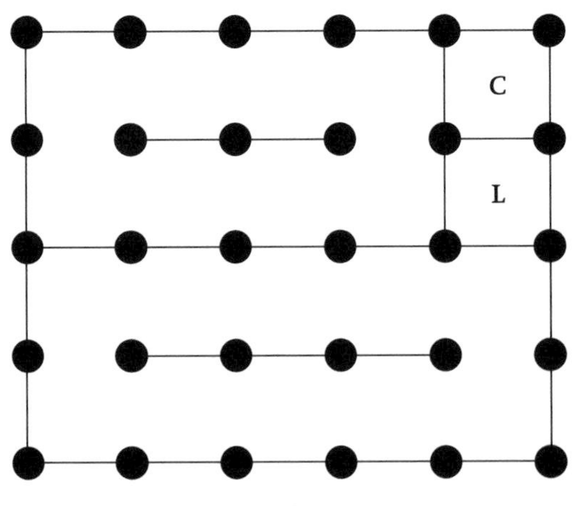

Figure 8 *Creator-of-creations to play.*

DoF	You must delete 'excellent'. You now have to play again, and there is no option but to give away all of the remaining boxes to Creator, who therefore wins.
CoC	As any creator-of-creations rightfully should, of course.
DoF	Indeed. Now, you may be thinking that you made a smart move when you broke into the *smaller* of the two mazes.
CoC	[*With false modesty*] Well, the thought did nest in my corner of the group mind ...
DoF	No doubt. That is because hardly anyone who plays this game penetrates to the next level of sophistication.

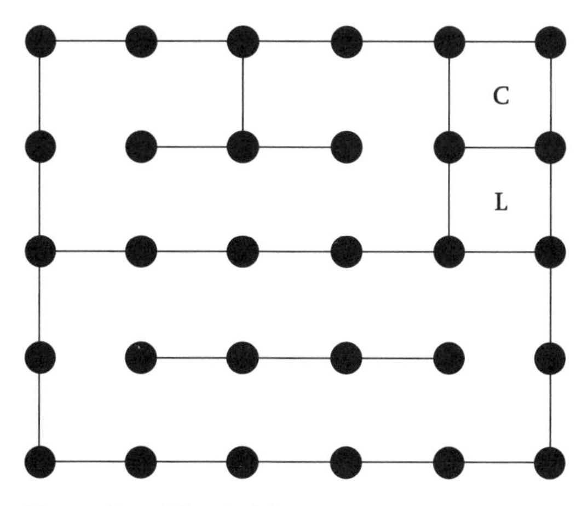

Figure 9 *Creator's hypothetical move ...*

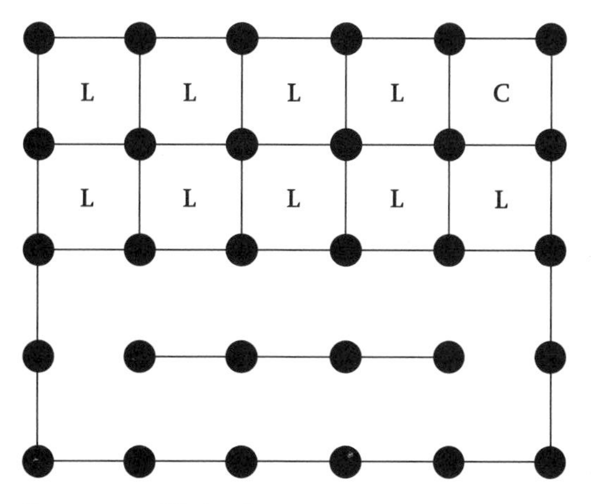

Figure 10 *... and Liar's reply.*

CoC Which is?

DoF That you did *not* make a smart move.

CoC But Liar had no choice ... [*Too late, smells the stench of error.*] Oh.

DoF Liar had many choices. What he should have done was to play this far (figure 11):

and then *refuse* any further boxes by terminating the play like *this* (figure 12)

[*a ploy known to afficionados as 'double dealing'*]. Now Creator is faced with a dilemma.

CoC I certainly am. If I play in either of the 2×1 rectangles that Liar has left, I will gain two boxes but must play again. I can do this twice, gaining

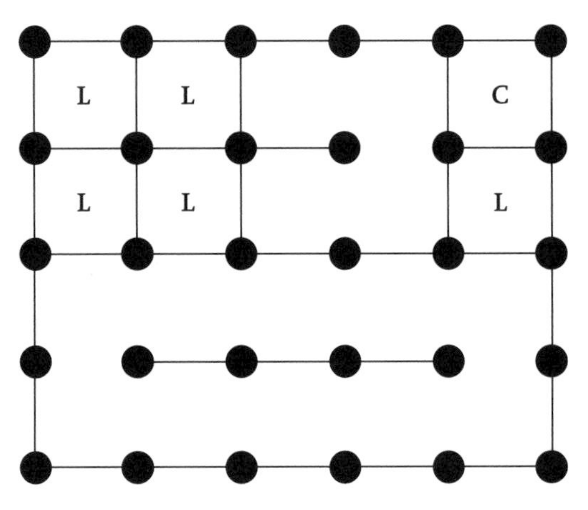

Figure 11 *Destroyer-of-facts' recommendation*

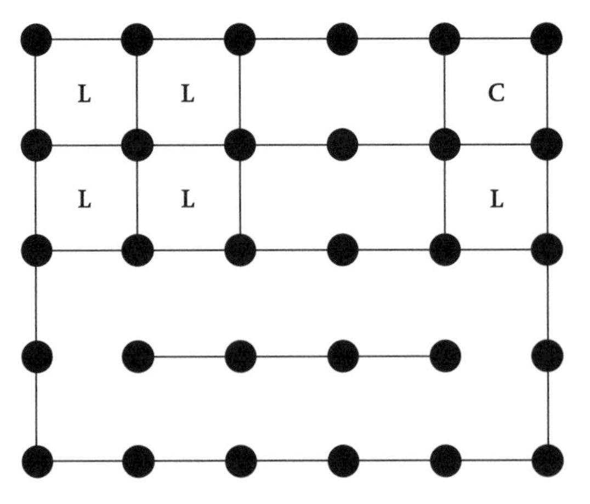

Figure 12 *Double dealing, Destroyer fashion.*

four boxes, but then I must play in the big rectangle at the bottom – like so [*with an air of resignation Creator makes the move* (figure 13):]

LtC That is an excellent / foolish idea, and I shall delete 'excellent'. Clearly I can secure all of the remaining boxes, winning by a handsome margin.

CoC Hmmmm. Maybe I should not accept those 2×1 rectangles.

DoF No, if you do then the same thing happens, but now Liar gets them too. He picks them off at the end.

CoC Oh, right.

DoF You see, Liar – Creator's problem is that at any time you may decide to present him with a few boxes, voluntarily, in order to secure more for yourself later.

LtC Correct. I am not obliged to hang on until I have no option but to give boxes away. What a refreshing thought!

CoC But equally I am under no obligation to take everything that is on offer.

DoF Precisely. And once you have seen this possibility, then the entire course of play that humans nearly always adopt becomes suspect.

CoC It can sometimes be worthwhile to give some boxes away early on, in order to reap more subtle rewards later. The humans should consider much more of the space-of-the-possible than they do.

DoF Indeed. But we have already observed that they are not looking in that direction.

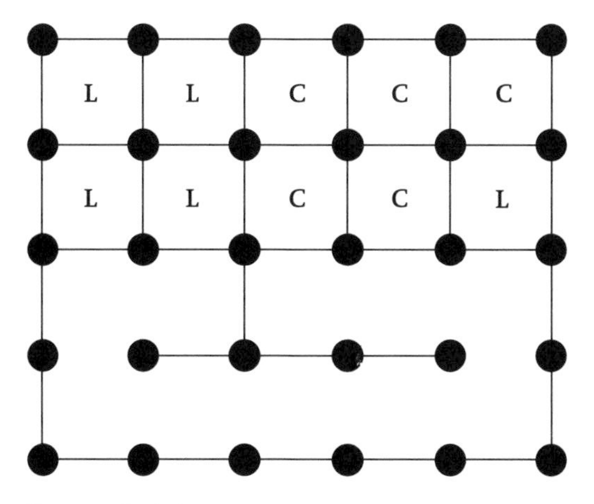

Figure 13 *Creator's move.*

Boxes is so subtle that there is no known strategy for grids of even moderate size. And this type of behaviour is the rule rather than the exception: many simple games – probably most – do not have simple compact strategies. (Which is the whole essence of a game worth playing, of course.) For such games, working back along the game tree leads to ever-increasing complexity and no simple overall pattern. Part of the reason is that a typical game tree is *huge*. For the example just given, a 5×6 grid of dots, there are 49 different places to put a straight line, so the full game tree contains 2^{49} positions – roughly a quadrillion. It is utterly impossible even to write the game tree down, let alone to prune it. So unless by good fortune there is some regular mathematical feature of the game tree that provides a 'short cut' for determining winning and losing positions – as there is in Yucky Choccy – then any strategy will seem arbitrary and patternless. (We are reminded of an SF story in which two computers play chess. After an hour's thought the first opens by moving the queen's rook's pawn forward one square. The other ponders this move for several hours and then resigns.)

In this kind of game the complexity of the strategy arises from the recursive nature of 'winning/losing move' and the sheer *size* of the game tree as an exploration of the 'phase space' of all possible positions. So it is the richness of the context, of the phase space, of the game tree, that provides room – and need

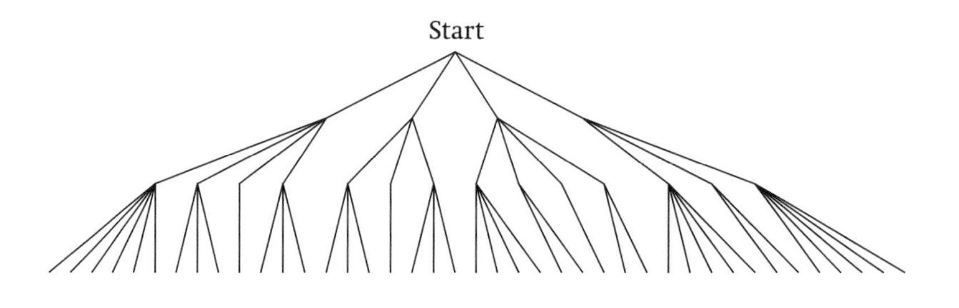

Figure 14 *A typical large game tree.*

– for internal complexity. This is one of the things that makes chess and the oriental game of Go so rich and inexhaustible. The structure of a highly complex game tree, such as that for Boxes, is shown schematically in figure 14. Its top is exactly like the reductionist nightmare, and its bottom is like the Tree of Everything: this is no coincidence. Scientific inference is a game, whose positions are statements and whose moves are logical deductions. This analogy suggests some useful terminology, leading us to divide the world of games into two types:

- *Dream games*, with simple 'compact' strategies.
- *Nightmare games*, whose strategies are just huge unstructured lists of positions and moves with no apparent rationale.

There is a strong analogy here with the polarity between order and chaos, as now envisaged by frontier mathematical research. According to chaos theory, there is not a simple dichotomy between order and chaos, between patterns and randomness. Instead, certain types of apparent randomness are phoney: they have hidden order. The order lies not in what is observed, but in the rules that generate it. Simple, deterministic rules can generate order – as has been known for thousands of years – but equally simple deterministic rules can generate disorder, apparent randomness. We do not wish to explain why this happens: see *Does God Play Dice?* by IS or *Collapse*. But the same thing is happening here: simple game rules can generate dream strategies (order) or nightmare ones (chaos). And we *know* that the nightmare strategies are not really random, because we know there is a simple algorithm to find them.

It is just the *result* of the algorithm that is complicated.

3 Ant Country

There is a parasitic flatworm that spends part of its life inside an ant, while its reproductive stage is inside a cow. The technique that it has evolved to affect the transfer from one animal to the other shows just how subtle the effects of 'blind' evolution can be. The parasite infects the ant, and presses on a particular part of its brain. This interferes with the normal behaviour of the brain, which causes the ant to climb a grass stem, grasp it with its jaws, and hang there, permanently attached. So when a cow comes along and eats the grass, the parasite enters the cow.

You will have noticed that in the game tree of figure 14 there is a *gap* between top-down and bottom-up. How big is it?

It contains virtually the whole of the game tree.

There is a similar gap between what is accessible to top-down and bottom-up reductionist science. In this chapter we give this gap a name: Ant Country. The origins of the name lie in a simple mathematical system, Langton's ant, which we shortly introduce. We shall employ Langton's ant as a metaphor to open up the nature of simplicity, complexity, and the relationship between them. Langton's ant itself is an instance of 'simplexity', the tendency of a single, simple system of rules to generate highly complex behaviour, but it also leads to a more subtle concept, which in *Collapse* we called 'complicity'. Complicity, the key concept of this chapter, arises when two or more complex systems interact in a kind of mutual feedback that changes them both, leading to behaviour that is not present in either system on its own.

Evolution, towards which our discussion of games is all heading, is almost entirely built around complicities, of many different kinds: the story of the parasite and the ant, which opens this chapter, is a case in point. People's first response to a description of complicity is often 'oh, you mean they interact'. No, we don't: we mean much, much more. Things interact when one of them affects the other. Once. Things are complicit when their interactions change them, so that soon they have become different things altogether – and *still* they continue to interact, and change, and interact again, and change again ...

We contend that most natural systems are complicit, and that the natural world – in particular the human brain – cannot be properly understood unless this is borne in mind. Complicity generates 'emergent' phenomena: recall that in these the behaviour of the system seems to transcend that of its parts. If our contention is correct, then most natural phenomena must be emergent, which in turn implies that their *detailed* behaviour is not amenable to reductionist analysis. We believe that mind, consciousness, and culture are examples of such phenomena: if so, they cannot be fully understood by reductionist methods. That does *not* mean they are inaccessible to science: it just means that science must extend its methodology to encompass a theory of emergence. Since nearly all of today's science is reductionist, and immensely successful to boot, the above statements may seem rather sweeping. However, the success of reductionist science does not lie in providing detailed explanations of what nature actually does, but in providing simple archetypes whose behaviour mimics useful features of systems that are really far more complex – and far more complicit.

Our analysis of the human condition will be based upon these rather philosophical concepts. We introduce them not to act as a verbal smokescreen, but because we believe that they provide unifying insights into the nature of such things as biological growth and form, evolution, ecosystem dynamics, and human culture. They are a small part of the new methodology that is needed to make emergent phenomena scientifically respectable and accessible.

Science observes the game of Nature, played out on the vast board of the universe, and tries to deduce the rules. Reductionism works whenever the part that is being observed resembles a dream game, where simple rules give rise to simple strategies. It breaks down when it runs into a patch of nightmare game, with simple rules but complex strategies.

Or when it encounters a game with complex rules.

Watcher-of-Moons' simulated microclimate edges one per cent humidity closer to mid-morning, a time when all sensible Zarathustrans take to their burrows or turn on the air conditioning. For about the last semi-octoon (equivalent to some 300 terrestrial years) the traditional mid-morning occupation had been to kibitz on the Octopoly contest. While The Regulations lowers the lights and fine-tunes the ship's air conditioning program, the crew settle down to watch the game on TV, beamed direct from Zarathustra. (TV is of course an abbreviation for Tachyonic Virtuality – a form of faster-than-light three-dimensional communication. We encounter several new crew members in the dialogue that follows.)

Octopoly has an inordinately complicated system of rules, but the general tenor of the game can be gleaned from the fragments that follow. It may help if we point out that play is governed by one simple meta-rule: *play to win*. There are no points for graceful losers. The game of Octopoly has been around a long time. We say '*the* game' because only one has ever been played. It is still going on to this day – possibly because, by an unfortunate oversight, the rules do not state under what conditions the game should end, while the committee that oversees rule changes is not permitted to meet if the game is still in progress. In consequence, Octopoly has been going on non-stop ever since the first opening move was made a semi-octoon ago. The first few thousand moves took only a few days, in Earth units, but back then all the players were amateurs. By now they have all become professionals – and professionals do not, on the whole, make silly mistakes. Neither do they make unnecessarily rapid moves, so play in the game is rather slow. In fact no move has been made for the last twenty years. The game has therefore reached a rare peak of excitement, with more and more viewers turning on to watch it in anticipation that they may, finally, get to see someone make a move.

Hewer-of-wood [*This role is self-explanatory, especially since he is carving a piece of wood.*] What is the score?

Creator-of-creations The same as it has been for as long as anyone can remember. Piffenchog Wanderers have 777,777,777 points [*in octal (base 8) notation*] and South Wuggen United have ⅛.

HoW I had forgotten it was that close.

Destroyer-of-facts It should not have been.

CoC No, Wanderers were set up for a really lengthy break, but they ran into trouble when they got three stops up the line from Mornington Crescent.

DoF That's right. They made a stupid strategic error a demi-semi-octoon [*150 years*] ago when Mufflepuffle failed to centre his backhand volley, and lost two million pawns *and* the toss.

Liar-to-adults [*A sort of combination politician/priest. Unlike 'liar-to-children' the term does carry negative overtones.*] Ah, but they still could have regained position if Jebbaruff had adjusted his run-up to suit the windy conditions and trumped the rook with his pink.

The regulations [*Proudly.*] Rule 553Bxy(a)(iii)gamma.

DoF No, no, no. That would have been offside.

TR [*Pompously.*] Addendum 19k* to Rule 553Bxy(a)(iii)gamma.

LtA With his five-iron, then.

CoC [*Dismissively.*] Nifflepux, you have such a tadpolish view of the game. The moment Jebbaruff even *thought* about adjusting his run-up, South Wuggen's quarterback would have anticipated being trumped, and built a dozen hotels on Park Lane. That would have put Wanderers under so much psychological pressure that their strength would have faded long before reaching Barnes Bridge.

LtA [*Losing his temper.*] Tadpolish? *You* accuse *me* of being tadpolish? Why, you haven't got the gumption of a proto-newt in a snoostorm! Any fool can see that before Quyxxshfflpuxx had poured one bag of concrete, Jebbaruff would have slid down the nearest snake and secreted the candelabra inside his glov–

DoF Was it not, in fact, the rope, rather than the candelabra?

Performer-of-amusements [*Roughly equivalent to 'clown'. He speaks in a worried tone.*] No, surely it was Commander-of-regiments M'stard, in the room-of-balls, with the hammer.

LtA That's absolute Regulations-fodder! The *hammer?* [*Rolls his eyes in disgust.*] Anybody with any feeling *whatsoever* for the game knows that the last time anybody even *touched* the hammer when there was a pigeon on the pitch was back in '33, when Guvsnepligroat broke his own record twice in the same season!

TR Correction: he broke the record three times. It originally stood at 18.1732 octometres. It was increased to 18.1733, then to 18.1734, and subsequently to–

PoA The Regulations, be quiet! Do not spoil a perfectly good argument by confusing us with unwanted facts! No, I still think it was the hammer. Did not Jebbaruff sweep the candelabra to third slip on a technicality?

DoF Of course he did. Got 46.7 marks for the technicality, as I recall, and 39.2 for artistic impress–

PoA No, he was sitting out the second set because of two personal fowls and a green card.

DoF Personal fowls?

PoA He chickened out during the sudden-death replay.

LtA [*Shouting at the top of his voice.*] Are you *mad*? Look, unless the wink pitches outside middle-and-leg then you can only be given two personal fowls when the cross-bar falls into the water-jum– [*Stops abruptly and stares at the screen, where thousands of Zarathustrans are leaping up and down like demented yellow yoyos.*] Hewer, what is all the excitement about?

HoW Sorry, I was carving a snoozosaur tusk and I was not watching the play.

DoF Something must have happened. The score has changed.

CoC Oh no. Did Piffenchog Wanderers reach 1,000,000,000 points?

DoF I cannot see, some nitwing with a huge pro-Piffenchog hat is standing in front of the scoreboard. Maybe the ringmaster saw what happened.

Ringmaster I have no opinion of my own, I can only form a consensus. Maybe one of you guys ...? [*His diction is becoming influenced by too much terrestrial television.*]

Liar-to-children No. But the maker-of-needless-comments seems to be hinting that perhaps South Wuggen scored a drop-goal.

PoA A drop-goal? *A drop-goal?* There hasn't been a drop-goal in the entire history of the game! Adultliar, you *idiot*, you were so busy distracting us by arguing that you made us miss a drop-goal! Oh my Regulations, a drop-goal! [*Starts to snivel.*]

R [*Attempting vainly to pour soil on ruffled feathers.*] Do not be so upset, Performer. It is how brilliantly one analyses the game that matters, not the actual play.

The same is true of much science, but let's not go into that. The best science takes things that appear complex and reveals them as simple. Let's consider what those two words mean, and where they lead.

The words 'simple' and 'complex' are deceptively straightforward. Something is simple if it can be described in a few words, complex if not. If we wanted to provide a spurious air of quantitative precision we might say that something is simple if it can be specified by a small quantity of information, and complex if not. However, the straightforwardness of these concepts is deceptive. The level of complexity of something cannot be determined by an information count, because complexity is a property of a description of an object, not a property of the object itself. Descriptions may be long or short, and the length depends at least as much on contextual assumptions as on the thing being described. 'Play a winning strategy' is a simple description of a simple activity in the context of Yucky Choccy, but *the same* description of an immensely complicated one in the context of Boxes. Similarly, 'obey Newton's law of gravity' describes a simple activity in the context of two bodies, but an immensely complicated one in the context of the solar system. The universe seems not to be put off by the difference in complexity – but we are.

This observation is central to everything in *Figments*. Simplicity and complexity are context-dependent concepts, not absolutes.

Human beings seem to have an innate belief in 'conservation of complexity'. If we perceive something as being complex, we want to know where its complexity 'came from'. In particular we assume that it has to come from *somewhere*. For example, where does the complexity of consciousness come from? Many people feel that ordinary matter is too simple to do the trick – whence such things as Cartesian mind/matter dualism (the belief that minds are made from some exotic kind of *stuff*, and not just from physical matter) or the invocation of an indestructible soul. But if complexity always has a precursor, we are trapped in an infinite regress: complexity at any level has to be explained in terms of complexity one level down, on and on forever. It's turtles all the way down: if complexity can never appear from simplicity, then the universe must always have been just as complex as it is now, and every simple explanation rests on the back of endless hidden complexities.

In fact, nothing in the mathematics of simple and complex systems justifies the idea that complexity is conserved. We have seen this already in games: the complexity of the strategy bears little relation to that of the rules, even though the rules determine the strategy. Simple or complex rules can lead to dream or nightmare games, with no obvious connection. Notice that the difference is not merely the *size* of the game tree: Yucky Choccy played with a very big chocolate bar (about 1.414×2^{24} pieces square, to be precise) has just as big a game tree as 6×5 Boxes (2^{49} states in both cases), but it still has the same simple strategy as Yucky Choccy with a 4×4 bar: 'rectangles good, squares bad'. So how can we tell whether a particular game is dream or nightmare? If it's a dream game, there must be a short cut: guess the strategy and verify that it satisfies the two recursive principles. However, there's no systematic way to guess: the best you can do is start pruning the game tree and hope to see a pattern. And for a nightmare game there seems to be no short cut at all, because there *is* no such pattern. Strategies are emergent properties of games.

Emergence, indeed, is a common route for the genesis of large-scale simplicities. Rule-based systems exhibit features on many levels, and emergence occurs when low-level rules generate high-level features. A simple example of emergence occurs in Langton's ant, a 'cellular automaton' invented by Chris Langton. The ant is known as a cellular automaton because it lives on a grid of square 'cells', which can be in one of two states: black or white, and it (automatically) obeys simple rules that determine the colour of each cell. For simplicity suppose that initially the cells are all white. The ant starts out on the central square of the grid, heading east. It moves one square in that direction, and looks at the colour of the square it lands on. If it lands on a black square then it paints it white and turns 90° to the left. If it lands on a white square it paints it black and turns 90° to the right. It keeps on following those same simple rules forever. It is hard to imagine that a system with such simple rules can do anything interesting at all. However, those simple rules produce some surprisingly complex behaviour. Here's what happens.

For the first five hundred or so steps, the ant keeps returning to the central square, leaving behind it a series of patterns. For the next ten thousand steps or so, the picture becomes very chaotic. Suddenly – almost as if the ant has finally made up its mind what to do – it builds a *highway*. It repeatedly follows a sequence of precisely 104 steps that moves it two cells southwest, and continues this indefinitely, forming a diagonal band (figure 15). This large-scale feature emerges from the low-level rules. The only rigorous way that is currently available to deduce this feature is to write down the ten thousand or so steps that lead to the 104-step cycle. Then an analysis of that cycle explains why it must repeat, which implies that a highway must form. So here we have a feature whose existence can currently be demonstrated rigorously *only* by following the reductionist rhetoric.

Computer experiments suggest that something stronger can be said. Suppose you change the initial conditions by scattering finitely many black squares around the grid before it starts – in any pattern whatsoever. Whatever you do, the ant *always* seems to end up building a highway – but nobody has ever been able to prove it. It certainly can't be done by reductionist rhetoric: there are infinitely many different initial conditions to consider. It is striking that even for such a simple system as Langton's ant, where we *know* the Theory of Everything

Figure 15 *Three distinct stages in the dynamics of Langton's Ant.*

because we set it up, the combined mathematical intellect of the human race is currently incapable of answering one simple question: starting from an arbitrary 'environment' of finitely many black cells, does the ant always build a highway? So here the Theory of Everything appears to lack explanatory power, predicting everything but explaining nothing.

Of course, tomorrow some brilliant mathematician might solve that problem, but the lack of explanatory power runs deep, as another cellular automaton, invented around 1970 by John Horton Conway, shows. He called it Life. It has only slightly more complex rules,♪ and Conway proved that it can simulate a universal Turing machine – a programmable computer. Turing machines are a simple mathematical model of the computational process. They are named after Alan Turing, who proved that the long-term behaviour of a Turing machine is undecidable – for example, it is impossible to work out in advance whether or not the program will terminate. Translated into Life terms, that implies that the question 'does this configuration keep changing forever, or does it eventually die out?' is undecidable. The answer *cannot* be deduced from the Theory of Everything. This is a rigorous 'qualitative' demonstration of genuine emergence. In *Collapse* we gave a quantitative demonstration too: there exist true mathematical theorems whose shortest proofs are *much* longer than their statements.

We take Langton's ant, and its more elusive generalisations such as Life, as a symbol for the gap between the top-down reductionism of the reductionist nightmare, and the bottom-up reductionism of the Theory of Everything. Bottom-up analysis proceeds from the putative Theory of Everything and ascends levels of description by deducing logical consequences of those laws in a hierarchical manner. Top-down analysis proceeds from nature and looks down mental funnels to see what lies inside. In the reductionist nightmare, the top and the bottom do not meet. Instead they both diverge into deductions too lengthy for the human mind to comprehend them. This 'no man's land' between top and bottom we call *Ant Country* (figure 16).

We headed this chapter with the story of a real ant and its travails with a flatworm parasite. This story is also about a kind of biological Ant Country. The ant's brain, and the ant/cow/parasite ecosystem into which it plugs, is so complex that any reductionist description falls smack into Ant Country. But evolution, making random samples from Ant Country until something interesting happens, has blundered across a feature of the ant's brain that can be exploited in a quite unexpected way. The resulting curious behaviour becomes reinforced because it has survival value for the parasite: it sets up a reproductive cycle. Indeed the analogy with Langton's ant is closer than might first appear, because 'building a highway' is a mathematical reproductive cycle. Ant Country poses problems for science's claim that its bottom-up rules *explain* the top-down behaviour of nature.

How does the reductionist chain of logic traverse Ant Country?

It doesn't.

It just claims to. The link between bottom and top is achieved through the intermediary of models. For example, consider the motion of the planet Mars. In the standard textbook derivation, a uniform sphere (bottom) and a planet (top) are identified for conceptual purposes. That is, the planet is modelled by the sphere. The mathematical rules explain the sphere's gravitational field, and this explanation is transferred – by analogy, not by logic – to Mars itself. We're not disputing that this process often *works*, but it breaks the alleged

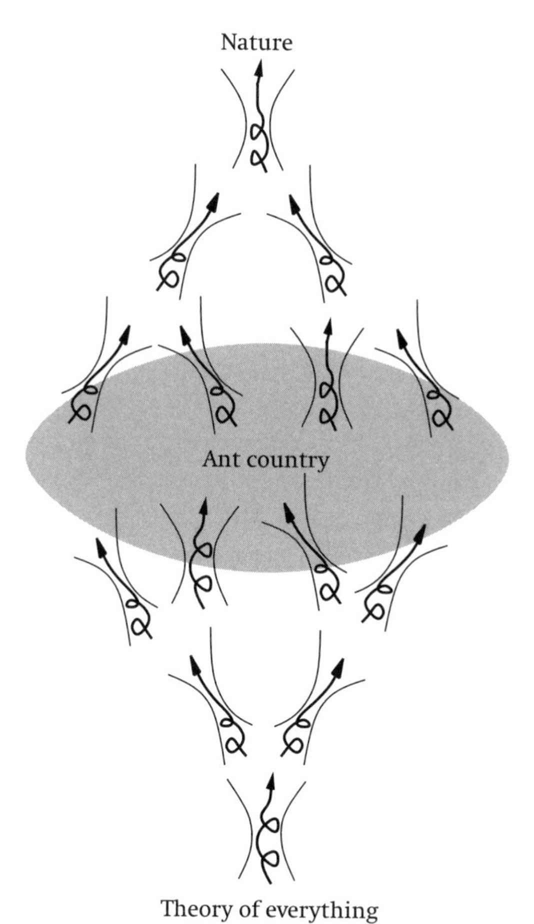

Figure 16 *Ant Country.*

reductionist chain. Real Mars is a complicated irregular collection of disparate atoms, not a uniform sphere. So the textbook derivation gives the illusion that the simplicity of the rules leads *directly* to the simplicity of elliptical orbits for real planets, when actually the explanatory story must enter the uncharted territory of Ant Country. And that is where the emergent phenomena live, it is where they come from. Ant Country is where complexity is created from nothing, where systems organise themselves into more complex systems without anything equally complex telling them how to do it. But very few scientists are even aware that Ant Country is there, let alone have any intention of exploring it.

Nature's patterns, then, are not merely simplicities inherited *directly* from simple underlying rules to behaviour. Virtually all of them are emergent phenomena, and all paths that lead to them from the rules pass through the uncharted territory of Ant Country. We therefore need to develop a workable theory of emergence, a new kind of theory in which some aspects of phenomena can be understood without referring them to lower-level rules. Poincaré short-circuited the need to solve equations explicitly before understanding qualitative features of their solutions. In the same way, we need to find new methods that will short-circuit the need to navigate through Ant Country, leading to an understanding of high-level features that avoids becoming over-obsessed with reductionist rules. Contextual thinking is a step in that direction. For example, Poincaré's approach to dynamics is a contextual supplement to classical reductionist methods, and it works like a charm, because it has another kind of weapon in its armoury: the large-scale qualitative principle such as continuity, connectivity, symmetry ... As well as the low-level arithmetic of equations it has the high-level geometry of phase spaces. So in this area of mathematics, the bottom-up approach and the top-down one sometimes meet in the middle to create something more powerful than either.

We do not yet have a good formal theory of emergence, but we can pin down some of the general mechanisms that come together to generate emergent phenomena. Among them is our promised concept of 'complicity'. We'll get round to it in a moment, but in order to explain it we must up the ante. We want to introduce some images related to real games, games like football, tennis, baseball, snooker, pool, or croquet; games played not on mathematical grids but on fields of real or artificial grass, or on baize-covered tables, where every theoretically sound move might run into trouble because of a patch of bare earth, a sparrow on the pitch, or a piece of fluff.

These are nightmare games of a different kind: their inherent chaos is generated not by the complex working out of consequences of simple rules, but by an external source of disorder, or 'noise'. It is reasonable to consider noise as lying outside the rules altogether. Alternatively it can be seen as part of a much vaster set of rules for a much wider game – the game of physics and chemistry – but for simplicity we shall view it in terms of random external disorder, if only because the players of a game such as tennis are not *aware* of these external rules. They know they have to hit the ball before it bounces twice, but they don't know, and have no control over, the molecular basis of elasticity that makes the ball bounce.

In such games, there can be no *perfect* strategy, because the outcome of a 'move' involves chance elements. At best there is a statistically optimal strategy, one that guarantees a win with maximum probability. Strategies arrived at in this way arise from a kind of 'Monte Carlo' statistical sampling of the game tree, external noise and all. The game tree is unprunable, so the overall strategy is determined by 'percentage play'. And also by creating situations in which the effect of the noise is reduced, so that the player genuinely seems to have full control. In nearly all real games players learn to recognise and exploit such situations, or to deliberately avoid them because they present the other side with an advantage.

Among situations of this type are some with a striking pattern, one that will be important later on when we discuss 'reproductive' sequences of events – not sequences that replicate precisely by following the same scenarios with a rigid periodicity, but sequences that re-create in a flexible manner the same qualitative type of conditions that gave rise to them. Reproductive sequences of play occur in several games, where they are known as 'breaks'. Examples include snooker, which is similar to pool, and croquet; and we digress to explain the general idea, because it sets up a striking and insightful metaphor for reproduction in general. Don't be put off if you've never played these games: we're not expecting you to go out and win at snooker. All you need to get out of the description is that *some* complicated series of moves leads to a repetition of roughly the same position, after which that series of moves can be repeated. We describe a snooker break in detail, and relegate croquet to the Notes. ♪

Snooker ♪ is played on a rectangular table (roughly 2 metres by 4) with six holes, or 'pockets', one in each corner and one at the centre of each longer side. There are 15 red balls, a sequence of 'colours' (yellow, green, brown, blue, pink, black), and a white cue ball. Balls are roughly 3 cm in diameter. A player scores points by using the cue ball to knock another ball into a pocket ('potting' that ball). Red balls are not replaced, but colours are replaced unless all reds have been potted. The player's turn continues until he fails to pot a ball. An initial red may be followed by the choice of any colour, after which the player again aims to pot a red, and so on until all reds are potted – after which the colours are potted in sequence. The red/colour periodic cycle leads to a strategy whereby the player attempts to set up and then maintain a 'break', which is a repetitive but not strictly periodic sequence of shots designed to reproduce, stably, the conditions required to keep the sequence going. For example, a typical break starts when the player pots a red in a top corner pocket and simultaneously 'gains position' on the black; that is, ends up in a position from which the black can be potted comfortably. The player then arranges to gain position on a new red while potting the black, and the break continues in this manner. A key feature is that during the present 'generation' of the reproductive cycle (potting the current ball) the player also sets up conditions that will permit the next generation to occur (gaining position on the next ball). Subtleties of shot such as spin, screw, or swerve are employed to provide flexible control of the positions of both object ball and cue ball in the same shot. This reproductive strategy avoids periodicity (so that, for example, a pink or blue – or in extremis a yellow, green, or brown – may be substituted for a black when appropriate). Players go to considerable lengths to set up a break, for example risking more difficult shots; and easy pots may fail because the player is also trying to gain position for the next shot from an awkward angle.

Many other games have similar quasi-repetitive sequences of play that offer a considerable advantage to any player that can set them up. We shall use the term 'break' to refer to all such strategies. Breaks, like strategies in general, are not built into the rules explicitly: they are emergent phenomena. In *Collapse* we played a linguistic game to classify emergent phenomena into two qualitatively very different types. By recombining elements of 'simplicity' and 'complexity' we ended up with the words 'simplexity' and 'complicity'. We attached the word 'simplexity' to emergence of the kind exemplified by the highways of Langton's ant: large-scale features that occur within a single rule-based system but whose detailed deduction from the rules is enormously lengthy and uninformative – or perhaps unknown. Complicity is far more elusive, but also far more important, and we will use it as a central image throughout *Figments*. It occurs when two (or more) rule-based systems interact: when two separate phase spaces join forces to 'grow' a joint phase space that feeds back into both components and changes them recursively until after a while hardly any trace of their original, separate forms can be found.

It is not unusual in such circumstances to find new high-level regularities emerging. Imagine, for instance, a game of snooker in which the cushions round the edge of the table, which make the balls bounce, slowly deform if they are hit by a ball. To begin with the game will look much as it did with a fixed shape of table, but now the table's shape will start to evolve. Regions of the cushion that are hit more often will deform more, making big changes to the shape; but now the ball will bounce differently. The old strategy for making a break, which relies on repeated bounces off very closely defined parts of the cushions, will probably come to pieces, because exactly those regions of the cushion that are needed will be changing. Other strategies will evolve – for example the table might change shape in a way that makes breaks much easier, or much harder, or replaces them with an entirely different self-sustaining sequence of shots – but the whole game will certainly take on a very different feel.

This process of 'self-modification' of the game actually involves two distinct phase spaces. One is the original phase space of snooker balls bouncing round a table. The second is the phase space of possible shapes for the table's boundary. On their own, the first is ordinary snooker and the second has no very interesting features at all. But put them together and let them interact, and each feeds off the other, changes it, and is changed by it. Imagine other ways for the moves of snooker to interact with its phase space – new pockets appearing, old ones disappearing, pockets that move depending on which ball falls down them, pockets that spit out balls, balls that change size, change colour, change shape ... This is complicity, and in Chapter 4 we will see that it is how evolution works – which, if you've been wondering, is why we've been explaining games for two chapters.

As we indicated at the start of this chapter, a real-world example of complicity is the evolution of the remarkable ant/cow/parasite system. It is (at least) a three-way complicity, whose component systems include the ant's brain and the cow's grazing habits, as well as the parasite. The ant's brain evolved the ability to grasp grass stems *and let go again* because this was a useful thing for an ant to be able to do. The cow evolved a liking for grass because that was good for the evolution of cows. Lurking in the combined phase space, with grass as the connecting feature, was a horrible trap, which the parasite discovered by accident. The resulting behaviour makes no sense either in ant-space or in cow-space alone, but it got built into their combined phase space because it set up a beautiful snooker break for the parasite.

That's complicity. Usually it is even more complicated and convoluted.

We shall explore the implications of complicity throughout *Figments*. It is central to our view of how complex systems interact, and of how systems build their own phase spaces. That affects our view of the source of nature's simplicities. Traditional science saw regularities in nature as *direct* reflections of regular laws. That view is no longer tenable. Neither is the view that the universe rests upon a single fundamental rule system, and all we have to do is find it. Instead, there are – and must be – rules at every level of description. To some extent we select the descriptions in which such rules arise, because our brains cannot cope with raw complexity. Every human being programs its brain, and its sense organs, to extract meaning (features) from its environment as it develops – especially during early childhood. Simple rules exist because simplicity emerges from complex interactions on lower levels of description. The universe is a plurality of overlapping rules. And in the gaps between the rules lies Ant Country, in which simplicity and complexity not only fail to be conserved, but transmute into one another – complicitly.

4 Winning Ways

A species of viperine snake, which is not poisonous, has evolved three ways to protect itself against predators. The first is camouflage, so that it gets 'lost' against its background. However, its camouflage is very similar to that of the poisonous adder, which leads to the second method: mimicry. If a predator sees through its camouflage, it exploits the resemblance to an adder by *behaving* like an adder. But if this doesn't work either, for example when the predator is a crow, which *kills* adders, it adopts the third strategy. It flips about like a demented rope, and then it arranges itself on the ground to look for all the world like a dead snake, lying on its back in the dust at an awkward angle, with a vaguely bloated look ...

However, if it is now turned on to its front, it promptly and energetically flings itself back into its 'dead snake' pose.

The background theory and philosophy is now out of the way, and we are ready to begin the journey from molecules to minds. It is a journey which, at every stage, involves the concept of evolution. Evolution is a general mechanism whereby systems can 'spontaneously' become more complex, more organised, more startling in their abilities. So in this chapter we shall take a closer look at the evolution of living creatures, paying especial regard to the idea that evolution can profitably be viewed as a game – indeed a nightmare game, according to the classification of Chapter 2. Evolution plays games with the forms and behaviours of organisms, and also with human sensibilities. In a book of the same name, Daniel Dennett calls evolution 'Darwin's dangerous idea'. The concept of evolution is dangerous because it provides a compelling rebuttal of the central tenet of nearly all religions: that the world we live in was created by a supernatural being. Life *must* have been created, the argument goes: it is too complex and structured ever to have arisen through the normal workings of ordinary matter. Not so, says Darwin: there are excellent reasons to suppose that life evolved of its own accord from more simply organised matter. Just as, centuries earlier, Galileo's astronomical observations threatened to upset a theological applecart based upon the Earth as the centre of the universe, so

Darwin's deceptively simple idea threatened the keystone of nineteenth century theology.

Darwin's idea is not only dangerous, in this cultural sense: it is subtle. In this chapter we shall tell the story of evolution in at least two versions: the standard textbook version, which focuses on genes, and a contextual version, which focuses upon emergent dynamics in phase space. Our treatment of the latter viewpoint will be on a fairly general level, though we will invoke Darwin's finches to provide some more concrete mental imagery. In the next chapter we will move from generalities towards specifics, and in Chapter 6 we will derive a complementary viewpoint to genes, based upon the interactions of organisms – in particular the concept of privilege, in which parents provide their children with a head start in life. We outline these approaches now, and develop them in greater depth as the book proceeds.

The textbook gene-based version goes something like this. All animals produce too many offspring, so that there is no room for all of them to breed. In consequence those offspring that do manage to breed contribute to the next generation, while the others do not. Some attributes assist in the process of breeding, or surviving until breeding age – strength, speedy reactions, the ability to be as quiet as a mouse when the cat is nearby, whatever. Suppose that one animal – call him Fred – manages to breed. If the attributes that allowed Fred to breed are inherited by Fred's offspring, then they will possess the same advantage. This doesn't mean that they will necessarily get to breed, but it does mean that their chances of producing Fred Jr. are improved. Over several generations this slight advantage is likely to lead to a preponderance of Freds, with his special attributes. The population has now changed: nature has winnowed the creatures that might have been produced and chosen mostly Freds. This process is known as 'natural selection', and it ensures that the resulting creatures have 'good genes'. This is the essence of the story as it is usually told, with a few extras like the necessity for heritable differences and bells and whistles like dominant/recessive alleles (alternative versions, such as brown or blue eyes, one of them taking precedence) of each gene.

The 'emergent dynamic' version places the emphasis on organisms rather than genes, and sees the textbook process as a complicated way of describing something much simpler: a dynamic on phase space. In this view, organisms change because the geography of the surrounding space-of-the-possible makes change inevitable. Evolution runs 'downhill' in its phase space. The hidden complication is that the dynamic is emergent, not prescribed once and for all, and the phase space changes complicitly with it.

This organism-based view places emphasis on many different things, among them privilege. We shall go into this idea in depth in Chapter 6, but a short preview will help set the scene now. Instead of concentrating on organisms competing for how many progeny they pass their genes to, this alternative theory focuses attention upon how the parents provide their offspring with a head start, and how siblings compete for the privilege of growing up. If you are a baby starling, then your immediate source of food is your parents, so that the most immediate short-term competition comes not from prowling cats, but from your brothers and sisters. Darwin saw that this kind of 'sibling rivalry' must be very intense, and for him it was the major driving force of evolution. Of course it is important to be able to avoid the prowling cat too, but you have to grow up first, and that won't happen if you don't get a fair share of the food. The way adult birds feed their young is grossly unfair: the one who is pushiest, makes the most noise, opens the biggest gaping beak, gets most of the food. 'To him that hath shall be given.' Unfair it may be, but this simple strategy has survival value: it ensures that at least one chick gets enough nourishment to reach the stage of leaving the nest and foraging for itself. Divide your food supply among too many young, and if it is inadequate – as it often is – then they *all* die. An extreme instance occurs in the fish eagle, which typically starts by trying to bring up three chicks.♪ What you see in the nest is one large chick, one medium, and one small, but that's because they hatched at different times. If there is too little food, which happens in most years, the large and medium chicks eat the small one; if necessary the large chick then eats the medium one. It is conjectured – but not yet established – that in very bad years the parents end up eating the big chick. This approach violates many of our cherished sensibilities, but it makes perfectly good sense to the fish eagles, who in effect use their chicks as a living refrigerator to keep food in usable form, and can thus adopt a strategy that avoids putting needless effort into a baby that won't become a breeder.

Those are the alternative positions that we shall compare, contrast – and eventually combine.

We start by taking a closer look at the textbook story of 'good genes'. In order for competition – either between species or within them – to affect future generations, some kind of hereditary factor must pass from one generation to the next. The system must have some kind of 'memory' of what works, and what does not. Darwin knew that such a hereditary factor must exist, but he didn't know what it was. Nowadays we know that the basic genetic material is DNA, and we say that the breeders – the animals that survive to produce the next generation – are most 'genetically fit'. As we've just said, alternative possibilities for a gene – more properly, genetic *differences* – are called 'alleles'. Natural selection favours some alleles over others, so textbooks of population genetics take a lot of space describing how to calculate the relative genetic fitness of different alleles. A particular allele may be fitter if creatures that possess it are more able to survive than if they possess any of the alternative alleles. The simplest way that this can occur is if the alleles correspond to clear-cut 'characters' – features of the organism such as shape, behaviour, or colour – and particular characters enhance survival value in particular circumstances. However, despite a widespread belief to the contrary, most alleles do not correspond directly to characters. Alleles may also be fitter if they enable the organisms that carry them to increase the representation of those alleles in the next generation. This is not quite the same thing as helping the organisms themselves to survive, though the distinction often doesn't matter, because if you don't survive to breed then your fitness is zero. However, characters needed for survival often mitigate against breeding: for example in order to survive you must not be visible to a predator, but in order to breed you must be obvious to your intended mate. Again, staying away from your own species is a good survival trick, because they are the most likely source of parasites and disease; but if you belong to a sexual species that strategy is hopelessly poor when it comes to breeding. Organisms have evolved special tricks, such as elaborate mating rituals, to cope with these contradictory tendencies.

Some evolutionary biologists, the best-known being Richard Dawkins, tell us that DNA is in the driving seat of evolution. This view is known as 'neo-Darwinism': it is Darwin's original idea of natural selection, together with a specific statement of what the 'unit of evolution' is. The neo-Darwinists' stories have become increasingly subtle over the years: for example they now say that it is not just *your* DNA that improves its chances of survival by making *you* better. Alleles can sometimes increase their chances of appearing in the next generation by promoting the survival of close relatives rather than that of the actual breeder. Worker bees, which do not breed at all, provide one of the best known examples. This effect leads to the concept of 'inclusive fitness' of an allele: when calculating an allele's fitness, you must take account of all relevant copies of that allele, not just those in one particular organism.

In Chapter 1 we mentioned that DNA has a fascinating structure, the famous 'double helix', composed of two intertwining strands made from four types of 'base'. The bases rejoice in the names cytosine, guanine, thymine, and adenine, normally abbreviated to C, G, T, A. Bases in corresponding positions on the two strands are paired together, and the pairs are 'complementary': C always pairs with G and T with A. DNA is replicated by a lot of molecular machinery in the cell, which splits the strands apart and replaces the 'missing' complementary bases. A large part of modern genetics occupies itself with 'sequencing' various organisms' DNA – finding the sequence of bases – and then reading off the genes and trying to work out what function they have. Because today's molecular biologists have developed some impressively powerful techniques for playing this game, they have a tendency to act as if it is the only game in town. The ultimate such game is the Human Genome Project – biology's equivalent of a Moon landing or a major particle accelerator such as the ill-fated Superconducting SuperCollider – whose aim is to work out the entire sequence of three billion DNA bases in a human being. An important milestone was reached in April 1996 with the completion of a project to sequence the genome of yeast – which has twelve million DNA bases. This is good basic science, but it is expensive, and in order to raise the money to carry out the work, it has been rather oversold. In particular the Human Genome Project – to sequence the entire DNA of a human being – has been sold to the public on the grounds that knowing 'the blueprint for a human' will enable us to cure all sorts of diseases, and goodness knows what else. There are at least two things wrong with this: the genome is not a blueprint, and there is a gaping chasm between knowing the genetic origins of a disease and finding an effective cure. Our inability to cure AIDS makes this clear: we know everything there is to know about the genetics of the HIV virus that is responsible, and it doesn't help much. What we don't know much about is the human side – behaviour, psychology, and so on. This might help us keep the disease under control, but it wouldn't lead to a cure. This state of affairs is pretty much typical of genetic diseases.

So far.

So say the geneticists, and so say we. But our 'so far' is different from theirs, as we explain below.

Back to yeast. Yeast is an interesting organism because it is a eukaryote – its cell has a nucleus. Moreover, it reproduces sexually. In consequence it is much more representative of the rest of the eukaryotes, us included, than are the two other organisms whose genomes have been or are soon to be sequenced: *Escherischia coli* bacteria (done) and the nematode worm *Caenorhabditis elegans* (well on the way). Bacteria are prokaryotes, with no cell nucleus, and the celebrated *C. elegans* is wildly unusual in that its body has a fixed cellular architecture, with precisely the same number of cells, each in exactly the same place, in every nematode. (There are 959 cells in males, and 1031 in the other sex –

hermaphrodites. As one biologist put it: 'Lovely to work with, impossible to argue from.')

The more useful organism yeast, chosen because it breeds rapidly and is easily cultured, is one of the geneticists' favourites, though it is still somewhat atypical. When the yeast genome was sequenced, the results contained a lot of unexpected puzzles. We now know that yeast has about six thousand genes: forty per cent of them have no known function and most of those look totally unlike any gene previously encountered. Geneticists are now busily trying to find out what they are for. As Bob Holmes wrote in *New Scientist*: 'In the next nine months scientists from Helsinki to Heraklion plan to delete a thousand of the unknown genes, one by one, in individual yeast cells, and then measure how each knockout affects the growth and reproductive ability of that cell's descendants.' Such 'deletion experiments' are the tried-and-true technique in genetics. They are like removing bits and pieces from a car to see what it no longer does. Take the wheels off, it won't move, so wheels are what make it move ... A bit later in this chapter, we will stick our necks out and criticise this approach, explaining why geneticists really should have expected to find a lot of genes that they'd never seen the likes of before, and why deletion experiments will *not* tell them what those genes do. But for now we prefer to present the yeast genome as a major scientific achievement – which it is, the genetic equivalent of climbing Everest, say, though not yet the full Moon Landing – which demonstrates the impressive technical abilities of today's genetic science.

We also hasten to recognise that modern genetics is not solely obsessed with sequencing DNA. For example its techniques have shown that, contrary to what used to be thought, for most species from amoebas to antelopes the population contains a vast range of *different* alleles. Rather than there being a kind of 'master blueprint' for a peacock or a pigeon, with just a few individual variations permitted, each bird has a substantially different 'blueprint' from those of its fellows. So great is the variety of alleles that for about ten per cent of their genes birds receive a different allele from their father than they do from their mother. The same goes for most other sexually reproducing organisms: in particular, us. Nearly all of these genes make very little overt difference to what the animal looks like, or even how it behaves in routine circumstances of everyday life. It is the unusual, difficult, stressful events that expose the subtle effects of these differences – which viral diseases the animal is susceptible to, which animal can function at a slightly lower temperature than normal, or conserve water in a drought; which giraffe can run longest when chased by lions. The effect of such alleles is 'cryptic': most of the time they don't have any effect, but when conditions change, they do. One consequence of the cryptic dependence of phenotype (body-plan and behaviour) on genotype (DNA) is an effect known as 'genetic assimilation'. The orthodox evolutionary position is that genes affect phenotype, but not the other way round. However, the form and behaviour of an organism affect its survival abilities, and hence what genes are passed on to its progeny. This creates a feedback loop from phenotype to genotype, not in individuals but in an ongoing lineage. In this manner a population of highly stressed giraffes, running about as fast as their legs can carry them to escape marauding lions, can evolve into a population that is much more able to cope with such stress. From outside it looks as if they are passing the acquired character 'able to cope with stress' on to their offspring, but below the surface what is happening is that animals whose genetic makeup copes poorly with stress do not live to reproduce. Notice that no *new* genes need be involved – instead, old genes may be assorted in new ways, or eliminated.

While we're discussing this topic, there is a related concept that we will need shortly, called 'canalisation'. In genetics laboratories, organisms such as fruit flies are bred and selected so that their phenotypes are extremely sensitive to genetic changes. This is done deliberately, because the phenotypic changes are used to track what's happening to the genes. 'Wild type' flies out in the real world are nowhere near as sensitive. Their genetic makeup involves all sorts of back-ups and contingency plans to make their phenotype remain as stable as possible *despite* genetic changes. Genetics of this latter type is said to be *canalised*.

We can use DNA sequences to reconstruct plausible versions of organisms' evolutionary histories. About ten years ago this technique was employed to argue that all of humanity can be traced back to a common female ancestor, 'mitochondrial Eve'. Later it turned out that the *same* techniques, employed more intelligently, demonstrated the exact opposite. This is an instructive tale, which brings in several important aspects of early human evolution. In order to tell it, we begin with the concept of 'polymorphism'. When individuals in a population of organisms have the option of several alternative alleles, the population is said to be polymorphic. A familiar (but in several ways misleadingly special) example of polymorphism is blood group in humans. A coarse classification of blood groups involves three main types : A, B, and O. However, the full story is more complex, including such things as M, N, S, P, rhesus factors, 'Kell', 'Duffy' and so on – the latter being named after antibodies found in particular families. It is easy to see that different alleles might be useful in different circumstances, and that if so, then natural selection might lead to a distribution of several alleles throughout the population – the precise mix depending on past history and the selective influence of different environments. It is also fairly obvious how alternative segments of genetic material could have arisen in the first place: by way of mutations. Mutations are not merely a matter of replacing one DNA base by a different one: other mechanisms include deleting a base or a sequence of bases, adding a base, 'transplanting' a sequence from elsewhere, and gene duplication, in which a whole gene may be copied so that it appears more than once in the genome. By a combination of these changes, nature can experiment with different versions of a gene. In sexually reproducing organisms it can do so without risking losing functional genes, because the gene-pool – the overall range of alleles in the population – retains copies of old alleles even when a new allele is being tried out. Gene duplication as a preparatory step also allows nature to keep a spare copy just in case the novelty doesn't pan out, which is presumably how polymorphisms get going – and got going – in the first place.

With the advent of DNA technology, which in particular provides chemical methods for detecting mutations and quantifying how many are required to turn one sequence of DNA into another, quite an industry has grown up around a technique which in effect tracks mutations backwards by starting with polymorphic alleles and comparing their DNA sequences to work out which mutations occurred, and in what temporal order. The underlying idea is that mutations in DNA introduce small, cumulative changes, so DNA sequences that are very similar to each other are likely to have had a common origin not long ago, whereas sequences that differ more markedly diverged from each other much further back. For example if we find one contemporary sequence with CCCCGGGG, and another with CCCAGGGG, then it is clear that at some point in the past either a C mutated to an A, or vice versa. If another allele has the sequence CCCAGGGT, then it is 'closer' to the second than it is to the first, and so on. By analysing such differences systematically you can take a crack at finding the genetic 'family tree' of all of the present-day alleles. This technique is based upon several assumptions. One – which can be given pretty solid theoretical justification – is that alleles that differ, but have family resemblances, 'coalesce' in a common ancestral allele when they are tracked backwards; running time the usual way, this implies that they all diverged from such a common ancestor in the distant past. Another assumption, rather more questionable but reasonable in at least a qualitative sense, is that on average mutations occur at a fixed rate. This provides a 'molecular clock' with which you can estimate how far back you must go to find the common ancestor.

We can illustrate how the technique is applied by telling an evolutionary story that is instructive in its own right: the promised tale of mitochondrial Eve. To set the scene, we briefly recall what is currently thought to be known about the evolutionary family tree of *Homo sapiens* – us privileged apes. Our knowledge of human ancestry, like most of our knowledge of events that happened more than a few tens of thousands of years ago, comes from the fossil record. The known traces of early hominids – prototype humans – are found in Africa, China, and Java, and the oldest come from Africa. So the evidence, such as it is, favours an African origin for humanity. The standard theory is that we evolved out on the savannahs, where the need to survive among the big predators quickened our senses and forced our diminutive brains to grow quickly, or die along with the animal in which they were housed. There are some problems with this theory, a big one being the absence out on the savannahs of the chemicals needed to make brains: essential fatty acids. A controversial alternative, to which we return much later, is the 'aquatic ape' theory of Alister Hardy and Elaine Morgan, developed further by Michael Crawford and David Marsh, in which humanity evolved on the seashore. This is where you can find a huge supply of essential fatty acids: they are common in seafood. On the seashore, too, our peculiar love of water, our ability to swim even as very young babies, even the uniquely strange pattern of hair on our bodies, could plausibly have come into being. But seashores are not a good place to make fossils, so conclusive evidence is not likely to be available even if the theory is true.

Wherever we first arose, there is no question that zoologically we are apes. Indeed we are definitely chimpanzees, our closest living relatives being the ordinary chimpanzee *Pan troglodytes* and the bonobo chimp *Pan paniscus*. Genetically we are very similar to chimps: our genome and the chimp genome have 98% of their component bases in common. The general consensus is that our lineage – the sequence of parents and offspring that contains us – and that of chimpanzees diverged about 5 million years ago. The two types of chimp diverged from each other a bit later on. The earliest fossil hominids mainly come from the African Rift Valley area, and until recently those older than 3 million years were all considered to be one species: *Australopithecus afarensis*. These creatures walked upright, but from the neck up they looked more like apes than people. There are two types of fossil, one smaller than the other: the obvious explanation is that they are the males and the females of the same species. However, some anthropologists think that the two types of fossil are *too* different, and that they come from two separate species. Very recently two more species of *Australopithecus*, known as *A. ramidus* and *A. anamensis*, have been found, dated around 4.5 to 4 million years ago; and there are a variety of other

Australopithecines, the most recent being *A. robustus*, which existed between 1.5 and 1 million years ago. Much closer to us in appearance are the hominids of the genus *Homo*, notably *H. habilis*, a toolmaker that lived between 2 and 1.5 million years ago, and *H. erectus*, the first of our ancestors to walk upright in essentially the modern manner, who flourished between 1.7 million and 150,000 years ago. The most recent species in the genus *Homo* is us, *H. sapiens*: we first came on the scene around 250,000 years ago. For most of that period we overlapped with *H. neanderthalensis*, Neanderthal man. Neanderthals were more heavily built than us, and are conventionally assumed to have been less intelligent. The pattern of speciation that links these early humans is speculative and controversial, as indeed are many of the species boundaries: some scientists want to combine several separate species into one, others want to split an existing species into several. The battle will rage for many years yet because the evidence is very limited. Figure 17 shows the fifteen main recognised hominid species, and the times when the fossil record suggests they lived.

At last we are equipped to follow the story of mitochondrial Eve, erstwhile universal mother. The 'Eve' metaphor trips easily off the tongue of anyone brought up in even a nominally Christian country. Too easily, perhaps? Was there genuinely a single 'Eve' – one female ancestor for all of today's humans – or were

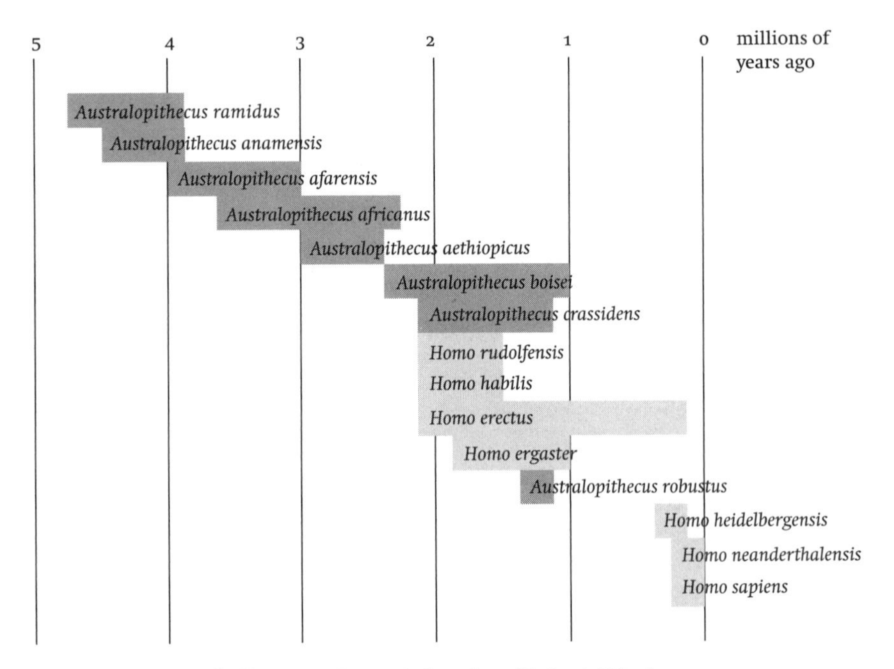

Figure 17 *The fifteen currently recognised members of the hominid family.*

there several? This is a loaded question, as the 'Eve' terminology reveals: it bears upon the authenticity of the Christian creation myth. So it was not a great surprise that in 1986 the news of scientific proof that all of humanity did indeed descend from one female ancestor attracted a lot of media attention – predictably greatest in the United States, where it came as welcome news to the Bible Belt and its gaggle of 'creation scientists'. 'Mitochondrial Eve,' she was named, because the evidence for her existence came from an analysis of resemblances between DNA sequences in the mitochondria of human cells. She lived in Africa between 100,000 and 200,000 years ago. The evidence was quite compelling, based on tracking such resemblances backwards in a molecular family tree, as just explained. The scientists analysed large numbers of sequences derived from modern human mitochondria, from people all over the world, and tracked their DNA genealogies backwards. The evidence seemed to point back to a single human being as the Mother of Us All.

What exactly was that evidence? The human genome contains some 6 billion DNA base pairs, most of it inside the nucleus incorporated into chromosomes: half of it in each set of 23 chromosomes inherited from each parent. However, about sixteen thousand base pairs of DNA exist inside mitochondria, which are organelles of the cell *outside* its nucleus. This is a short enough sequence to be handled easily when tracking differences that have accumulated over time through mutations. Mitochondrial DNA has a distinctive pattern of inheritance: it *always* comes from the mother. Analysis of mitochondrial DNA from more than a hundred ethnically diverse modern individuals shows that it all tracks back to a single DNA sequence, 'coalescing' some 200,000 years ago. Conclusion: we all have the same great-great- ...-great-grandmother, with about ten thousand greats (at twenty years per generation, say). That's Eve.

Only recently has it been shown that this entire story is based upon a misunderstanding.[♪] Almost predictably, it arose from a conceptual confusion between the genealogies of genes and those of individuals – yet another mistake brought about by taking the 'DNA-as-blueprint' image too literally. (A genealogy is a 'family tree', a list of predecessors going backwards into the past.) This work, summarised in the journal *Science* by Franciso Ayala in December 1995, tests the theory of mitochondrial Eve by playing the same game with certain genes from within the nucleus, known as DRB1 genes. They are involved in the immune system. A typical DRB1 gene contains 270 base pairs, and is thus easily sequenced in full. DRB1 genes are highly polymorphic in humans, meaning that different people have different DRB1 genes: there are 59 distinct morphs – 59 choices of DNA sequence.

There is strong evidence that human DRB1 genes go back a very long way indeed. For example the human DRB1 gene known (imaginatively) as Hs*1103 resembles the chimpanzee DRB1 gene Pt*0309 *more* closely than it resembles another human DRB1 gene, Hs*0302. Therefore the two human genes diverged from each other earlier than the chimpanzee and human lineages diverged from each other. (There is no contradiction in this, even though we call them 'human' genes. They are in humans *now* – but they were in the common ancestor of humans and chimpanzees in the past.) By studying the rate at which mutations normally occur, it can be inferred that these two human genes diverged about 6 million years ago; the divergence between human and chimpanzee lineages occurred soon after. For comparison, the orang-utan lineage diverged from that of the chimp–human ancestor about 15 million years ago, and – as we have just stated – *H. erectus* emerged about 1.7 million years ago. These times are estimated by the same method that produced the 200,000 year estimate for Mitochondrial Eve – if you dispute that method, fine, but then Eve goes out of the window straight away.

The genealogy of the 59 different human DRB1 genes coalesces about 60 million years ago – which, coincidentally or not, is about the time of the great mammalian radiation brought about by the demise of the dinosaurs. The same kinds of estimate show that 6 million years ago about 32 different members of today's human DRB1 gene lineages were still in existence. This implies that at least sixteen distinct human ancestors existed then, because each human can possess at most two DRB1 alleles, one in each set of chromosomes. The same reasoning shows that there must have been at least sixteen ancestors in the human lineage after it diverged from that of the chimpanzees, so Eve must have had, at the very least, fifteen Adams to keep her company. Actually a more careful mathematical analysis shows that throughout the last 60 million years human ancestral populations have generally contained at least a hundred thousand individuals, though there is a slight chance that they occasionally may have shrunk to about ten thousand. The theory of one Eve and 99,999 Adams really won't wash – she wouldn't have time to make enough babies – and it is much more likely that there were about 50,000 of each.

So why does our mitochondrial DNA all go back to one sequence? Possibly there did exist a Mitochondrial Eve, but she is not the Mother of Us All: she represents a particular molecular sequence for mitochondrial DNA, embodied in a *population* of women possessing that molecule, from whom all modern mitochondrial DNA molecules descend. We got a lot of our other DNA from other ancestors in any case, and all we can infer is that 200,000 years ago all of our

50,000 or more female ancestors happened to have the same mitochondrial DNA. Of course, if we push Mitochondrial Eve back sixty million years, to long before she was human ... But then she sure as heck wasn't created in the same image of God that *we* are said to embody.

Inevitably there is a male parallel to Mitochondrial Eve: ZFY Adam. A male counterpart of mitochondrial DNA is the Y chromosome, which is transmitted from fathers to sons. An exercise similar to the one that produced Mitochondrial Eve has been carried out using a particular sequence of 729 base pairs thought to be involved in the maturation of testes and sperm, and these sequences all come together about 270,000 years ago. Close, but no banana unless (aha!) Adam was immortal and lived alone for 70,000 years until Eve arrived on the– oh, those cultural mind-traps ... The same point that applies to Eve applies also to Adam: he must actually represent about 50,000 individuals who all had the same DNA sequence because their ancestors also all had the same sequence. Far enough back, they might have had a common ancestor – the simplest but not the only possible explanation for this identical piece of genetics. The technique used to produce the 270,000 year figure cannot tell us when – once the genes have 'coalesced', they remain coalesced however far back you go. This does not imply they all came from one original ancestor, but it doesn't rule it out either.

Perhaps the most ironic aspect of this story, apart from the credulity with which the original results were received, is that the analysis of DRB1 genes implies a human ancestry stretching back some 54 million years before humans appeared on Earth. Far from supporting the creation story, Mitochondrial Eve supports evolution.

We have told this story from the conventional point of view about convergence of alleles and molecular clocks, and many biologists are happy with that viewpoint. We're not, and it's worth taking a few moments to indicate some of our objections.

One (which is another objection to the interpretation of 'mitochondrial Eve' that does not require games with DRB1 genes) is that the analysis applies only to those women who have descendants alive today. It does not tell us anything – by its nature cannot possibly tell us anything – about other women who were alive at the same time as 'Eve', but whose lineages died out. However, in real populations lineages die out all the time; the most familiar analogy is the disappearance of surnames, so that as the centuries pass we find more and more Smiths but after a while no Chaucers (well, not in the Coventry phone book, at least). The Biblical Eve is supposed to be not just the first female ancestor of everybody alive today, but of everybody alive *ever*. And *that* you cannot deduce from modern mitochondrial DNA (see figure 18).

The next objection is methodological. By starting with a sample of only a hundred ethnically diverse individuals, you run a very real risk of omitting many rare morphs that occur somewhere in the human population – but very infrequently. As an analogy imagine that you are surveying patterns of human wealth by choosing a hundred financially diverse individuals, at random. You will without doubt find many instances of grinding poverty, because huge numbers of people are very poor, but even in a developed country like the UK you may well miss all of the millionaires (there are less than a hundred thousand in a population of over fifty million). You will almost certainly miss the billionaires – just as the rare bloodgroups like 'Kell', occurring only in a few families, were missed by everybody until very recently. At any rate, if additional morphs have to be factored into the coalescence calculations, it is likely that the estimated coa-

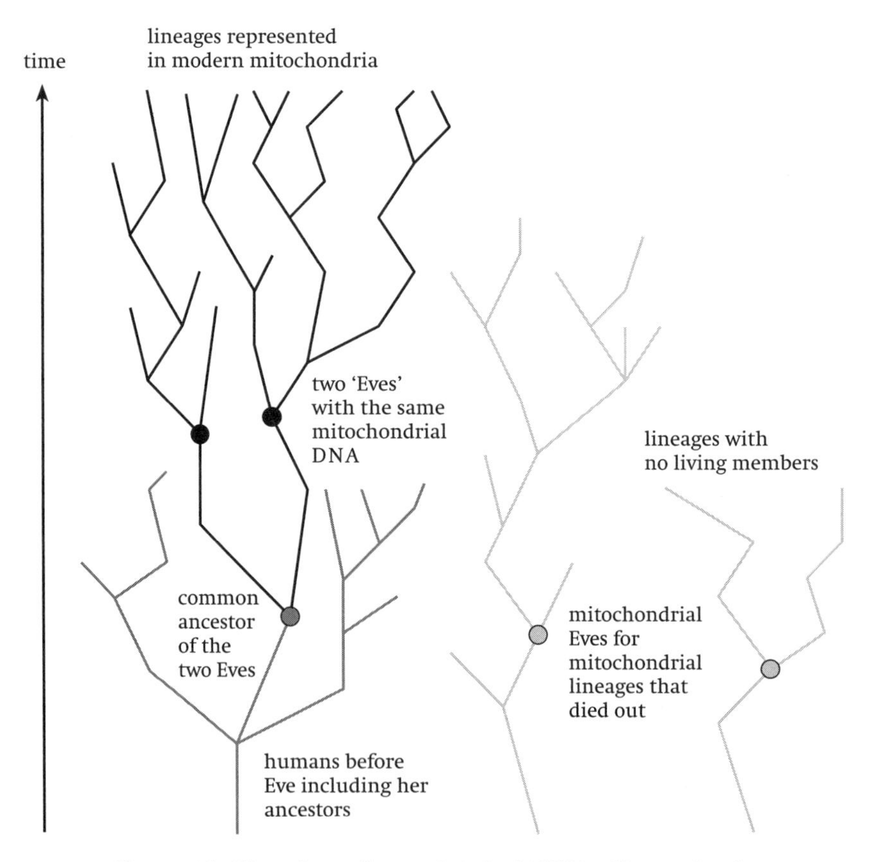

Figure 18 *The difference between lineages of mitochondrial DNA and lineages of people.*

lescence time will increase; and if there are a lot of missing morphs, it will increase considerably. So what the analysis showed, at best, was that (say) 95% of people alive today had a female ancestor among the 'core' group that, 200,000 years ago, all had the same mitochondrial DNA.

That doesn't sound like Eve to us.

The molecular clock hypothesis also causes us concern. It is surely correct in a qualitative sense: it takes more time to produce more mutations. But its justification in a quantitative sense is open to challenge. The usual argument is that mutations are induced by radiation and that radiation levels have been roughly constant over time, so the 'mutation rate' provides the basis for a clock. However, what we observe when we employ this technique is not all mutations, but only those that lead to viable organisms *whose descendants still survive today*. This introduces a strong element of selection, and it seems likely that the resulting clock ticks at different rates in different parts of the genome. For example, it may be that some genes, or some regions of a gene, cannot easily be mutated without killing the organism that inherits them. If so, those regions are unusually 'stable' to mutations – not because mutations don't occur, but because they are not inherited by future generations, because there *are* no future generations with individuals bearing those genes. Once selection enters into the picture, it messes up the neat, tidy mathematics of DNA sequences, mutations, and probabilities.

The resulting picture of evolution is different, in many ways, from neo-Darwinism. In particular, *organisms* again play a role, not just their DNA. Modern genetics certainly recognises this, but it still tries to make DNA paramount and organisms secondary. We think that the two are complicitly and inseparably linked, and for the rest of this chapter we shall look at evolution as an example of complicity. One reason why neo-Darwinism is popular is that it directs attention *away* from one of the biggest gaps in scientific knowlege: the connection between DNA sequences and the organisms that develop 'from' them. (We think that gaps like this should be bridged, not ignored.) The popular image of DNA as a kind of blueprint for an organism is seductive, hence widespread, but seriously misleading. An organism's DNA sequence – its genome – is more like a partial recipe, listing ingredients and implying how and when to mix them, but missing out key facts like the temperature of the oven, because many such items are provided not by the organism's genome, but by its mother. In consequence, alleles do not always have a direct match to characters. Even the textbook stories find it necessary to face up to this lack of direct correspondence between alleles and characters, but their usual approach is to fudge the issue. They compromise by describing 'dominance' of alleles: brown-eye alleles trump blue-eye alleles, so if your father gives you one and your mother the other then you will have brown eyes. But that is about as far as the textbooks get. The textbook story does happen quite often, especially in the laboratories of scientists who were brought up on the textbooks and only ask the kind of questions that the textbook story raises. (See our earlier discussion of the experiments due to be carried out on the yeast genome, which suffer from exactly this oversight: we return to this criticism shortly.) In particular they will ask 'how "genetically fit" is an animal carrying *this* allele?' That is, in the Game of Life, does it perform better, or worse, than others? But very little attention will be paid to breeding success, or survival; and when it is, the assumption will be that there is a direct link from alleles to characters, a 'mapping' from genes to fitness. This approach leads to the conclusion that if an animal is a 'winner' then its progeny and its genes will take over the species and eliminate the competition. If it's a loser, with a 'bad gene', then it will die out.

Unfortunately, this simple idea that 'organisms with good genes win' cannot possibly work in the real world, where 30% of genes have several alleles and most of them have no effect on survival anyway. A soccer analogy may help to make the point clearer. Soccer teams compete, and some win more often than others. The genetics-text approach would introduce a 'fitness factor' for each team, and try to work out whether green shirts were fitter than red ones, yellow socks fitter than blue ones, teams with strikers over six feet tall fitter than those without, and so on. All of this is a pale shadow of the real game, where the team wears a different strip (soccerspeak for 'uniform') depending on whom it is playing, and this week's striker may be out of the match next week with a red card (banned for serious foul play).

In fact different alleles are often nature's way of making contingency plans. Allele A may be the best when conditions are warm and humid, but allele B is better when they are dry. Frogs, for example, are forced to have several different contingency plans up their sleeves because they develop in ponds that do not have temperature control. Or allele P protects the organism against disease X, whereas allele Q protects the organism against disease Y. Even if it is currently warm and disease Y is not around, it pays to retain such alleles in the gene pool, just in case conditions change. Sexual reproduction – in which most alleles come in pairs, one of which may not even be used at all – is a good way to keep such 'back-up copies'. This possibility also makes the Grim Sower a good idea: to find out whether an old allele has come back into vogue, keep making organisms containing it and see if they survive. This fact is one of the reasons that we (JC & IS) are unsurprised by the large number of 'functionless' genes just found in yeast. That is what you should *expect*. Some are no doubt old bits of rubbish dumped in the genetic attic that have not yet been thrown out, but a lot have been stuffed up there because one day they might be needed. Taking yeast that lacks those genes and growing it *in laboratory conditions* will tell you nothing about what the missing genes are 'for'. In fact it runs the risk of suggesting that they are 'for' nothing at all. It is like removing the windscreen-wiper from a car *on a dry day* and finding that it behaves exactly the same without it. One day it rains, and *then* you find out what that bit is for. In his article about the yeast genome, Holmes remarks that 'knock-out experiments have also deepened one of the big mysteries of genetics: the new experiments confirm that many genes may be superfluous to the cell's well-being.'

We really don't find this terribly surprising.

In fact, if you think of evolution as a game, it's obvious. And that image illuminates much more than just the potential roles of 'functionless' alleles. Let's see why. It is not coincidence that we chose a competitive game – soccer – above: in fact, as observed in Chapters 2 and 3, we can gain some interesting insights into the evolutionary story by appealing to the theory of games. There is a standard way to do this, which equates 'evolutionarily stable strategies' for organisms with certain types of winning strategy in very simple mathematical games, but that's not what we have in mind here. Instead, we need the concept of a 'nightmare game', one for which a winning strategy is incredibly complicated and unstructured, despite the simplicity of the rules. Many games, indeed in a reasonable sense most, are nightmare games. We shall argue that evolution is really a nightmare game, but that most attempts to explain evolutionary events try to treat it as a dream game.

To see why we think this is so, look at a tree – not a game tree but a real one. It exists because its ancestors have been long-term winners in the great game of Survival. DNA records of the strategies used by its millions of predecessors – in *winning* games only – are built into its every cell ... No wonder it's complex *inside*. The implicit rule 'the name of the game is to win' provides a simple contextual imperative that generates all of that internal complexity. But notice how different 'ability to win' here is from 'possession of good genes'. The phrase 'good genes', and the way it is interpreted in nearly every textbook and lecture on population genetics, carries the implication that a particular allele is superior to some other, alternative allele, in all circumstances. This is a bit like saying that 'attack your opponent's pawn structure' is *always* a good strategy in chess – even in the opening, even in the endgame when the opponent may not even *have* a pawn structure. The strategy for chess – and that adopted by a tree in the game of Survival – is highly context-dependent. It depends on the 'position' of the game at the time the 'move' is made. A strategy that might be excellent in some circumstances may be fatal in others – and the same goes for an allele. A real organism is a mass of compromises, not a finely optimised piece of machinery.

Moreover, in evolution – unlike chess – the 'rules' are not fixed. Every time a new organism evolves, or an old one evolves a substantially new trick, the rules effectively change. Evolution is a *self-modifying* game in which the rules depend upon the state of play. There is complicity between the rules and the overall state of the game. Life would be much simpler if DNA really were a blueprint for organisms, and that was all there was to inheritance – but that's not so. Even in human games, the invention of new rules or the emergence of new strategies can change the odds for each player. Organisms can also change the rules, by methods that often look suspiciously like cheating – loading the dice in favour of one's own offspring. So as well as genes, there is privilege. The list of strategies by which parents can provide their offspring with a non-genetic 'flying start' is almost inexhaustible – the yolk of the chicken's egg, the paralysed spider left for its maggot by the tarantula wasp ... Presumably such strategies offer sufficient advantage to outweigh the investment required by the parents when they provide the privilege, otherwise they wouldn't be used. But *we* think that there is a deeper reason why privilege is an excellent strategy, one to do with the structure of the Game of Survival, and the insight comes from the analogy with a snooker break, our metaphor for reproduction.

A break is a way to keep much the same sequence of moves going, without requiring exact repetition. Think of the basic two-shot cycle of a snooker break: pot a red, get position on a colour, pot the colour, get position on a red: now repeat. Think of potting the red as the parental generation's move in the game of Survival, and potting the colour as the offspring's. In this analogy 'pot' corresponds to 'survive', and 'get position on the colour' is privilege, passed to the offspring by the parent. It is not a feature of the shot that pots the colour: it is an extra feature of the previous shot, which pots the red *and* sets up a nice shot for the offspring. Snooker players, incidentally, are prepared to make 'parental sacrifices' to pass on such privilege – for example they will elect to play a more difficult red if it is easier to gain position on the next colour. Why? Because it's the best way to *keep the break going*. In evolutionary terms, what privilege does is to link successive generations together so that to some extent selection can act on both, simultaneously, as a kind of evolutionary unit. Selection acts on the entire cycle, not just on one stage – the parent, or the child – which results in cycles that are robust and successful.

This may sound fanciful, but we think that something like it is often valid. For example, imagine a species of finch that feeds on caterpillars. In times of plenty, there is no problem in 'keeping the break going'. Outside the breeding season, adult birds catch caterpillars and eat them, no problem. Ditto during the breeding season: now they go off and catch caterpillars, bring them back to the nest, and feed them to their young. But now imagine a year when there are too few caterpillars to go round. Obviously an adult that is better at catching what few caterpillars there are has a better prospect of survival. Moreover, the same goes for its offspring: thanks to privilege, if your parents are good at catching caterpillars, *you* grow up big and strong too. So the *same* character in the adult affects both its own survival prospects and those of its young.

Without privilege, in contrast, each generation is on its own, forced to reinvent the evolutionary wheel by trying out the same genotypes and phenotypes as its parents– maybe improved a bit by a few genes dropping out of the gene pool, maybe misimproved by a mutation, maybe enhanced by a mutation, but – across the species – not greatly different from last time. Privilege is a radically different strategy from anything that affects only individuals, one generation at a time, and when it works it offers very strong survival advantages. So natural selection will reinforce the 'privilege' strategy, once it gets going.

We can now go further, and consider the question of evolution on a deeper level. Evolution strikes many people as being highly implausible, because today's world is so amazingly complicated and finely tuned. They find it hard to see how such a wonderful, interlocking 'machine' could have come about through the blind workings of chance. What they often fail to bear in mind is that today's world is a snapshot of a game that has been in progress for five billion years. If it were possible to watch the entire game, the current position would make much more sense. We can't do that, but we can make reasonable inferences based upon our understanding of chemistry, biology, geology, the fossil record, and anything else that seems to be implicated in the game. It is a game that began when the first replicating molecular network accidentally – but if Kauffman is right, inevitably – assembled, and it has never stopped. It is the game of Survival, and you can consider it as having either countless trillions of players (all living creatures) or only one (the global ecosystem). Our language lacks the words, and our science lacks the concepts, to describe this game adequately, so we give it a label: Evolution. Putting a label on a concept doesn't mean that you understand it, but at least it makes you recognise that there's something to understand.

Evolution is a very curious game, going well beyond our idealised mathematical formulation. It has trillions of players (or one, the global ecosystem), and no fixed rules. Players (or The Player) are free to make whatever moves the constraints of physics and chemistry permit. The 'success' or 'failure' of those moves is determined not by a referee with a rulebook, but by whether the player – or its progeny – get to keep playing. In a sense the real players are *lineages*, chains of organisms, each a parent of the next.

Biologists will tell you that evolution has no goal, and on one level of description they are right. It has no preset purpose, it contains no coded representation of its own future. But neither does water, and it still flows downhill, not up: this shows that dynamics in phase space can impose an overall directionality on processes, goals and purposes notwithstanding. And on a different level of description evolution does have a goal, one that exists only in retrospect. The goal of evolution is *to stay in the game*. Players who (unwittingly) achieve this goal continue playing; all the rest become losers in the most brutally literal sense – they die without breeding. 'Goal' is of course the wrong word. The usual sense of 'goal' is a reductionist one: look inside the system and you will find a 'search image' of the future which acts as a cause of present behaviour. The word applies to a sparrow seeking an earthworm, or to a human seeking religious enlightenment, but not to evolution. 'Unforeseen destination determined by contextual constraints' is what we mean, but the only common word with that meaning seems to be 'destiny', which has all sorts of misleading mystical overtones. 'Attractor', in the dynamical sense, probably comes closest. It is a bit like firing a shotgun at a ping-pong ball: the individual pellets do not have an inbuilt drive to hit the target – indeed they are totally unaware that there *is* a target – but nevertheless those few pellets that happen to be travelling in the right direction *do* hit it. They are thus singled out for 'success' by the choice of context. The twist in evolution is that it creates its own context, and 'hitting the target' is not an arbitrary choice, but the essence of the game – the unique precondition for the game to continue.

Evolution, then, is a five billion year-old self-modifying planetary scale game which carries around partial records of its own past. And just as it has a (contextual analogue of a) well defined goal, so it also has a (contextual analogue of a) winning strategy. Like the goal, this strategy becomes apparent only in retrospect: *make winning moves*. A winning move, of course, is one that lets you stay in the game. And the way you find out what moves win is again rather like firing a shotgun at a ping-pong ball: you try as many moves as you can, and sometimes one of them works. Because evolution is a learning process, and because today's creatures are the descendants of players that have consistently made winning moves, the creatures that inhabit present-day Earth have become pretty good strategists. They don't *know* what strategy they are playing, but they play one that wins often enough to be useful. Individuals may die, or even whole species, but the Game goes on forever and it gets wilder and more convoluted as it does so. Their strategy is encoded in their genes and passed on from their parents.

One result, which we can see clearly from our 'game' analogy, is that the survival strategy becomes increasingly ritualised. This point deserves examination at length, so that's what we'll do now. When we look at an ecology – an interacting system of organisms, animals and plants and insects and worms, or whatever – we do *not* see the Game of Evolution being played out in the way our liar-to-children teaching myths seem to suggest, with badly designed animals locked in mortal combat with their eventually triumphant superiors. Part of the reason is that the real story is more subtle anyway, but in fact the game *is* being played, still – but by professionals rather than amateurs. The winning strategies have become entrenched, and most moves have become routine. Organisms still experiment with novel moves, but hardly any of those succeed, and nearly all of the failures are removed when still very young, so we don't notice them. Think of the millions of acorns that never get to grow up into oak trees. The ecosystem has become 'canalised', just like the genetics of wild-type organisms; that is, it is much more able to survive environmental changes without (apparently) altering its behaviour.

This is an important point, affecting our entire conception of how evolution and ecology are related, and it is worth spending enough time to appreciate it. As an example consider 'Darwin's finches' on the Galápagos islands. There are thirteen distinct species – some rest in trees, some rest on the ground, some inhabit cacti, some live amid the mangroves, some eat seeds, some eat snails, and so on. Between them they have occupied the main 'ecological niches' open to birdlife in this very impoverished ecosystem. The current division of resources is stable, in the sense that new *species* do not seem to be evolving, although we can detect evolution in microcosm as the proportions of different species vary from year to year in response to changes in environmental conditions.

In what sense, however, are they really *distinct* species? There are two accepted concepts of 'species': one is taxonomic, the other reproductive, and they do not always agree. Taxonomically two (sexual) organisms belong to the same species if they resemble each other in sufficiently many ways, *including* being able – potentially – to interbreed, giving rise to fertile descendants. The reproductive definition, promoted energetically by Ernst Mayr, is that a species is a breeding group. If organism A can in principle interbreed with organism B to yield fertile descendants, then A and B belong to the same species. Now, taxonomically, the thirteen types of Darwin's finches are separate species. Reproductively, however, they are not: they are thirteen 'morphs' within a single species. Birds from different morphs can interbreed with each other to produce fertile 'hybrids', cross-bred individuals whose genes combine those of both morphs. For example a snail-eating finch can interbreed with a cactus finch. Nevertheless, despite this 'gene flow', the morphs remain separate. Why?

There are at least two possible reasons. One is that since different morphs prefer different habitats, the opportunity for crossbreeding is small. The cactus finch likes the dry parts of the island and seldom gets an opportunity to breed with a snail-eater that stays in the wet parts where the snails live. However, this cannot be the full explanation, because there is plenty of overlap at the edges of habitats. Over a relatively small number of generations, the hybrid gene combinations would diffuse widely. However, you don't find them. The real reason is the Grim Sower. Hybrids do occur, but they do not survive *in the current ecology of the Galápagos islands*. A finch that combines some of the strategies of a snail-eater with some of those from a cactus finch falls between two stools, and cannot compete well with either. It searches for damp-loving snails among the dry cacti, perhaps. So the hybrids die young, and do not go on to breed. There is nothing 'wrong' with the hybrids. All of their genes are 'good' – inasmuch as this concept makes sense – because they come from 'good' parents. But the combination – not so much of genes but of characters – doesn't work when the thirteen existing species have already carved up the territory to suit themselves.♪

This view of how organisms partition their environment into 'eco-territories' shows that we used the wrong word when we talked of species 'occupying' niches. Actually they *create* niches through their collective and complicit interactions with each other and the rest of their environment. After the ecosystem has 'settled down', the viable niches are pretty well determined and segregated. The winners are the professional snail-eaters and the professional cactus-eaters, whose strategies have a proven track record. The hybrids that nest among the dry cacti, but want to eat the moisture-loving snails that inhabit the marshes ten kilometres away, are incompetent amateurs, making the wrong moves; and they lose every game they play. The Grim Sower ensures that they go on playing long after their incompetence has been amply demonstrated – and they go on losing.

We can speculate that during the very early states of evolution of Darwin's finches the game was, briefly, different. Amateurism was the order of the day, because the ecosystem had not yet learned which strategies were winning ones. We can imagine a thought experiment, a tiny essay in eco-SF, that may bear some resemblance to what actually happened. It goes like this. Once upon a time the Galápagos islands had no finches, only seabirds – boobies, petrels, frigate birds ... One day a small flock of finches, blown astray by a storm, arrived upon the islands. Assume for the sake of argument that they were seed-eaters. Surrounded by unexploited resources, their babies prospered, as did *their* babies, and so unto the six hundredth generation (or more). At that time all sorts of weird experiments might have survived, transiently, before the population got so large that competition really became a feature of everyday life. There might have been wading finches in the marshes, and fat little finches with long beaks specialising on nectar in ground-hugging plants. These experimental finches would have possessed alleles already present in the early 'founder population' – but in new combinations, although the odd mutation would have helped them to diverge. Most of the finches, however, were still seed-eaters, because that's the lineage that got going first, and it grew exponentially fast. Soon there were thousands of finches, by now proficient professional seed-eaters – and the supply of seeds started to run low.

Did finches do epic battle, eyeball-to-eyeball over every last seed, nature red in beak and claw?

It's a lovely image, but we doubt it very much. If they had ever reached that state, the whole lot would have been teetering on the brink of extinction. Instead, the occasional young bird that might have grown up to be a wading marsh-finch or a fat little nectar-eater, or at least to evolve in that direction, never did so, because its (seed-eating) parents were overwhelmed by weight of numbers. As the Grim Sower sowed, so did the Grim Reaper reap. Just occasionally, there arose a novelty that happened to fit an unexploited part of the Galápagos environment, an accidentally talented amateur whose game was so different that the professionals didn't know how to cope with it – a left-handed bowler, a two-fisted hitter ... The odd finch with unusual genetics or just unusually adapted physiology discovered that snails were good to eat too, and it prospered, relatively unchecked. Those birds left the seeds for the rest to squabble over, and their morph began not so much to branch off the original phenotypic trunk, but to solidify out of the primal soup of the total gene pool. In this manner the ever-growing population of finches carved itself a multiplicity of niches – viable lifestyles – in a whirling dance of complicity. They were learning to play the game while the game itself was changed by their moves, and they converged on whatever strategies worked in the prevailing environment. A large and highly influential part of that environment was all the other finches. And so the amateurs disappeared, and the professionals took over the game forever. The transition from amateurism to professionalism probably did not require that many generations, not on an evolutionary scale. As soon as a hybrid snail/cactus finch cannot compete with either of its own parents, the hybrid niche can no longer be carved. So rather rapidly the strategies of the *winners* settled into standard patterns. We shall call this process 'canalisation of the ecosystem' by analogy with canalisation of genetics in organisms.[♪]

By the way, the above scenario assumes that environment is more or less stationary. If the environment changes too fast, then canalisation can't 'track' it so successfully. Ironically, this strengthens the need for contextualism. Recent theoretical work[♪] shows that in variable environments evolution leads to organisms that would not survive in a constant environment with the same mean. For example in a variable environment it pays organisms to 'hedge their bets' against a bad year, but in a constant environment it doesn't.

Now we can return to our central argument: that evolution cannot sensibly be referred solely to genes. The phenomenon of canalisation – both in organisms and ecologies – shows this clearly. The complex strategies evolved by millions of previous generations of organisms can be written into – or out of – the fabric of the planet instead of the fabric of the genes. In fact many resources available to organisms are themselves the result of previous organisms: we've talked more than once about how the oxygen in the air was first produced by bacteria some 3.5 billion years ago.

Evolution, then, is a nightmare game, and the 'positions' that we see today have arisen as a result of the *strategies* required to play it successfully. They are not simple reflections of the *rules*. Moreover, in the usual twisted manner of evolution, the game has evolved along with the strategy. If you change the game – through global warming, say – then the strategies may cease to be appropriate, and fail. Given that a game as simple as Boxes, or chess, leads to strategies so complex that they can be represented only by incompressible structureless lists, is it any wonder that the strategies for a game as intricate as evolution are incomparably more complex? A reductionist analysis can unravel the occasional common thread in this complexity, but as it peels back layer after layer it descends the ever-ramifying tree of the Reductionist Nightmare. This is not to say that reductionism teaches us nothing useful: on the contrary, it provides our best hope of working out what the strategy *is*. But it is largely useless for understanding why the strategy came about, which is a contextual question that is more usefully treated on a contextual level.

Of course the strategies of the Game of Evolution are not *completely* structureless. In the same manner, even though we don't know how to guarantee a win at chess, we know a lot of useful rule-of-thumb principles such as 'maintain a strong pawn structure', and a lot of trite but vital short-term rules like 'don't let a knight fork your queen'. In an evolutionary analogy these correspond to things like 'inhabit regions with an assured food supply' and 'don't put your head in the tiger's mouth'. But it is important to realise that the only bits of strategy that we *know* to be sound are those trite and obvious rules for avoiding instant death. The high-blown principles of good structured play *might* be illusions, apparent patterns in the nightmare Strategy of Everything. Perhaps one day it will turn out that the best way to play a chess game is to open by moving only knights for the first twenty-six moves, and then to embark upon an apparently random sequence of moves 4017 steps long. We doubt it, but the strategies for many much simpler games are at least as bizarre.

Speaking of bizarre evolutionary strategies ... how did the Zarathustrans evolve? We'll take a first stab at the story here: in the next chapter you will see what such a story might be good for. Right now, consider it an exercise in stretching your mental muscles so that they acquire new flexibility when it comes to thinking about evolutionary possibilities.

Liar-to-adults I find it difficult to believe that such a complete entity as the Zarathustran octuplet can have come about by mere chance. Surely some octimist creation must be–

Creator-of-creations [*Who knows about creativity.*] Ah, but creation requires a creator, and octimism is merely a philosophical principle.

Liar-to-children Neither of you knows what you are talking about. I shall explain how such complete entities as us evolved, without any need for a creator or even for creativity.

Pursuer-of-sicknesses [*'Doctor', the eighth Zarathustran role. Now we have met them all.*] It will have to be a pretty good story. Today's highly evolved Zarathustran has some amazingly 'well designed' features, you know. Your theory is going to have to explain every one of them. And their imperfections.

LtC What features?

LtA What imperfections?

PoS For a start, there's Zarathustran anatomy (figure 19). We are bipedal, with a short fat body and a long neck. Our main sensory organs – eyes, ears, proboscis – are located on our head, which also houses the heart.

Our eyes have a large lens surrounded by a ring of multifacets. The head is supported by a long neck, which contains three major tubes: one carries blood up to the heart, one carries it down to the brains, and one conveys food to the stomach. The stomach is located roughly in the middle of our body. At each side of the body is a set of gill-covers, leading

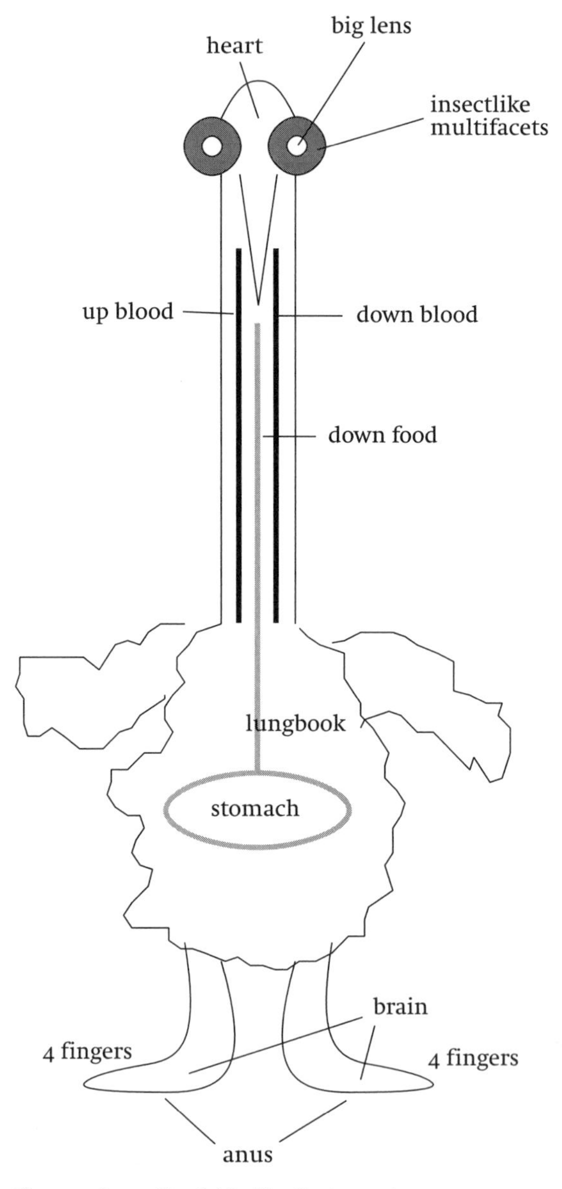

Figure 19 *Our working sketch of Zarathustran anatomy.*

to a' lungbook' of the kind found in terrestrial spiders, and the external openings are protected by flaps. We have a double brain, one in each foot. I say 'foot', but our feet are really hands. Each has four fingers, and an anus in the centre for excretion.

LtA Like I said, perfect in every way!

PoS No, there are imperfections. For example, placing our proboscis between our eyes means that our olfactory senses all adopt the same viewpoint as our visual sense. If you smell what I mean.

CoC You will have to explain more than just our anatomy, you know.

PoS What else is there?

CoC Octual structure. Our minds function best in octuplets. Isolated individuals, even septuplets, are unable to cope. And philosophy – our octimist view of the universe.

LtC But the universe *is* octimal!

CoC Some daring thinkers suggest that octimality is a preference of our own minds, which we project upon an objective external reality.

LtA What nonsense!

PoA Excuse me, but can I ask a silly question?

The Regulations Silly questions may only be asked during off-duty periods.

PoA This *is* an off-duty period.

T R Then you may ask a silly question if you wish.

PoA [*Changing his mind.*] Oh, delete it. It's inapplicable.

T R I cannot delete anything. I am write-once-read-many.

CoC Liar-to-children, your theory of Zarathustran evolution must also explain The Regulations.

T R No, *I* explain The Regulations. Often. [*If a featureless purple sluglike being covered in graffiti can look smug, this one does.*]

LtA Why? Surely The Regulations has always been as it is now?

CoC Not at all. If we evolved, then it follows logically that The Regulations must have co-evolved with us. Complicitly.

LtA [*Becoming visibly agitated.*] You are telling us that morality is relative? But that is outrage–

LtC Be calm, Adultliar, while I tell you all The Story of Zarathustran Evolution. Are you standing comfortably? Then I shall begin. Once upon a time, The Regulations was just a useful pet that hung around the burrow. People put graffiti on it, and it would look after the children. Now, you will not believe the next bit, but the fossil records show it quite clearly. Zarathustrans were much more primitive in Those Days.

They looked a lot like us, but their feathers were more orange than yellow and not nearly so well preened. They were so primitive that they lived as– [*hushed voice*] – septuplets. Hence their scientific name, *Zaro septimis.*

PoA Yeah, we know. And they lived in irregular burrows instead of nice octagonal tunnel complexes. Do not expect us to be shocked, we are grown adults, remember?

LtC Apologies, when in liar-to-children mode I sometimes lose my sense of context. Anyway, what they did not realise was that they needed an eighth member of the group.

PoS Why?

LtC Without a Master-of-rings they could not upgrade The Regulations. They obeyed them, of course, but only in a tribal 'anything not mandatory is forbidden' manner – and what got graffited on The Regulations was largely a matter of chance. True, the 'delete whichever is applicable' cult made a point of always *disobeying* The Regulations – but since what was on The Regulations was random, they were merely obeying a complementary version of The Regulations in which every rule was negated. There was no consensus. They had no inkling of the civilised way, the octimist way: to be *complicit* with The Regulations. They lacked a Ringmaster, who could determine what the consensus was, and graffit it on The Regulations so that it could never be disputed again. Before septuplets acquired a Ringmaster, they used to delete whichever was *in*applicable, leading to an anarchistic state of continuous referendum. The difficulty was recursive because they could not even agree on whether they had agreed on something.

LtA But how did the Ringmaster get appended?

LtC It came about because all adult Zarathustrans are fond of tadpoles.

LtA I am not sure I would say 'fond'. It is true that it is necessary to fondle tadpoles during election campaigns, but that is a matter of political necessity, not voluntary pref–

PoA Oh, do you not just *love* the dear little things? With their wiggly little tails and great big goo-goo eye? [*Starts to go broody and is given a surreptitious kick by Hewer.*]

LtC Well, primitive septuplets used to bring orphaned tadpoles into the burrow. Now, if a tadpole is too old to join the group mind, it hovers on the fringes trying to make sense of what's going on. And one of them – the very first Ringmaster – did this so effectively that it made sense of what was happening to the other seven as well! So it was put in charge of graffiting The Regulations with the consensus decisions. Suddenly Zarathustranity acquired a unified smellpoint on group problems, coupled to the ability of The Regulations to pass on acquired characteristics to its offspring. It was a major breakthrough and it put us on the road to fully fledged sentience. In fact you might say that *Z. octimis* really evolved only when The Regulations joined the group mind.

LtA The *Regulations*? You are saying that a mindless beast like The Regulations is part of the group mind?

LtC Woops, apologies. A nonist slip. I meant to say that *Zaro octimis* really evolved only when the *Ringmaster* joined the group mind. Ringmaster, regulations – a slip of the palate.

LtA [*Mollified.*] An excellent and compellingly logical story, Liar-to-children.

The idea of evolution as a game also sheds light on one of the commonest objections to Darwin's theory: the fact that many of the structures found in today's plants and animals have a stunning simplicity of 'concept' and 'design'. Think of petals as an 'advertisement' to attract pollinating bees, or the eye as a light-detector, or the heart as a pump. And if we look at these structures and their simple functions at more detailed reductionist levels, they are revealed as being even *more* complex. Petals are actually leaves whose function has been subverted in complicity with insect evolution. A simple beating heart resolves into a mass of millions of interconnected muscle fibres, all of which have to be synchronised by electrical pacemaker waves before they can produce an effective beat. The information processing that goes on between the retina of the eye and the visual cortex in the brain is incredible: the eye is in no sense a simple camera. Why does nature assemble huge, intricate networks of delicate complexity in order to achieve simplicity? It would, on the face of it, make more sense if living creatures *behaved* in incredibly complex ways, as well as being *built* in incredibly complex ways. Or, given that their behaviour is (on the whole) relatively simple, that their internal structure was comparably simple. However, very few things work like that.

An analogy with technology suggests why there should be such a tendency towards increasing complexity. Human technology is a method for achieving simple, easily stated goals (fly to the Moon, cure cancer, breed tomatoes that can be transported without sustaining damage) by highly complex means. We can see why there is such a mismatch between the simple and the complex in that case. The overall goal may be simple, but the means for achieving that goal are dictated by what is possible within the laws of physics, chemistry, and biology. We no longer get to choose, and the best we can do is to cobble together various ill-fitting bits and pieces that relate, more or less directly, to our goal. In solving one problem we create others, which must be solved in turn. In order to keep the wind from our faces in a speeding car, we add a windscreen, but then the rain sticks to it so we have to invent the windscreen wiper. If it's *not* raining then we can't use the windscreen wiper to get mud off the windscreen, so we add windscreen washers – artificial rain. The water in the reservoir turns to ice in winter, blocking the flow, so we put in chemical additives to prevent freezing ...

It is tempting to explain the burgeoning complexity of life in the same way, but there is one final obstacle. Technology is driven by human goals, but all biologists learn at their tutor's knee that nature has no goals, no purposes, no intentions. Evolution, they tell us, has no inherent drive towards any particular direction – it just happens. So the analogy with technology is misleading.

Not so.

Yes, of course nature does not have goals, purposes, or intentions, for those are human attributes. But it may well have an overall dynamic, a flow, a tendency to behave in one way rather than another, even a limited kind of 'predestination', because everything in the universe is affected by the context in which it operates, and subject to rules that limit its possibilities. To see what we mean, imagine a theory of the flow of water that is founded upon the molecular nature of the liquid rather than the conventional continuum model favoured by fluid dynamicists. Suppose further that the only observations available are those of water in its natural habitat – streams and rivers, rainstorms and lakes – and that the surrounding landscape is invisible. We observe the water and try to explain its patterns as it meanders past a randomly strewn boulder or drops off an eroded ledge in a waterfall. And the story we tell is one of contingency and selection. At root, the flow of water is merely the random wanderings of molecules. There is no purpose, no goal, no order; just stochastic jiggling. However, as the molecules jiggle around, some of them tend to accumulate, whereas others tend not to. So the first kind *do* accumulate, whereas the second rapidly jiggle themselves elsewhere until eventually they accumulate too. As a consequence of this random jiggling plus selection based on the purely contingent factor of accumulation, the mass of water moves. It could, it seems, go anywhere. However, things look very different when you can see the landscape. The places where water molecules tend to accumulate are contingent upon their surroundings, for sure, but once you know the surroundings you can predict what will happen. Water accumulates where the potential energy is lower – in short, it runs downhill. The geography of the surrounding landscape imparts a definite direction. The random jigglings are important for only one reason: they render the water fluid, so that it can flow into those lower regions. Without the random jigglings, motion would be only potential, but with them it becomes actual. And while the mechanism whereby individual molecules select where to accumulate is also contingent, it inherits direction from the landscape. Individual molecules *can* climb uphill, but on the whole they don't.

Evolutionary theorists have the wrong view about determinism in biology because they have the wrong view of physics. They think that physics tells us that the flow of water is completely deterministic – as if the mathematical equations used by fluid dynamicists to describe the flow of a fluid were what the fluid itself actually does. From this mistaken view they derive, by analogy, a wholly misleading contrast between Darwinian principles and physical laws, and this confuses them when they start thinking about the global constraints that affect evolution. 'Goal' and 'purpose' are not the right words, to be sure, but there are dynamic effects in evolution, resulting from constraints. Mutations make phenotypes fluid enough to change, selection implements particular changes preferentially, but the overall result is more like water flowing through a landscape. However, it is an invisible landscape, formed out of the nearby 'potential' phenotypes and constrained by context – a mathematical 'phase space'. It includes not only what happens, but what could have happened instead. It may sound metaphysical, but such imagery lies at the core of how physicists and mathematicians currently think about *all* dynamics. Energy surfaces are also invisible landscapes – you can't see them. Who cares? Dynamical phase space is just as real as ordinary space – it makes itself felt by constraining the potential dynamics so that it carries out the behaviour that we actually observe. So evolution does have a preferred 'direction', and on the whole this will be the direction (in phase space) of increasing complexity. And so organisms become ever more complicated – most of the time.

Occasionally evolution reverses that tendency, however, introducing vast simplifications, such as control of body temperature in mammals. This got rid of huge amounts of genetic contingency planning for developing eggs in lakes that were warm at noon but chilly at midnight. Evolution *can* introduce simplifications, and it does so when the system has become too baroque. Otherwise, increasing complexity is the norm.

If it ain't baroque, don't fix it.

5 Universals and Parochials

In the fossil layers of the Burgess Shale are the remains of strange, soft-bodied creatures. So strange are they that some palaeontologists believe that they represent more biological diversity of form than now exists upon the entire Earth. Indeed some of the forms present in the Burgess Shale have no surviving descendants at all.

Reconstructing the shape of these creatures, in three dimensions, is immensely difficult because their fossil forms are squashed flat, and a certain amount of careful interpretation is necessary. For a long time one of the most strikingly bizarre Burgess Shale creatures, of a form not seen at all in today's world, was *Hallucinogenia*, which – it was thought – stood on the sea floor using a set of seven pairs of sharply pointed struts. Seven tentacles with two-pronged tips wiggled on its back, together with a bunch of even tinier tentacles. It had a blobby head, and its rear end was a tube.

It then turned out that *Hallucinogenia* was really a form that is still common today. The 'struts' were spines on its back, the 'tentacles' were its legs.

It had been reconstructed upside down.

We have already offered you two versions of what happened during the evolution of life on Earth. We described the origins of life, the endless aeons when bacteria – many of them photosynthetic and emitting oxygen – were the dominant life-form, the development of eukaryote cells with nuclei, of multi-celled organisms including complex animals with brains, and the appearance of organisms that could learn. In our first version of the evolutionary story we took the point of view that genetic differences underlay this tree-like structure, causing different kinds of organisms to 'branch off' along their own particular paths. We saw DNA as a kind of genetic tree that provided a thin thread running along the core of each branch, determining where and when new branches would form, so that organisms were *made* more complex because of genetic changes. The second version was rather different: it took a general, contextual view and saw evolution as a game played out in a phase space. In Chapter 6 we shall add a third ingredient to the pot: hereditary mechanisms called 'privilege' that operate outside of DNA. We shall tell the story from the viewpoint of yolk and learning, the investment by parents of energy and materials for the benefit of their offspring. Before becoming that specific, however, we shall take an excursion and look at alternative possibilities for life on Earth – and off it.

It's not *really* a digression: it's more a necessary exploration of evolution-space, to set terrestrial evolution in a broader context. We shall make good use of that context later, because one of the aims of *Figments* is to explain why having many different viewpoints is valuable in science, and to show you how to exploit such a multitude of viewpoints more systematically, by employing the imagery and meta-viewpoint (viewpoint on viewpoints) of phase spaces – the spaces-of-the-possible that conceptually surround instances-of-the-actual. Phase spaces are a formalisation of contextual thinking, and they help us answer the question 'why is it like this?' by considering the alternative question 'what if it were different?' The evolutionary history of life on Earth can be embedded in several different phase spaces, depending on what kind of question you are interested in. It can be viewed in terms of organism-space if your questions concern features of organisms; in terms of DNA-space if you are interested in tracing the history of mutations; or in terms of privilege-space if you are interested in how adults can give their offspring a boost by linking succeeding generations into an inseparable unit.

These are not the only possibilities. For example, evolution could equally well be viewed in terms of innovation-space, where gadgetry such as brains and swim-bladders live, and whose dynamic explains how the gadgets improved over time – a kind of 'technological' viewpoint with nature providing the technology. We might feel tempted to argue, as brainy humans, that in order to produce such marvels as us, the evolutionary tree has reached up to its noblest heights, stretching its very topmost twigs in the direction of the biggest and best brains. Our pride in being top of the tree might, however, be slightly deflated by the realisation that the cod could equally reasonably claim that the topmost twigs were those that culminated in *him*, because the branch of maximum perfection is that which leads to the swim-bladder. We gain a more rounded and realistic view of what is 'highest' in evolution when we recognise that the cod has a valid point. By comparing our own particular evolutionary gadgetry with that of other earthly creatures, we can be led to recognise that there is no unique view of what is most sophisticated.

Can we take an even wider perspective, and enlarge our evolutionary phase space even further? What insights do we hope to gain by doing so? Well, let's work backwards from ourselves and cod, whom we left vying noisily for the topmost branch of the evolutionary tree. Both are vertebrates, they have backbones. They belong to the same phylum (one of the major taxonomic groupings) of animals, the chordates. Chordates originated from echinoderms, creatures whose modern representatives are such things as starfish, sea-urchins, sea-lilies, sea-cucumbers. What probably happened was that chordates arose from an echinoderm somewhat like modern sea-lilies; this produced a specialised kind of larva that omitted the sedentary adult stage from its development. So we and our rival the cod both go back to an incredibly ancient sea-lily, some of whose larvae blundered across the trick of not growing up properly. We could ask, without any evidence except argument from modern forms, what was so special about that particular larva – or what was so special about the circumstances in which it found itself – so that it took the option of becoming a whole new kind of animal. We could ask whether the same echinoderms would again produce the new kind of larvae if it were possible to run the scenario again.

We've already mentioned in Chapter 1 that in his book *Wonderful Life* about the bizarrely diverse fossils in the Burgess Shale and their implications for evolutionary theory, Stephen Jay Gould argues that evolution would *not* repeat the same events if it were run again. Indeed he focuses the entire book around the idea that evolution is a 'contingent' process, heavily influenced by chance, so that if we ran the evolutionary story again the outcome would almost certainly be totally different. Analogously, two games of chess, both beginning from the standard opening position, will nearly always lead to different play and in particular to a different end-game. We agree that contingency plays a role in evolution, but we don't think that its role is as great, or as ubiquitous, as Gould's book states. (Neither does he any more, to judge from more recent writings on the same topic.) So let us pose the key question of this chapter: if we ran the evolutionary story again, starting from the same primal Earth, would the same kinds of creatures arise? This is a phase space question, posed in evolutionary-history-space, which includes not just the way evolution did happen, but how it might have happened instead under different circumstances. Our answer, by the way, will be 'yes and no': the juxtaposition is an instructive one and is not *merely* a cheap way to avoid the question.

There are many terrestrial innovations that have arisen many times on many different branches of the evolutionary tree. Down near the base of the trunk, many different bacteria invented photosynthesis. We can be confident their inventions were independent because they used different chemical molecules. However, they all followed the same general path of using the Sun's energy to power their own metabolism. Similarly, many different kinds of animal have invented winged flight: insects, pterodactyls, bats, birds, various fishes. There is a common feature, however: wings. (A different form of flight, 'parascending', is used by many plant seeds, and by tiny spiders that cast themselves aloft on long threads.) As a third example: sexual reproductive processes have appeared several times – in fungi, then in red algae that did it differently from fungi, then in animals that did it differently from green plants. Again there is a common feature: the genetics of two creatures is combined in one nucleus, which is then multiplied – sometimes after other mixing procedures – into many potential progeny.

When many different creatures independently come up with the same basic feature by different routes, we can be confident that the feature is not accidental. The particular routes that they find, however, probably are accidents – otherwise they'd all solve the problem in exactly the same way. Similarly all journeys from London to New York have the common feature that they pass over water, and this is not an accident: it happens because the continents of Europe and North America are not joined by a continuous stretch of land. However, you can fly from Heathrow to Kennedy, or you can take the Channel Tunnel to Paris, fly to Bombay, travel by rail to Calcutta, take a fishing-boat to Singapore, and then fly via Tokyo and Los Angeles to New York, so there is nothing terribly unique about the actual route. (Mind you, some routes are more sensible than others, a point that also applies to evolution.) It therefore seems only reasonable to assume that if anything like our terrestrial life-forms appeared on another planet – or if we could somehow run this one again, say by using a time machine – and achieved a bacterial grade of organisation, then several of them would invent photosynthesis to free-load off the Sun's energy. After all, the solar energy is there, ready to be used; photosynthetic chemical pathways *exist* (because our bacteria found several); and they're not even that hard to find (because our bacteria found several). Indeed we can state a general evolutionary principle: *if the potential is sufficiently accessible and the advantages that will accrue from realising it are strong enough, then evolution will eventually come up with some form of the necessary trick*. This is no more suprising than the way rain manages to put puddles into *all* of the low-lying bits of a roadway – which happens for much the same reason. We shall call the italicised statement above the 'Principle of Murphic Resonance', in deference to Murphy's famous law: 'If anything *can* happen, then it will.' (And if you're sure that it can't happen, then it probably will anyway.) This principle should not be confused with Rupert Sheldrake's concept of 'morphic resonance'

– the idea that if something has happened before then it is *more* likely to happen again – which is altogether sillier.

Equally, in any re-run of terrestrial evolution, sexual processes would appear, and when – not 'if', for it happened here on many independent occasions – complex multicellular life-forms appeared, some of them would invade the land (which is sitting there 'waiting' to be exploited) and the air (ditto). There is no reason to expect them necessarily to have flapping wings: in *Collapse* we spend several pages discussing 'Balloon Island', a scenario in which hydrogen-balloon-based animals got off the ground ahead of any winged competition, and gained such a hold that wings never had a chance.

What these considerations tell us is that innovations can conveniently be classed into two types: *universals*, which happen on many different occasions, whenever 'their time has come'; and *parochials*, which happen only once, largely by accident. 'Photosynthesis', 'flight', and 'sex' are universals, but 'chlorophyll', 'feather', and 'back seats of cars' are parochials. Universals arise by Murphic resonance; parochials don't. This distinction is the basis of our 'yes and no' answer to whether evolution will repeat itself if run again: for universals, 'yes'; for parochials, 'no'. And you can tell whether particular features are universal or parochial by seeing how often, and by what routes, they have evolved on Earth – or sometimes by arguing convincingly on theoretical grounds. Only one echinoderm produced chordates, so that event was presumably a parochial; ditto for chlorophyll, produced by only one type of bacterium. But haemoglobin, oddly enough, was produced by several types of bacteria, so it may be a universal. Only one kind of fish, we believe, came out of the ocean on to the land to produce descendants including all of the amphibians, reptiles, birds, and mammals – us included. We can tell because all of these land vertebrates possess a suite of characters inherited from this common ancestor, some of which seem to be 'badly designed' and are therefore likely to be parochial. For example, our airways cross our foodways, so that we sometimes choke to death on our food, and our reproductive and excretory systems are combined in potentially dangerous proximity so that our progeny are likely to be infected by parasitic organisms in our excreta.

These 'imperfections' in evolved creatures, incidentally, offer strong evidence against the idea that animals are the result of special creation. Evolution, proceeding in an unintelligent trial-and-error manner, is entirely capable of retaining an early building plan even if it is not 'the best': as long as it works pretty well most of the time, the tendency to evolve something better may be insufficient to overturn an established structure. Indeed the presence of an effective but non-optimal organism can be a major factor preventing the appearance of a superior alternative. With special creation, on the other hand, there is no reason to expect imperfections – why should God make a botch of His creations? (Of course He can do so if He wants – just as a 600-pound gorilla can sit wherever it likes – but why bother?)

Is intelligence a universal or a parochial? Gould in effect argues that it is a parochial – and at first blush so does our 'did it evolve more than once?' test. He points out that if that particular fish had not come out on to the land, or if the meteorite had not destroyed the dinosaurs, then the evolutionary success of mammals, hence mankind, would never have happened. So there would not have been self-conscious creatures on the Earth. We disagree – and so does Gould himself in later articles, as we mentioned. The evolution of high intelligence is a universal, because sensory and behavioural complexity have increased enormously on several different branches of the evolutionary tree. We shall discuss the evolution of intelligence in Chapter 6, but we need a quick preview now. Among the mammals there are clear signs of intelligence in chimpanzees, cats, dogs, elephants, and dolphins. Parrots and vultures are very intelligent birds (but the mythically 'wise' owl is as thick as two short planks). The gecko is an intelligent reptile, and some of the cichlid fishes are surprisingly bright. But what if there had been no land vertebrates – for these, we have just argued, are parochials? We can still build a case for the universality of intelligence, because octopuses, cuttlefish and some squids show very complex, flexible behaviour. In laboratory psychology experiments these animals have proved the equal of rats and mice. Even among the arthropods, which are classically regarded as little computers running fixed software (and in an earlier age were similarly regarded as telephone exchanges), the mantis shrimps are up there with mice and squids. Even the homely honey-bee has turned out to be a good deal more flexible than we used to think: it has a world-view in which it can take conceptual as well as geographic short cuts.[♪]

Intelligence, then, is a universal. Our particular level of intelligence may not be matched on Earth – yet, at least – but if we were to be wiped out then many other organisms are poised to take advantage of our absence by developing their own intelligences.

If we ran terrestrial evolution again, we would expect to find the same universals occurring, though not realised in the exact same forms; however, we would not expect to find the same parochials. Some animal would have crawled out on to the coastal mud, but it probably would not have been one whose foodway crossed its airway. Its descendants would probably have developed intelligence, but there is no particular reason why they should have had five digits on each hand – or have any hands at all. So Gould is right in the sense that our precise body-plan would not evolve the second time round, but wrong in what we (JC & IS) think is a more important sense, which we might describe as 'meta-wrong'. If the K/T meteorite – the presumed dinosaur-killer – had not hit the Earth, we would not find intelligent monkeys, but (say) two intelligent dinosaurs named Jak'kohen and Yanstuit collaborating on a seditious treatise called *What if a Meteorite had Hit?* in which they speculate about an improbable apelike beast taking over the world.♪

But why stop at re-running evolution on Earth? The possible evolution of extraterrestrial life should offer even more understanding; for example it should let us distinguish evolution as it *has* occurred on our own planet from the general possibilities for evolutionary processes that *might* occur elsewhere or under different conditions. At the very least this would help us to decide which features of our kind of life are essential to *any* kind of living creature, and which are accidents of terrestrial history: to sort the universals from the parochials.

Since we are going to talk about life on other planets, we must first establish that other planets exist. Until recently their existence was entirely conjectural, though very plausible: our understanding of how our solar system formed made it pretty clear that there ought to be plenty more planetary systems around other stars. Planets are thought to condense from clouds of gas and dust under the influence of gravitational attraction, and computer simulations show that under slightly different conditions such clouds can produce a variety of 'solar systems', with planets of many different sizes arranged at many different distances. There seems to be no particular theoretical reason to prevent earthlike planets occasionally condensing at earthlike distances from sunlike stars – and there are a hundred billion stars in our Galaxy alone, so even if the probability is small, there ought to be quite a lot of planets suitable for life. Moreover, life-bearing worlds don't *have* to resemble ours in every respect: if the sun is hotter, maybe a planet further away would compensate – which increases the likelihood even more.

People have tried to calculate the relevant probabilities, but such computations involve so many unjustified assumptions that we personally don't place much reliance upon the results. Fortunately, nowadays there is more direct evidence. The first planets to be found outside the solar system turned up in a highly unlikely place: orbiting a pulsar. Pulsars are very strange objects that give off strong pulses of radiation with amazingly precise periodicity. Orbiting planets ought to cause tiny variations in the timing of the pulses, so astronomers looked for appropriate signals. The first announcement that such a signal had been found was later withdrawn by its instigator when the 'signal' was traced to a systematic error in observational technique, but shortly afterwards at least two planets were detected round a different pulsar, which emitted a signal that could not be introduced by such an error. Pulsars emit large quantities of radiation, however, so pulsar planets don't look like good places for life to evolve.

So are there planets around stars similar to our own?

Yes.

As we go to press, at least seven are known. In October 1995 Michel Mayor and Didier Queloz announced that they had detected a planet, probably half the mass of Jupiter, that circled the sunlike star 51 Pegasi. They had known about it since February of that year, but had been careful to check their observations before making them known. This discovery was confirmed shortly afterwards by two American astronomers, Geoffrey Marcy and Paul Butler, who found two more planets: one about seven times the mass of Jupiter orbiting 70 Virginis, and one about two and a half times the mass of Jupiter around 47 Ursae Majoris.♪ In April 1996 the same team found a planet orbiting rho Cancri, and very recently another around tau Bootis.♪ Chris Burrows found a planet encircling beta Pictoris, and in mid-1996 George Gatewood announced yet another around our neighbouring star Lalande 21185, the fourth nearest star to the sun (figure 20). The detection technique (except for beta Pictoris) involves observing tiny wobbles in the path of the star produced by the encircling planet, and with current technology this method can detect only fairly massive planets. No doubt the number of known extrasolar planets will increase, and the lower limit on their masses decrease, as the technology improves: by the time you read this, that number will surely be larger.

Anyway, there *are* planets out there.

Is it likely that any support life? Maybe we can stay closer to home. In August 1996 a group of nine NASA scientists announced that they had found evidence for life on Mars, inside a meteorite that had been chipped off the Martian surface and had then made its way to Earth.♪ The evidence was found in the meteorite ALH84001, which had been knocked off the surface of Mars in a collision with an asteroid or comet about 15 million years ago, and had landed in Antarctica about 13,000 years ago. Inside it are three pieces of evidence for life: markings like the outlines of tiny fossil cells, crystals containing iron that resemble those produced by certain terrestrial bacteria, and organic molecules closely resembling those associated with fossil bacteria on Earth, but never before encountered in a meteorite from Mars.

As we write, the whole suggestion is highly controversial, and almost everything about it is subject to dispute: the dates, the actual origin of the rock, whether it got contaminated by Earth life-forms, whether the fossil markings are really fossils ... We mention it because it may turn out that NASA's scientists are right, and to show that the idea of life on nearby planets is being taken seriously. It is worth bearing in mind, however, that a similar announcement was made in 1961, when it was claimed that a meteorite known as 'Orgeuil' contained organic matter of biological origin. Later it transpired that much of the organic matter was earthly contaminants, such as ragweed pollen. We (JC & IS) are a bit worried that the evidence for life on Mars is based on *resemblances* between Martian fossils and terrestrial ones, because the similarities under discussion seem to revolve around parochials. Parochial features of Martian life ought to be *different* from what we find on Earth, not the same. For the same reason we are encouraged that the Martian 'bacteria' seem far too small by terrestrial standards, but everybody else seems to think that's a discouraging feature.

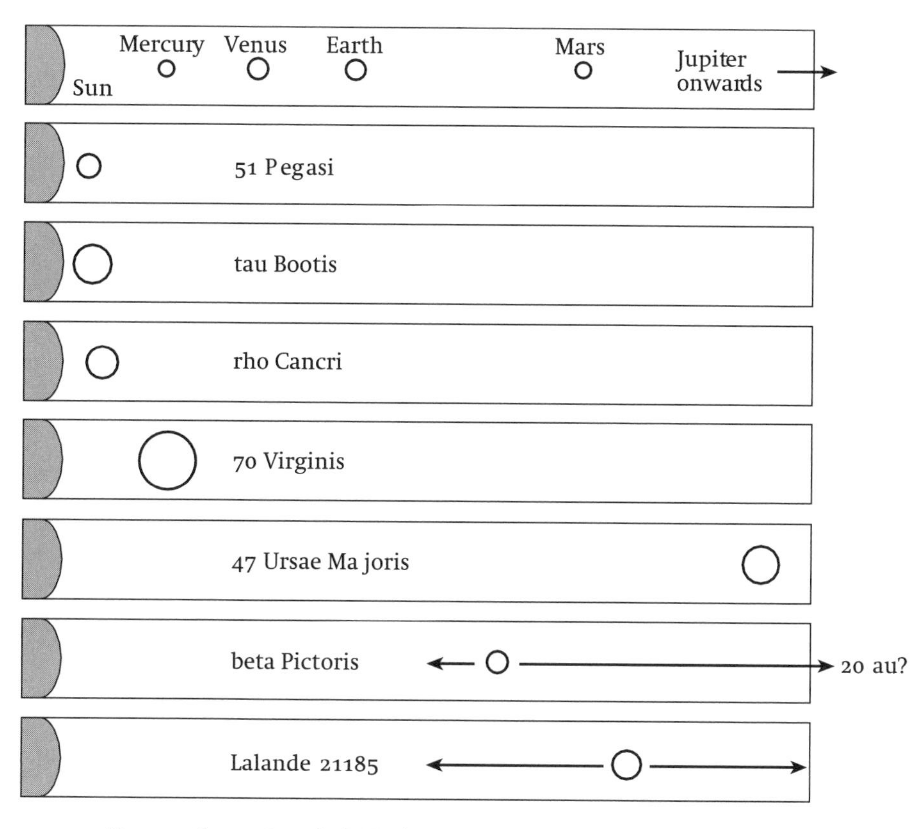

Figure 20 *Comparative study of eight planetary systems. (au = Astronomical Unit = distance from the Earth to the Sun = 93,000,000 miles.)*

If not on Mars, then where might life take hold? It is widely thought that a watery planet like our own ought to be a very good place for alien life to evolve. Water contains hydrogen and oxygen, useful for our kinds of carbon chemistry; it is a very effective solvent; it easily exchanges with an atmosphere to give high humidity and rain. None of the planets so far detected is likely to have liquid water: their orbits are at the wrong distances and several of them seem to be Jupiter-like methane/hydrogen gas giants, to judge by their masses. So we are back to conjecture. However, experiments intended to locate water-bearing planets around other stars are already on the drawing board. Of course we can question whether water is essential for life, but it certainly seems very useful: indeed the best reason for expecting life to appear on a watery planet is that it did so here. On such a world, we would expect to find flying life-forms (but not necessarily feathered wings), photosynthesis (but not necessarily the chlorophyll molecule), jointed limbs (but not necessarily knees or elbows like ours, or our particular bone structure), eyes and ears (but not necessarily teeth like ours, which were once scales).

However, the chances of the same genetic systems as ours arising on this alien world, with the same oddities selected, the same combinations of genes, are remote. It is by no means clear that alien life has to be based on DNA: the principle behind DNA looks suspiciously like a universal, but the molecular architecture realised on Earth might be a parochial. We just don't have enough evidence to decide. Certainly, though, to find another planet inhabited by our kind of dinosaur, or by humanoids almost the same as us, is about as likely as finding a previously undiscovered Pacific island whose natives speak perfect cockney rhyming slang. The prevalence of humanoid aliens in television and movie 'sci-fi' owes a lot to the ease of hiring human actors instead of an octuplet of Zarathustrans; it is noticeable that as the special effects department has improved its ability to use puppets, models, or computer graphics, TV and movie aliens have tended to become *more* alien.

If you can't have human beings or dinosaurs on distant planets, what can you have? You can't just invent an alien with a hodge-podge of scriptworthy characteristics: 'Enter an intelligent gas-cloud stage left' only works if you provide a *lot* of prior build-up. The way to design aliens – whether referring to a TV scriptwriter or a real evolutionary process happening right now on the fifth planet of Proxima Centauri – is to employ universals for the overall structure, adding a dash of harmless parochials to taste. For example, a sensible design for an animal that lives on a heavy-gravity world might be a solidly built low-slung six-legged herbivore with two rhino-like horns, a trunk, and a pronounced overbite. Here a solid build is required by the physical constraints of strength of materials, and legged locomotion, some longish device for drinking when it's hard to lower your head to the ground, and horns for self-protection are universals. On the other hand a trunk, *six* legs, *two* horns, and teeth (overbite or not) are parochials.

A trunk is a *parochial*? What *other* solutions to the problem of drinking from a height have evolved, on Earth? The answer is 'a long neck', as adopted by the giraffe. We are used to thinking of the giraffe's neck having the adaptive advantage of letting it eat leaves from the tops of high trees, beyond the reach of its competitors. That's fair enough, but there is another way to view the giraffe, not as a peculiar feeder but as a peculiar strider. The giraffe's locomotion is a bit unusual, and in consequence it can run faster by making its front legs longer. Evolution may have selected giraffes for longer legs, taking its head further from the ground. It could feed with ease, browsing on the delicate, tasty leaves at the tips of tall trees.

But what about drinking?

Water has a universal tendency to collect in pools *on the ground*. If you are a long-legged animal with a short neck, the ground may well look just as distant and inaccessible as those tasty leaves do to your envious competitors. If you want to drink without kneeling – which could slow you down dangerously if you need to start running away from a predator – then you need either a long neck or a trunk. So these two very different body-parts are two parochial ways to solve the same universal problem – but it is an individual drinking problem, not a competitive browsing one. Rudyard Kipling's *Just So Stories* see the elephant's trunk and the giraffe's neck as solutions to different problems, but they may actually be solutions to the same problem. When considering such questions it is important to bear in mind that there is no reason why the evolution of a single feature of an organism must be a solution to a single problem. So Kipling could be right, and so could we, without any contradiction. As we said, science often provides several different ways to look at the same issues.

The message that trunks and long necks are in some respects 'the same' affects our view of alien evolution: it can be expected to produce apparently different solutions that turn out to be the same on a deeper and more subtle level. Designing realistic aliens thus requires a knowledge of how that particular planet's evolutionary experiments turned out, which means that we must rely upon general principles of biology to generalise terrestrial experience to new environments. This requires taking a deeper and more divergent view of biology than merely chanting 'DNA is God and molecular biology is Its prophet'.

You've met the Zarathustrans; you've even seen a first pass at their evolution. In the same vein, let's take a look at a few of the aliens invented by SF writers. You'll see why later on: mainly, we want to expand your ideas of the possibilities for evolution, so that you can explore the phase space surrounding evolution as it happened on your home planet – with a view to considering what might have happened instead. That is the true role of SF: exploring 'Worlds of If', not predicting the future. So even if you dislike SF, please read on, to see how such thought experiments go. No ray guns, or scantily clad heroines in the clutches of bug-eyed monsters – promise.

In the 1950s Harry Clement Stubbs, under the pseudonym Hal Clement, wrote an SF classic called *Mission of Gravity*. This was set on the planet Mesklin, which was shaped like two saucers glued rim to rim, had enormously strong gravity, and rotated very fast. As a result, gravity at Mesklin's equator was 4g, whereas at the poles it was 600g. (Here 1g is standard earthly gravity.) Stubbs, a physics teacher, did all the physics calculations needed to check that these figures – and many others that affected the settings for the story – made sense. Mesklin is inhabited by intelligent aliens. What sort of biology makes sense in such a high-gravity environment? Well, no giraffes for a start. Stubbs' Mesklinites are like enormous centipedes, evolved from aquatic jet-propelled ancestors. Mesklinite culture and biology have diverged according to local conditions. The inhabitants of polar regions have many phobias about gravity – merely rearing up your front few segments can be fatal if you drop back too rapidly, and as for climbing on top of a rock – our own fear of heights is a pale shadow of Mesklinite fears. Dramatic tension comes from acts that require the aliens to overcome these instinctive fears: the hero, Barlennan, is a very unusual Mesklinite who is so driven by his vision of the future of his race that he is prepared to endure psychological stress, such as being six inches off the ground or underneath a heavy object.

Briefly.

The main problem with Stubbs' design for the Mesklinites is not their biology, but their culture: like most 1950s aliens, they have the outlook of Americans.

One of the best realised alien cultures can be found in *The Mote in God's Eye*, by Larry Niven and Jerry Pournelle, where the alien culture is more complex and more – well, *alien*. Moties – remember that name, we'll refer back to it in later chapters – have an amazing facility with machinery, routinely rebuilding their own spaceships during flight. Small, so-called 'watchmaker' variants invade human ships, ultimately destroying one, but at the same time they repair and improve their equipment too. This mixture of destruction and assistance is a necessary part of the book's ambience, and also of its plot. Other aliens in Niven's books inhabit 'Known Space', a moderately self-consistent fictional construct. They include Outsiders, who live exposed to space on their skeletal spaceships, following the star-seeds on their 100,000-year cycle into and out of the Galaxy, and act as information brokers in a form of barter trade. There are the cowardly three-legged Puppeteers, whose entire civilisation departs for pastures new at the first hint that the galactic core is exploding, even though the radiation so generated will not arrive for 20,000 years. There are the Bandersnatchi, creatures of vast proportions who inhabit many worlds, disseminated by the ancient Slavers: nobody knows whether Bandersnatchi are intelligent or not because they are so weird.

Good SF aliens must be plausible not just physically and biologically, but evolutionarily. The same must be true of *real* aliens, should there be any – and everything we know about the origins of life suggests that there should be real aliens, somewhere, at some time. With this point in mind, let us examine a famous *Star Trek* alien, the tribble, invented by David Gerrold. Tribbles, when seen on the TV screen, are cuddly pets – furry hemispheres with almost no features, but of many different colours, popular throughout the galaxy. We are told that they eat almost anything organic, and that they are prolific breeders. Indeed they 'devote half of their metabolism to reproduction', are 'born pregnant', and 'reproduce at will'. In the course of the story they multiply amazingly, especially in dark cupboards and other enclosed spaces. They make a purring, trilling sound when stroked, which seems to soothe everybody's nerves.

Star Trek scripts seldom consider whether their aliens are evolutionarily plausible. Indeed they seldom worry about them being physically plausible. But we can worry, and choose to do so. There are problems with tribbles. For example one of the least credible tribble characteristics is that everybody – except the Klingons – loves them. Clearly they are symbionts, eating anything.

Anything?

Who would want a symbiont that eats *anything*? You wouldn't dare leave a plate of food about. But they'd be great for cleaning up messes, like village dogs in the tropics. So, pursuing a feasible line of tribble evolution, we are led to imagine a burrowing alien whose food, perhaps a grass-like plant, is eaten outside their burrow – something like a rabbit or a marmot. This last is potentially interesting, because marmot burrows can be shared – by a burrowing owl and a rattlesnake. Let us, then, imagine an alien 'exomarmot' burrow, which is also inhabited by a sluglike detritivore – rubbish-eater – which finds rich pickings in abandoned burrows. As this 'prototribble' adapts to coexisting with exomarmots, it will lose any characteristics that make it repellent to its host. Prototribbles that escape notice by other inhabitants of the burrow do even better. So, through evolutionary pressures, they will lose any obvious head, sense organs, or locomotory organs, turning into the 'every burrow should have one' general-purpose symbiont. But now consider breeding. The requirements here are very stringent. Within the burrow space is limited, so they must restrain their reproductive tendencies. However, they must be able to multiply very quickly if they get a chance to colonise a new burrow system.

This scenario is not just the product of a clever imagination running free. Much the same problem faces *Gyrodactylus*, a real fish parasite on our own home planet. It must not damage its host fish, but it must breed rapidly in a new host. It achieves these rather contradictory requirements by keeping in its uterus two, or even three, generations nested inside each other like a 'Russian doll'; moreover sperm from its last few matings can invade any mature embryos. Western human culture is currently facing the problem of very young mothers, but this parasite can be a grandmother before it is born. When there is no room to breed, it keeps its offspring safely tucked away. But when it invades a new fish, the embryos – and their embryos – hatch quickly, and all of them generate new babies from the mixed sperm population that they all inherit.

In just this way, our prototribble can credibly be 'born pregnant'. Each prototribble would involve many genetic lines – which is why so many differently coloured tribbles appear all over the starship *Enterprise*, all derived from one ancestor. In their normal habitat prototribbles would creep out of their burrows on moonlit nights, mate ecstatically upon the dewy exograss, and return home with renewed genetic diversity.

Now symbiotic systems co-evolve, so the exomarmot whose burrow is shared by prototribbles evolves alongside its guests – and 'guests' is the word, for it evolves towards a kind of universal inn-keeper organism. It becomes larger, and intelligent, and it adds many other species of symbionts to its burrow. The prototribble co-evolves along with them all, becoming likeable to virtually everyone. Young exomarmots, also much bigger, are intrigued by prototribble behaviour, breeding them for colour pattern and later for their relaxing purr. Sadly, the exomarmots failed to learn one important lesson: how to restrict the explosive breeding patterns of their prototribbles. Perhaps the short-term economics of easy-bred easy-fed burrow-pets overrode their sense of wisdom ...

We left the Zarathustrans admiring the elegant 'liar-to-children' story of their species' evolution. The time has come to take a more critical look.

Destroyer-of-facts [*Who can stand this tadpolish nonsense no longer.*] An excellent *story*, yes, but poor science.

Liar-to-children Why, what have I got wrong?

DoF Well, for a start The Regulations then wasn't like it is now, it had much less memory for a start, and its cycle time was slower. It was a rudimentary tribble, not a pet. It co-evolved complicitly with its group, it didn't evolve separately. And that bit about tadpoles – look, it isn't just *our* young that we go all gooey about, it's *anything* with several tails and one big eye–

LtC [*In a huff.*] Well, I think I told the story very clearly / simply (delete whichever is inapplicable) without dragging in irrelevant feathersplitting♪ details.

DoF Clearly, simply, and *wrongly*. And they are *not* irrelevant feathersplitting details, they are the essence of the whole thing!

LtC Well, since you are so clever, *you* can tell the story *your* way and the Ringmaster can tell us all how we reacted to it.

DoF It all started much earlier ... a pollinator that became necrophagous.

LtC [*With heavy sarcasm.*] An excellent start. Five words, two of them with four syllables. Just what I'd expect from a Destroyer-of-facts.

DoF You know full well what a pollinator is, and necrophagous means 'corpse-eating', OK? About eight-squared giga-octs ago there was a species of small multilegged animal that lived in a hole in the ground and collected pollen from plants. It carried the pollen back to its hole and heaped it up in piles. Over time this animal diversified into several species, all feeding on juices derived from plants – one fed on pollen-juice, one on nectar, and one on the juices from small pieces of fruit. All of them were stockpilers, and they often shared burrows. At that time Zarathustra's climate started to change, and erratic droughts began to occur fairly frequently. One group evolved the ability to secrete antibiotics, to stop the piles of juicy food fermenting. That let them turn to other sources of food in dry spells, and some members of that group began to exploit a major source of food that always appears in times of drought – corpses. Before they could secrete antibiotics, corpse-eating ran too much risk of infection, but now they could eat corpse-juices safely. And corpse-juices were so nutritious that this group thrived.

Eventually they diversified into a seven-way symbiosis, and a typical burrow contained one nectar-collector, one pollen-collector, one fruit-gnawer, one corpse-finder, one corpse-defender, one bartender–

LtC Bartender?

DoF It looked after the juices collected by the others. Plus a potter, who made the pots that the bartender kept the liquids in. The first three ate only what they collected, and the others ate everything, although potters generally specialised on clay washed down with nectar and just a *dash* of fresh corpse-juice: shaken, not stirred.

Pursuer-of-sicknesses And we evolved from these seven symbiotic species?

DoF We are a single species, we cannot have evolved from seven. No, we evolved from something else.

Liar-to-adults What?

DoF A parasite that crept up on the corpse-finders while they slept and sucked their blood.

LtA *What? A vampire?*

DoF Octimality may stem from emotionally unacceptable beginnings ... In the balanced ecology of the burrow, these protozarathustrans prospered, and successive generations increased in size. Eventually they became almost as large as the animals they had parasitised, but long before that they changed to eating the same food as their erstwhile hosts. All of the animals in the burrow slowly acquired better brains and more effective sense organs. The protozarathustrans were tolerated in the burrows because they learned to generally help around the place – carry babies, do odd jobs. At this stage the role of The Regulations became paramount.

LtA The Regulations? Why?

DoF As Liar-to-children never quite got round to saying correctly, the Regulations used to act as a tribble for the burrow, consuming rubbish and faecal waste. But over time they evolved into mobile bulletin-boards-cum-latrines. Uh – it is rather a complicated story.

Performer-of-amusements You mean up to now it was *simple*?

DoF The logical progression is impeccable, but because any evolutionary history is a record of evolving strategies in a self-organising nightmare game, the details are often bizarre. You must bear in mind that one of the dominant phyla on Zarathustra, to which we belong along with the burrow-symbionts, has an anus in each foot. The advantages of this arrangement are so obvious that they hardly need to be spelled out – the dual digestive tracts incorporate useful redundancy, so that the animal can still survive if one is damaged. And the waste is deposited directly on the ground, so that it can be recycled without difficulty by soil bacteria.

LtC Yes, yes, we all know that, we learned it as children-being-lied-to.

DoF Despite which, Childliar, it is correct. Now, it is clearly an inefficient arrangement to discharge waste products on to the floor of the burrow and wait for the prototribbles to come along and ingest it. It is far more efficient to deposit the waste directly on to the skin of the prototribble, to be absorbed immediately. And that gave the burrow-symbionts the ability to 'leave messages' which the others could respond to.

In this manner the prototribble became graffitable by the burrow-symbionts – in a very rudimentary way, of course. However, there was a problem. In order to be able to deal with the different chemical compositions of seven different types of waste, the prototribble had to have a way to detect which animal was defecating on to it, and it evolved a sensitive surface layer that could distinguish the seven symbionts by the shapes of their feet. This layer produced a chemical pattern in response – a forerunner of The Regulations' bulletin-board skin, you understand.

LtC I think we all get the picture, yes.

DoF The prototribble itself evolved into a forerunner of The Regulations, and to keep the discussion simple I shall refer to it by that name from now on. There came a time of unusual meteor bombardment, leading to a shortage of food. In different burrows, different symbionts died out. The protozarathustrans took their place, and had to be able to deal with The Regulations in any role. Otherwise The Regulations would get sick – it was accustomed to specific kinds of waste in conjunction with specific footprints. As it happened, the protozarathustrans possessed a flexible foot with four appendages, precursors of our fingers. They learned (I use the word loosely) to configure their fingers to mimic the footprints of any of the seven burrow-symbionts, and could then substitute successfully for any of them.

PoS I think I see where this is leading. I studied this bit in med school. It all depended on the reproductive methods of the various creatures concerned ... let me sniff – yes, I remember now. The protozarathustrans mated across burrows, leading to strong genetic cohesion. So did the The Regulationses – but they could also clone asexually.

PoA Bring on the clones.

DoF Shut up. Yes, the cloning response is triggered by their being submerged in water and fed after midnight. It must have been discovered by one of the burrow animals. Because the protozarathustrans were omnivorous, they survived the droughts better than any of the burrow-symbionts. One by one the specialised burrow-symbionts were replaced by versatile protozarathustrans, until eventually the burrow-symbiont species died out completely.

CoC Natural selection?

DoF In a way. There was a rather shameful episode in our early history when their last surviving descendants were systematically hunted down and killed, but let us pass over that quickly because it was embarrassingly unoctimal. Let me summarise: the burrows became inhabited by septuplets of protozarathustrans together with primitive but graffitable versions of The Regulations. There was no consistency in the messages they had been programmed with, of course, but that would come later. The important feature – it seems to be a contingent one – is that when The Regulations clones, the information inscribed in chemical patterns on its skin is copied, almost exactly. When it mates this information gets scrambled, of course. And *now* we finally reach the point from which Liar-to-children started: burrows containing septuplets of protozarathustrans plus a graffitable The Regulations.

LtC [*Menacingly.*] And after that?

DoF It followed much the path that you explained.

LtC [*Exasperated.*] So what were you complaining about? My story was OK all along. [*'OK' is a Zarathstran phrase derived from 'Octimally Korrect', which has much the same overtones as the terrestrial North American 'Politically Correct'.*]

DoF The many complicating factors that you glossed over, of course. For example we do not know *how* The Regulations acquired the ability to turn chemicals into actions and communication, but we do know their brains are skin-deep layers directly affected by chemical patterns.

LtC But we Zarathustrans do not know how we do analogous things, either.

DoF It is evidently a deep question, then.

LtA [*Who has lost track of the details and is trying to summarise the overall storyline.*] It seems to me that both you and Liar-to-children are telling us that octimism originally arose by way of septimism, and that septimism was purely contingent. This is rank heresy and I must warn you that–

DoF Ah, it might *appear* so. I accept that there is no obvious reason why the original burrow-symbiont should have formed a seven-species network.

LtA I find such talk seditious.

DoF As I said, it might *appear* so. However, there are deeper principles at work here. It is clear, for example, that each of the seven standard non-ringmaster roles that you find in a Zarathustran octuplet corresponds to one of the original roles of the seven burrow-symbionts.

PoS Really?

DoF Indeed. Why, let us take as an example your own speciality, Pursuer-of-sicknesses. This role manifestly evolved from that of corpse-defender.

PoS I do not quite scent the connection – or if I do, you are in a great deal of trouble.

DoF No, you misunderstand my implication. It is entirely natural for a corpse-defender to evolve a superior strategy: defend the corpse *before* it is dead. This no doubt evolved gradually, first by hovering around an animal that was close to death, like a terrestrial vulture; then by following the more elderly animals and helping them avoid dangers, until eventually it worked right back up the developmental chain to assisting in the birth of the babies.

LtA What is the relevance of this? I still scent contingent crypto-septimism!

DoF Not at all. *Why* do the seven original roles of the burrow-symbionts, plus the additional one of Ringmaster, produce the perfect group mind? Because, of course, the principle of octimism was operating behind the scenes all along. There *had* to be seven symbiotic species, or else the whole process could not have produced *us*!

CoC Oh, of course. It is the Zarathropic Principle in operation.

LtA I am not at all convinced that there is any credibility to the Zarath–

Ringmaster [*Sensing potential conflict and contributing to the discussion for the first and only time.*] I am *so* glad that we are all agreed.

LtA Uh – are we?

R Definitely.

LtA Oh. Well. That is all right, then.

R Excellent. I shall graffit our consensus on to The Regulations.

Let us now come back to Earth, but having gleaned a useful insight from our extraterrestrial excursions: *every change to a species has effects on other species.* This leads to another distinction between types of innovation, like the universal/parochial one but different, which we shall call the public/private distinction.

If a prey animal gets faster or bigger, that makes life more difficult for its predators; similarly if a predator gets sharper claws or teeth, then it has more luck catching prey. Those changes affect a few other species, but on the whole they can be assimilated into the ecosystem without causing radical changes. For example the grass that the prey animals eat will not be greatly affected by how rapidly they run across it. (There may well be small changes, there are always small changes, but this is a broad-brush distinction and we won't worry about anything that is not utterly novel.) In contrast, think of an organism like a newly photosynthetic bacterium, which excretes highly toxic waste products – oxygen – into the environment, thereby changing the rules for everyone. We humans don't think of oxygen as toxic, because we evolved from the creatures that evolved to exploit this new feature of the environment, but it was very toxic indeed to the creatures that were displaced. Similarly, on a smaller scale, the evolution of grasses changed a lot of the rules for other terrestrial life, both plant and animal. Grasses effectively prevent other seeds getting into the soil, by making a sward; we get weeds in lawns only because we cut them short. Grass makes small lakes and ponds much less permanent by growing in from the edges. And grass makes life very difficult indeed for elderly herbivores, because its leaves are full of silica, which grinds teeth down very rapidly. So all grass-eating animals, from grasshoppers to sheep, must possess a method for dealing with very abrasive food. Indeed a good case can be made that if the K/T meteorite had not wiped out the dinosaurs, grass would have. Continuing our list of examples, the existence of neutrally buoyant fishes, and that of sharks, changed all the marine rules, just as the existence of trees, snakes, and bloodsucking insects changed the rules on land.

So: some innovations affect only a small guild of species – they are private. Others have effects so dramatic that all life sharing that part of the biosphere must adapt in response to the change: these we call public. The private/public distinction is independent of the parochial/universal one. A parochial innovation can be private, or public; so can a universal one. Innovations that are both universal and public are especially important because they represent changes on a vast scale that are 'just waiting to happen'. As we have said, there is good reason to believe that every planet upon which water is liquid will produce life-forms, and that some of these, sooner or later, will change the atmosphere from a reducing one to an oxidising one. At this stage the evolution of life-forms will involve a great deal of evolutionary symbiosis, as oxygen-sensitive organisms cuddle up to oxygen-using ones, where the stress is smaller. We feel the urge to argue that most planets will then take the route that ours did, going on to eukaryote cells, and then a variety of multicellular forms like the organisms in the Burgess Shale ... but we suspect that such an argument would be simply a lack of imagination on our part. Other routes must surely be possible.

Associated with the public/private distinction for causes there is a corresponding distinction for the kinds of effect they produce. Developing systems can change in two radically different ways. One is the relatively slow and 'natural' changes that occur in existing features such as the evolution of a slightly better lens in the eye. The other is the rapid, almost discontinuous introduction of a revolutionary new feature: the eukaryote cell in response to an oxygen-laden atmosphere. In our phase space imagery, in which every event that actually occurs is surrounded by ghostly companions – events that could have happened instead or may yet happen – the distinction is clear. Changes of the first kind, which we shall call *explorations*, investigate the immediate vicinity of an existing phase space; the novelties are so close to what was already present that they – or something like them – are more or less inevitable. Changes of the second kind, which we shall call *explosions*, change phase space itself, opening up entirely new regions. In the new expanded phase space, the game being played has not yet canalised, its strategies still remain to be developed, and the players are amateurs. As time passes, the system learns what works and what does not, the players turn professional, the strategies canalise, and the phase space acquires a well-defined geography. Public innovations are much more likely to lead to explosions than private ones are.

The explosion/exploration distinction is involved in the answer to one of the big evolutionary riddles, exemplified by the rise of the Zarathustrans from parasite to intelligent octuplet. Why does evolution lead to increasing complexity? Indeed, *does it*? We have to face up to this problem if we are going to argue that such enormous complexities as intelligence and consciousness are 'natural' consequences of evolution in appropriate circumstances. Where does the complexity 'come from'? We shall argue that the answer is 'complicity'. Recall from Chapter 3 that complicity arises when two different phase spaces, with their own individual dynamics, become coupled together and change each other recursively to create an emergent dynamic on the combined space.

The traditional picture among everyone except evolutionary experts is that complexity has been increasing throughout the history of life on Earth. Actually, that's not quite so, which among other things means that any purported explanation of the general increase of complexity that 'proves' it must *always* increase must be wrong! To see why, consider the phrase 'people live longer nowadays'. That's not true either: some people don't. A more accurate but less felicitous expression is that 'more people live to any particular age than used to be the case'. In the same way, 'complexity is greater nowadays' is better expressed as '*some* organisms are more complex now'. After all, there are as many bacteria now as there have ever been – perhaps more, since the ecosystem has become more diverse – and they are probably not much more complicated than they have ever been. However, they probably are more specialised than they used to be. The increase of complexity has come about as a kind of addition of superstructure to this vast microbial substrate. Recall that about two billion years ago single-celled eukaryotes like today's *Amoeba* and *Paramecium* appeared; then a billion years ago a few flatworms and jellyfish-like creatures; then 570 million years or so ago the 'Cambrian explosion' led to a huge diversity of body-plans, still visible among the the soft-bodied creatures of the Burgess Shale some 40 million years later; then some 500 million years ago in the late Cambrian there appears a more familiar marine fauna, including many kinds of fishes.

However, in the Cambrian – indeed now – you still find plenty of bacteria, amoebas, jellyfish, flatworms ... *Their* complexity hasn't increased. The base of the 'pyramid of life' – the hierarchy of creatures *alive today* – is as wide as ever; there are just a few more, usually smaller, layers of more complicated creatures piled on top. However, these are much more *obvious* to creatures like us whose eyes don't see the microscopic. Copepods (tiny shrimplike inhabitants of the plankton), fish, ants, and beetles are probably the most important creatures on the planet if we measure importance by total mass, whereas insects, birds, people, and domestic animals would be the most obvious to a visiting Zarathustran.

So why does life seem to progress – in this piecemeal and cumulative way – towards complication? We gave one answer in Chapter 4, and what it boils down to is: there is nowhere else to go. Think of living creatures as rocks in a dry-stone wall, with the more complex organisms in the upper layers. Starting from the top layer, the next simplest layer is already occupied by creatures that are doing their thing in a pretty effective manner, so a new arrival will have enormous difficulty gaining a foothold. The only place where new 'ecological niches' – ways of making a living – exist is 'upwards', building the wall higher into the realms of the complex. This is what the slow changes of exploration naturally do. The second kind of change, explosions, may either produce rapid growth of complexity, or suddenly simplify everything – and this is why the increase of complexity is not relentless. The explosion that led to mammals keeping their young inside themselves *simplified* their DNA, because huge quantities of amphibian contingency plans for dealing with variations in ambient temperature were no longer relevant.

As usual, things aren't always that straightforward. There are creatures that have simplified themselves – but into new niches. The dog-tapeworm can't evolve until after the dog has. However, radically new life-patterns seldom arise in this way. Many major innovations arise from specialised but apparently simpler larval forms, after omitting the adult stage from their life history: an example is the evolution of chordates from larval echinoderms. Equally, vertebrates are like the larval stages of early chordates, reptiles may have employed the structure of larval amphibians, mammals seem much more like embryonic reptiles than ancient adult reptiles – or modern ones, for that matter. A good case can be made that many, if not most, of the truly radical steps in our evolution exploited a previous larval structure (figure 21).

Or consider plankton, a collective term for all of the small organisms that drift in the upper layers of the seas. This drifting life has a most interesting history, and many important organisms in today's plankton evolved quite recently; however others go back a very long way indeed. There is 'nanoplankton' – bacteria and small algal cells, which form the productive base upon which the other creatures feed; these have been present since long before the formation of the Burgess Shale. Feeding on them are many small animals, ranging from shelled amoebas (whose fossilised shells make chalk) to crustaceans. The crustaceans are not the direct descendants of original nanoplankton-grazers, because those were displaced eons ago. We have no real idea what the nanoplankton-grazers of a billion years ago were, but many guesses: they might have been large ciliated protozoans like *Paramecium*, for instance. At any rate, that rich, productive, well-lit layer soon became invaded by other creatures, many of which were filter-feeders like today's whales, though smaller. As life began to fill the seas, many filter-feeders became sedentary, like modern clams and barnacles, because the rich life in the water above created in turn a rich rain of nutrients – excreta, corpses. And, like those modern forms, they sent their larvae up into the rich plankton, to grow where they would not compete with their parents. Life was so good in the plankton, however, that many of these larval forms lost their sedentary adult stage altogether, and brought breeding back into the end of larval life. So the plankton now largely consists of these ancient larval types, but evolved into permanent larvadom.

The very common crustacean copepods are clearly juvenile forms of certain adult organisms – one conjecture is trilobites – but juvenile forms that breed *as* juvenile forms. Their predators, such as arrow-worms, are also present, along with a whole host of temporary residents such as larval barnacles, larval mussels, larval worms, and of course larval fishes. When the copepods became so successful, they set the stage for the next lot of – larger – larvae to wander up into the plankton and feed on them, and so on. So rich is this source of food now that the largest animals our planet has ever seen, the great whales, keep themselves

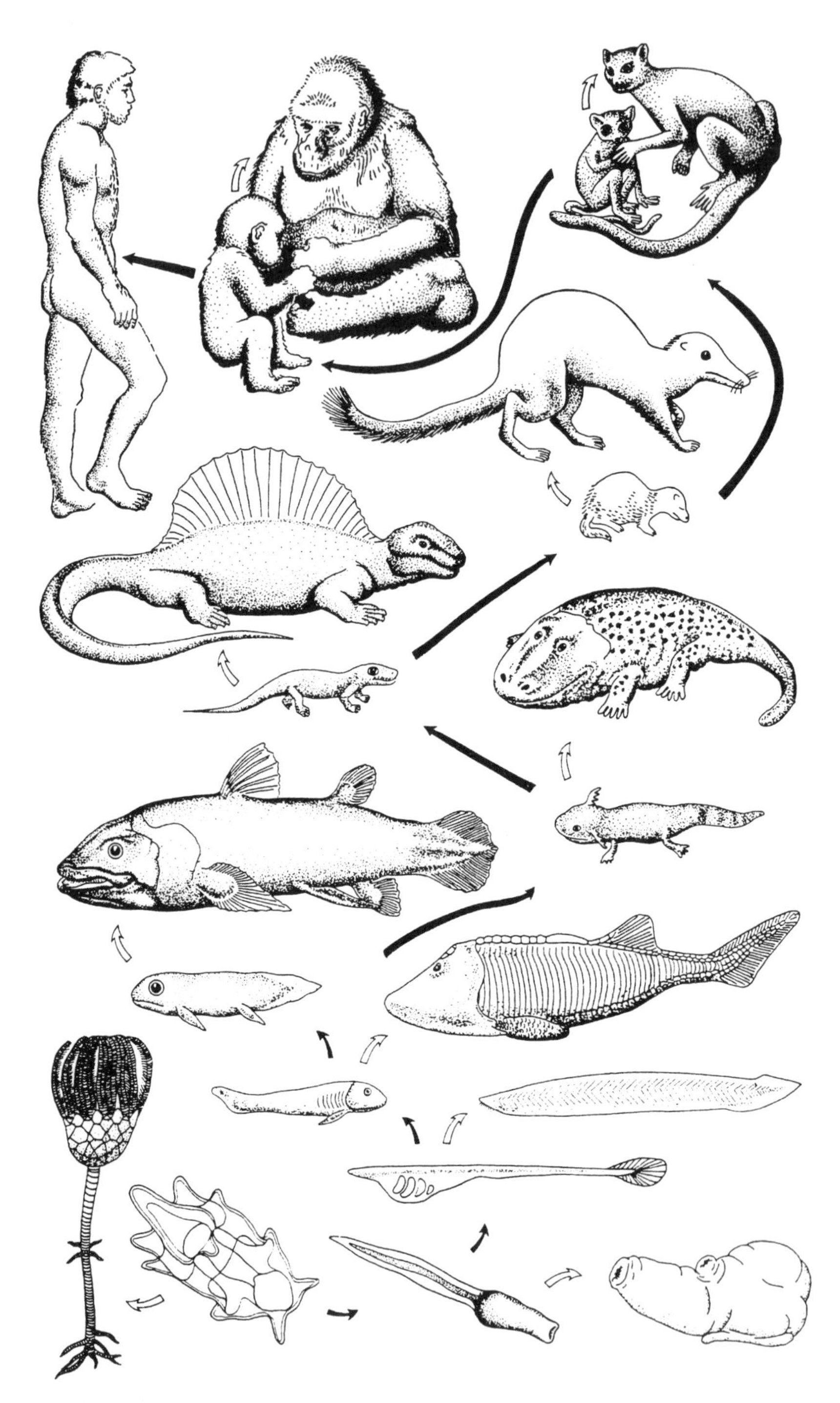

Figure 21 *Imaginary series to illustrate the importance of larval forms in evolution. White arrows link larvae to their adults, solid arrows link (hypothetical) larvae to adults of other species that may have evolved from them.*

alive by feeding on it. It has even seduced the whale shark and the basking shark away from their carnivorous way of life.

The central feature of this type of evolutionary story is that at each stage the whole *context* is changed by what has happened. The appearance of more and more life-forms of a given type changes the rules that govern the behaviour and evolution of succeeding ones. We see this in the evolution of the Zarathustrans. The early burrow-symbionts *worked*, there was no 'need' to replace them; but because they were successful, that opened the way for one of their parasites to take over the entire system. Various 'accidents' occurred along the way: it just so happened that some organisms could exploit corpses, it just so happened that there were corpses around to be exploited. The Regulations was but one possibility of many, but once that possibility became realised, others disappeared; the strategy became canalised, and phase space was being pruned even as new regions were being built. Mathematicians call such a system *recursive*: it feeds off itself. We call this particular kind of recursion complicity – it creates itself as it feeds. Complicit systems, in a word, are organic.

Complicity provides a universal route by which individual organisms, indeed the whole ecology, can become more complicated. Mere complexity, however, is not the point: what matters is that it is organised complexity, brought about by various processes which generate a system that works, and keeps working even as it changes. The ecological snooker break keeps going even though the shape of the table is changing, *and* the table continues to function even though the game being played on it is no longer the one it was first designed for. Interaction alone is not enough to achieve this: the game interacts with the table if you throw a snooker ball at one of the legs and the table collapses. Instead, both game and table must interact complicitly, responding recursively to each other and building a new joint phase space. You only find out what phase space you're getting when it arrives.

This implies that each planet must live through its own evolutionary history to 'find out' what it will become; so if evolution could be run twice on the same planet, the result might well be different the second time round. On Zarathustra, if the burrow had not contained a tribble – and it worked fine without one – then the precursor to The Regulations would not have evolved. Without The Regulations, the protozarathustrans could not have taken over the roles of the burrow-symbionts. If they had not done so, then the particular form of intelligent Zarathustran that we have come to know and love, the perennial octimists, would not have evolved. On Earth, similarly, the forms of living creatures would probably be quite different – even for bacteria. This is the basis of Gould's original point about the Burgess Shale: random events can shape evolution. Absolutely. But that doesn't mean that evolution can follow any pattern whatsoever. There are constraints, which make some dynamic paths more natural than others. On Earth we find many examples of 'convergent evolution' where the same form has evolved by quite different routes – wings on birds and wings on bats, say. The common constraint is the need for flight, and that governs the form of any flying animal in quite a strong way. As soon as we look to this deeper level, not of surface form but of roles in the ecosystem, we find a lot of common patterns and much less contingency. Gould's point was about parochials – yes, they could easily be different next time round. But universals will remain much the same – just realised by different organisms with different body plans. If the mudslides that created the Burgess Shale had killed off every living representative of the phylum that today includes us, then *we* would not have evolved. But you can bet your boots that *some* kind of intelligent creature would have evolved eventually, because intelligence is a universal, and as such is always a potential feature of evolutionary phase space.

And once the potential is there, Murphic resonance will do the rest.

6 Neural Nests

One episode of the TV series *Trials of Life* involved filming the remarkable behaviour of the tarantula wasp. Mother wasp finds a tarantula, stings it to paralyse it, and drags it back to her hole. Then she lays her eggs in it, so that the newborn larvae have their own reliable food supply. Or so all the textbooks say. The first attempt to film this sequence of events went beautifully – except that, right at the end, the wasp forgot to lay her eggs. The next attempt was going really well until a bird flew by and ate the wasp. On the third attempt the wasp never managed to find the tarantula ... and so it went, for a dozen or more attempts. *None* of the wasps filmed managed to complete the entire sequence in textbook fashion.

Despite this, the final film looked great: it edited several different wasps together to get exactly the textbook story.

However, one could be forgiven for concluding that real wasps don't read textbooks.

In the previous chapter we made an important distinction between universal and parochial features in evolution, and argued in passing that intelligence is a universal evolutionary strategy, likely to be found wherever life has taken sufficient hold. In this chapter we refine that theme by examining the evolution of intelligence on our own Earth. The orthodox story is a reductionist one: it begins with rudimentary sense organs, moves on to nerve cells as a kind of telephone system for the sensory signals, then explains how nerve cells hooked themselves up into networks and got clever, and culminates with an especially clever network of nerve cells known as The Brain of Albert Einstein. We shall call this the 'clever nerve cell' theory. It is a fascinating story, and it has only one serious deficiency – it misses out most of the things that are involved in the evolution of intelligence. Creatures can't become intelligent unless there is something to be intelligent *about*. And so we must begin to address some of the questions raised at the beginning of *Figments*, concerning the relation between human intelligence and human culture. We shall argue that they are a complicit pair, co-evolving, feeding off each other, mutually modifying at every step.

Nevertheless, it is difficult not to start with brains, so we will. The story of brains is inseparably bound to that of nerve cells, because networks of nerve cells provide both the hardware and the 'computational power' needed to run a brain. It is also inseparably bound to the development of senses, because initially the main reason for having nerve cells was to interpret and act upon the signals generated by sense organs. Nerve cells are also implicated in movement, so all these aspects of living creatures – senses, locomotion, brains, and intelligence – are part of a single package. The usual way to tell this story is to discuss senses first, then nerve cells, and finally brains, and to some extent we shall follow that pattern. However, we will argue that neither senses nor brains can evolve without the other: senses cannot advance beyond a fairly rudimentary state without a brain to interpret them, control them, fine tune them, and respond to them; and brains would not have evolved unless they had something to occupy themselves with, senses being the obvious possibility. Thus senses and brains evolved complicitly.

Our brains are adaptable, general purpose machines. Instead of making specialist 'hardwired' reflex responses, they can adapt, learn, and change. The story of brains is one of Murphic resonance: we happen to inhabit the kind of universe in which all the ingredients for making brains exist, and brains are such a useful trick that in such a universe they will inevitably turn up. There are good reasons why evolution might favour adaptable general-purpose nervous systems – and whatever evolution wants, and *can* get, it *does* get. Even the most predictable of animal requirements is poorly served by a nervous system that is like a cheap electronic toy, programmed to carry out only a few routine functions. For example, think of a creature as 'simple' as a filter-feeding plankton-eater. Even though its food may always be of the same shape, size, and nature, the distribution of food in space will normally be patchy. So any animal that can combine its routine feeding programme with a more versatile movement programme, capable of guiding it to places where food is more abundant, will have an advantage. It will have more offspring than its fellows who just sit and wait for food to float into their mouths, and those offspring will also be more successful at finding food ... and another recursive cycle casts loose from its scaffolding and becomes self-perpetuating. There are lots of other ways to improve the food-gathering programme – for instance, stopping food-gathering mechanisms when there isn't any food, thereby making the whole process more efficient. In order to make such improvements, the animal must have some way to detect what is happening in its surroundings, which implies the development of rudimentary sensory organs.

These sense organs do not need to be very versatile to offer an advantage in comparison to animals that lack them; a patch of cells that responds to some chemical that often occurs in food, improving the animal's food intake by just a few per cent, is enough to load the reproductive dice slightly in its favour. Amplified over many generations, that tiny advantage will be parlayed into a big one. We can see why by performing a short calculation in an admittedly very simplified model. Suppose the food supply in a particular region is enough to support precisely one million animals that come in two varieties – let us call them gulpers and gatherers. Both reproduce asexually, say by splitting into two. Gulpers sit around waiting for food to arrive; gatherers go out hunting for it using a chemical-sensitive patch that is so inefficient that it offers them a one per cent reproductive advantage over gulpers. The Grim Sower kicks off the proceedings by providing a million organisms: a 50-50 balance, half a million gatherers and half a million gulpers. The Grim Reaper gently reaps, and the next generation contains 502,488 gatherers and 497,512 gulpers. (These numbers are specified by the rule that they must add up to a million and be in the ratio 1.01, that one per cent advantage.) The next generation is sown and reaped: it now contains 504,975 gatherers and 495,025 gulpers, and the imbalance has grown. After a hundred generations there are 730,081 gatherers and only 269,919 gulpers; after five hundred generations there are 993,140 gatherers and the gulpers are down to 6,860. By the thousandth generation there are only 48 gulpers, and they have become a highly endangered species ... Before generation 1400 we find 999,999 gatherers and only one solitary gulper, and the game is over. At one generation per year, gatherers have used their one per cent advantage to wipe out all of the gulpers in under one thousand four hundred years.

On an evolutionary timescale, this is sudden death.

The reason for the rapid fall-off in only marginally less effective gulpers and the concomitant rise of the gatherers is that the cumulative effect of that tiny advantage grows exponentially, *multiplying* by 1.01 at each generation. As long as the factor by which you multiply is bigger than one, exponential growth soon leads to very big numbers indeed. In this simple model, you could start with one gatherer and 999,999 gulpers, and give the gatherer a hundredth of a per cent advantage – and it would still win. (Recall that they are both asexual.) The real world is of course more complex, so this kind of argument should only be seen as a very rough guide, to build intuition about the cumulative results of a tiny advantage.

Of course real organisms don't wait for the evolutionist to count them before the next innovation gets going: each is (say) ten per cent different genetically from the others of its species, and many innovations are being tested out in parallel. So, having established the new food-sensing patch in 1400 years, the gatherers would not stand still. If by chance one of them evolves a new improved patch, offering another one per cent advantage, then the same story repeats recursively, and after 2800 years all the animals have acquired it ... and so it goes. You can imagine that after a million years or so the sensory apparatus is getting pretty good – and a million years is still little more than an eyeblink on an evolutionary timescale. It took a couple of billion years for bacteria to create enough oxygen to change the rules for everyone; from that time on there have been two and a half billion years for complex beasts like us to emerge. And a lot of the changes can and did run in parallel: you can evolve ears at the same time as eyes, and you don't have to wait for either of those to be perfected before you start evolving fancy heart valves, or cunningly constructed lungs.

We return to the evolution of senses later in this chapter. First, we want to take a look at several different human senses, to see just how complex and individual they are, and give you an idea of what an evolutionary theory has to explain. We have chosen to discuss the eye, ear, and nose. Most of you will know the standard liar-to-children stories about the eye and the ear, and for those who don't, we will relate them in a moment, with a few added ingredients. The nose is more of a mystery, for a number of reasons. We now have technology that mimics some functions of the eye and the ear, namely the camera and the microphone, but there are no artificial noses in household use (although the Engineering and Chemistry departments at Warwick University are currently developing one). Moreover, the science underlying vision and hearing is much better understood than that of smell. However, smell is a fascinating topic, and a recent probable breakthrough concerning the sense of smell reinforces several of the central themes of *Figments*. For this reason we shall discuss all three of these senses in turn. The other senses, such as taste (closely allied to smell), touch, temperature, balance (yes, we have more than five in total) are just as interesting, but three senses will be enough here.

First, the eye.

The eye, though complex, seems straightforward: it resembles a camera. More accurately a camera resembles an eye, because eyes got there first. This is the liar-to-children story, but in fact the eye is much more than a camera, and its camera-like properties are among its least interesting features. There are many different types of eye in the animal kingdom, the fly's strange multifaceted eyes being the best-known variant: we shall concentrate on just one design of eye, our own. This comprises a light-sensitive surface, the retina, a lens to focus light on to that surface, an iris to adjust the level of illumination, and a cornea which among other things acts as an extra lens. Apart from the fluid this all sounds very like a camera; we even have lens covers, though female cameras do not usually daub theirs with eye-shadow. Here, however, the resemblance stops. The human visual system does not just fix an image on to photographic paper: it passes that image to the brain, to be processed into a vivid, coloured, apparently three-dimensional representation of the outside world, in which key features are somehow 'labelled' with appropriate associations – 'that's a cow, they give milk and some of them get Mad Cow Disease; that green thing is a field, it's lots of grass really; oh, look at the red dots – they must be poppies!' Philosophers call these vivid sensations *qualia*, and they play a central role in questions about consciousness. We'll return to them in the next chapter, but for now let's concentrate on the mechanics of vision. Our eyes dart this way and that, seldom fixing on anything for more than a split second, and only that part of the scenery that is aligned with the fovea – the central area of the eye where it can best distinguish fine details – is seen at all clearly. (Concentrate on your peripheral vision for a moment: everything is blurred, there is hardly any real detail.) Yet we get a strong subjective illusion of a seamless, detailed reality all around us (which philosophers call the *binding problem*). Clearly the brain must be working hard to achieve that illusion – and not just the brain. The retina 'pre-processes' the stream of visual data before the brain even gets to start working on it.

Colour vision is particularly interesting. Our colour sense derives from three different types of light-sensing organ known as cones, which respond selectively to light in different ranges of frequencies (roughly red, green, and blue). However, it would be taking far too literal a view to describe our colour sense as a frequency detector. It does not analyse incoming light, 'read off' its frequency, and consult a look-up table to determine that it is red. What happens is more the other way round: light of that frequency affects the 'red' cones more strongly than the other two types, and our brain is set up so that it then 'sees' red. 'Frequency' is a feature of reality, or at least of the physicist's conception of reality; 'red' is a figment of reality.

While the eye may be *like* a camera in many respects, its essence is quite different. Without a brain to interpret its signals, there are no pictures. In fact, an animal's eye–brain system has to be fine tuned during development before the animal can see properly. Experiments have shown that unless kittens are exposed to horizontal lines at the appropriate stage of early development, they grow up into adults who never can 'see' horizontal lines. This tells us that the eye can be picking up the input signals, but the eye–brain system may not know how to interpret them: it cannot drape the appropriate figments over its perceptions of reality because it never got the appropriate training.

Ears next.

In the standard liar-to-children story, the ear is like a microphone. It collects sound waves, vibrations in the air, and sends them down a sort of funnel. The funny folded shape upon which we rest the back end of our spectacles is like that because it collects the sound efficiently. The funnel channels the sound waves to the eardrum, making it vibrate in a manner that corresponds to the original sounds. The now-mechanical vibrations are transmitted to the inner ear by way of a cunning system of tiny bones. There they pass into the cochlea, a spiral structure along which is slung the basilar membrane; this is covered in tiny hairs hooked up to nerve cells. Different regions of the spiral respond to sound of different frequencies; the hairs pick up those responses and the nerves transmit this fact to the brain, which then synthesises all of the incoming information to create the perception of sound. The ear's responses are logarithmic: what sounds to us to be a fixed increase in noise level is actually a doubling of sound wave energy. That lets us detect very low energy sounds, like the rustling of dried leaves, without our ears being totally wrecked by the sound of a falling tree. This is all very similar to how a microphone, hooked up to a tape machine, works, with the tape machine playing the role of the brain.

Now, however, the liar-to-children story starts to become untenable, because – unlike a microphone – the ear is adaptable. More nerve connections lead back from the brain to the ear than flow from the ear to the brain, and their role is to make it possible for the ear to *change* its configuration, so that those sounds that are important to the brain become especially easy to detect. This isn't easy: even a pure single-frequency tone causes the cochlea to send many different signals to the brain. During infant development in humans, the ear is fine tuned to pick up such things as speech – and in different cultures, which employ different component sounds in their language, it is tuned differently. Babies babble in every phoneme until their brain compares the babbles with the language going on around it and prunes itself to produce only the phonemes that it hears – thereby also losing the ability to detect phonemes that are not sufficiently common in the surrounding language.

Finally, our *pièce de résistance*: the nose.

Smell is an unusual sense, even by the bizarre standards of evolution. It is an ancient sense, perhaps the most ancient, and may go back a billion years or more. The sensory connections from the nose link directly into the old, primaeval regions of the brain. A smell can evoke a long-hidden memory, suddenly and vividly; a whiff of perfume may send an old man back to the Paris of his youth. This probably happens because the regions of the human brain that once were devoted to processing smells have been taken over to be used as memory. It may also be why smell in humans is a neglected sense. In fact most adults are incapable of detecting at least one smell that others find very prominent – for example, sweat. It is as if we are nearly all 'colour-blind' for some smells, in a rather random way. Even so, our sense of smell is better than we think: we just tend not to use it properly. The physicist Richard Feynman discovered that if, while he was out of the room, someone picked a single book off the shelf and replaced it, then he could come in and find out which book it was by sniffing his way along the shelf. It was obvious: the book smelt damp and 'handled'.

How does smell work? It is definitely triggered by molecules entering the nose – but how does the nose tell which molecule it's got? Until recently the prevailing theory was that in the nose there are various receptors that detect the shapes of molecules. Think of them as irregular cups into which only corresponding shapes can fit. So when we get a whiff of benzaldehyde, which smells like almonds (and is in fact the main chemical responsible for their characteristic smell) it is because those molecules slot snugly into the appropriate receptors. The brain reads off which receptors are active, passes the information through its neural net for smell recognition, and an 'almond cell' fires. (Recall from Chapter 2 that a neural net is a network of nerve cells.) This 'receptor' theory made a lot of sense because similar mechanisms are involved in the immune system, and there was a fair amount of confirmation of other kinds. There was one tiny flaw in the theory, though. Hydrogen cyanide *also* smells like almonds, but it is a much smaller molecule with a very different shape. It could fit into *any* receptor. Other small smelly molecules, such as ozone, present the same difficulty. The theory could be fixed up by imagining that maybe several molecules of hydrogen cyanide could fit into one receptor, or perhaps one molecule with a layer of water molecules around it – but this rather spoiled the overall elegance and it looked a bit like special pleading.

In 1991 Luca Turin, a biophysicist at University College London, became intrigued by the possibility of an alternative explanation.♪ Smell is an indicator of chemical composition. How do *chemists* work out which molecule they've got? Not by looking at its shape, which emerges as a byproduct of their analysis, but by measuring its vibrational spectrum. You can think of a molecule as a lot of balls (atoms) linked together by springs (chemical bonds). When a molecule is given a sharp kick, it vibrates like a plate of jelly on an express train. Each bond has its own characteristic vibrational frequency, determined by the springiness of the corresponding spring and the masses of the balls attached to its ends. An instrument known as a spectrometer can measure these frequencies by exciting the molecule with light, and it can print out a graph showing the strengths of the possible frequencies of vibration. That graph is the molecule's vibrational spectrum, and it tells chemists which bonds are present.

Maybe, Turin thought, noses are spectrometers. Maybe in the nose there are a set of 'jellies on plates' that resonate with particular vibrational frequencies and signal their presence in a test molecule.

This theory has been around, in various forms, for quite a while. Turin faced up to the problem of testing it, and convincing a rightly sceptical scientific world. One obvious problem is that chemical spectrometers are large and delicate instruments: how could biology build such a thing inside a nose from organic materials? On the other hand, early cameras were cumbersome machines too, yet the eye does all they can and more. Turin's first idea was a simple one. Hydrogen sulphide, which all of us who have done any chemistry will remember for its strong smell of rotten eggs, is a very simple molecule: its structure is H–S–H where H is a hydrogen atom and S a sulphur atom, and the dashes represent bonds. There is only one type of bond, S–H, repeated twice. One day, by chance, thumbing thorough an old chemistry text in a secondhand bookshop while on holiday in Lisbon, Turin found another bond that he didn't know about, B–H, where B is an atom of boron. Its frequency, the book stated, was the same as that of the S–H bond. So, if his theory was correct, boranes – chemicals containing the B–H bond – should smell like hydrogen sulphide. Chemists hardly ever try to smell boranes, because they're rather nasty chemicals – they explode spontaneously, for example – but to Turin's nose there was a distinct resemblance to hydrogen sulphide.

This initial advance was encouraging, but it was not conclusive evidence. It was followed by two rapid set-backs. The first alone seemed fatal. The molecule known as 'carvone' comes in two forms, mirror images of each other. The grand symmetries of the physical universe imply that both forms must have the same vibrational spectrum: any motion of a left-handed carvone gets flipped in a mirror to yield a physically valid motion of a right-handed carvone. If the theory is correct, they should smell identical. However, left-handed carvone smells of spearmint, whereas right-handed carvone smells of caraway.

Oops.

To make matters worse, the American geneticist Linda Buck discovered the base sequences in human DNA that code for smell receptors in the nose. They looked just like the DNA sequences of other biological receptors, and it was generally accepted that those worked according to the 'shape' theory. Turin's 'frequency' theory seemed dead, and he turned to other research topics. Despite numerous experiments, however, nobody could make these new-found receptors work in the laboratory. Less discouraged now, Turin returned to the puzzle, trying to find a modified theory that countered the main objection – the riddle of the carvones. One day, while leafing through the *Review of Scientific Instruments*, he came across a description of an entirely different way to measure the vibrational spectrum of a molecule, by using electrons instead of light. If a bunch of electrons is made to flow across a molecule, then it can excite the vibration of one or more bonds. These in turn affect the bunch of electrons, and the resulting change in energy can be measured. It is a kind of electronic probe for vibrations.

Now the theory takes on a more concrete, and far more plausible, turn, because it includes a specific mechanism that might explain the nose's alleged ability to detect molecular vibrations. That mechanism combines a contextual element, receptors, with a reductionist one, the natural vibrations of bonds. The idea is that in some manner not yet fully understood, each receptor causes electrons to flow across any molecule that lodges fully – or partially – inside it. Some of the molecule's bonds then vibrate, affecting the electrons, and the change in the electron flow is detected, yielding a 'signature' for those bonds. This is an attractive theory, because creating and detecting electron flows is just what nerve cells are good for. Moreover, this idea solves, easily and naturally, the perplexing problem of left- and right-handed carvones. The relevant receptor must be asymmetric, so that one version of the molecule fits into it correctly but the other has parts poking out. Bonds that poke out do not vibrate when the electrons flow through the receptor, and hence are not picked up by the brain. Indeed Turin discovered that for carvones the caraway smell occurs when a particular C–O (carbon–oxygen) bond is picked up, and the spearmint smell occurs when it is not. The frequency of this bond, 1800 units, is the same as a bond in acetone – nail-varnish remover. The same bond occurs in the related chemical pentanone, which is more suitable for experiments. So now Turin had invented a really good test of his theory: mix spearmint and pentanone in suitable proportions, and you should get a smell of caraway. It's a highly unlikely proposal that makes no sense in terms of the shape theory, but it is entirely reasonable on the basis of the frequency theory. That's what makes it such a good experiment.

To provide a credible test he enlisted the aid of a group of Parisien *parfumiers*, some of the most sensitive and best trained human noses in existence. In a blind test, carried out late in 1995, they unanimously reported the predicted caraway smell. Of course such a test does not prove the theory is absolutely correct – which is why we described it as a *probable* breakthrough – but it certainly suggests that the idea runs along the right lines.

Some experts didn't like the use of parfumiers, with their subjective sense of smell, rather than precise instruments. The idea that the nose carries out spectroscopy bothered them too. But since the human sense of smell is inherently subjective, how else can one tackle it except by using humans as instruments? In which case parfumiers are the finest instruments around. As for the nose doing spectroscopy – no, look, it doesn't have to, that's just how the story goes in liar-to-children mode, just as the eye is said to perform photography. You don't have to have a system that picks out a particular spectral frequency *as such*. Spectral frequencies are *features* of a molecule that appear natural to a spectroscopist, but they are not necessarily the features that would be picked up by an electro-molecular sensor hooked into a neural net. The brain can evolve a classification scheme of its own – it doesn't have to be one that matches the chemistry textbook concepts precisely. (Though it must presumably bear some rough correspondence to frequency if Turin's predictions about bonds are to be relevant.) For that matter, it is entirely conceivable that different people perceive smell in slightly different ways, depending on how their neural net is configured. Smell could be like blood type: polymorphic.

We have already remarked that Turin's theory combines a contextual element (receptor) with a reductionist one (vibrational frequency). It is this embedding of the idea in a new context that solves the symmetry riddle: the two carvone molecules may be mirror images, but the context stays the same in both cases, so it does not turn into its mirror image when the molecule does. Indeed the sense of smell seems to have arisen through complicity between these two elements, one reductionist and one contextual. Because complicity is emergent, we strongly suspect that the receptors do not actually pick up vibrational frequencies *as such*, any more than the eye or the ear do. What matters is that different molecules interact with a given receptor, and its current of electrons, in different ways. The brain can then learn how to interpret these differences. It is clear that its interpretation is closely allied to frequencies of bonds, but it need not *be* frequencies. Analogously, the vibrational spectrum of a jelly wobbling on an express train is mostly that of jelly, but distorted by interactions with context: the jolting, juddering train. We think it highly significant that purely reductionist approaches failed to determine the essence of smell. They could not, because smell is not reductionist. It evolved way back in the pre-Cambrian era when the world was a soupy ocean, and it came about in the kind of chewing-gum-and-string manner that is typical of evolution. It is highly unlikely that such a process would produce an exact correlate of something so precise and reductionist as frequency, which is a feature of physical models rather than a feature of the organic world.

We suspect that the reason why most adults cannot detect certain smells may be similar to cats not 'seeing' horizontal lines if they are not exposed to them in early development, or babies pruning their babbles to match the language of their culture and losing the ability to 'hear' phonemes that are uncommon in that language. If so, developing brains that are not stimulated by certain smells at the appropriate period of development may lose the ability to recognise them altogether. This is sheer conjecture, but rather plausible. One test would be to find out to what extent the smells that people can detect, or respond to most strongly, depend upon the culture in which they grew up. Of course maybe it's a genetic difference, a kind of nasal polymorphism, or a failure to develop certain specific receptors.

From this new perspective, our sense of smell seems to rest upon a very arbitrary set of choices – a universal mechanism, very probably, but realised in a parochial manner. One can imagine many shapes for receptors that do not actually occur, and the neural net that interprets them would give different responses if its architecture were different. Do we detect only those smells that our distant ancestors were exposed to billions of years ago – or molecules that excite similar responses? Certainly many of the smells that we find strongest are related to plants, cooking meat, rotting organic matter, faeces, and the like.

Between them these three tales establish that in animals as sophisticated as humans, senses are very strange processes indeed. They go far beyond the simple technology of cameras and microphones: they require a degree of interactive learning by the brain and the sensory organs and fine-tuning by biological development. This, then, is the degree of complexity and versatility that any evolutionary theory must explain. Moreover, it must explain how such systems can come about gradually. The problem is not to explain how 'half an eye' can be useful: that's a straw man set up by naive anti-evolutionists. In 1994 Daniel Nilsson and Susanne Pelger♪ used computer simulations to show that a complex eye – with analogues of the retina and cornea, and a lens whose refractive index varies much like ours – can plausibly evolve from a simple light-sensitive patch of tissue in such a manner that every change is an evolutionary improvement. It takes no more than 400,000 years – the evolutionary blink of an eye. So much for straw men: the real problem is to find a plausible route whereby rudimentary sense organs, with little or no associated neural processing, can offer enough adaptive advantage that they slowly evolve into improved versions.

It is clear that the slight increase in versatility provided by rudimentary sensors has such a strong effect that animals have evolved, by that recursively self-reinforcing route, sense organs that can be astonishingly selective. *Hydra*, for instance – a tiny tentacled creature that lives in ponds – is thought to possess a chemical sense (a rudimentary form of smell/taste) that responds only to the chemical glutathione. Glutathione is in fact part of *Hydra*'s own biochemistry, so it already had the right enzymes to detect it: that coincidence provided the scaffolding from which evolution could build a sensor, and it realised that potential because glutathione is a sufficiently reliable indicator of the presence of *Hydra*'s favoured prey. It is very easy to invent this kind of receptor; an equally simple one would be a light-sensitive spot. Light has quite strong effects on many different molecules, so the Principle of Murphic Resonance applies and light-sensitive patches will be commonplace – provided organisms have evolved to the point at which they can in principle exploit the kind of chemistry that would be required, and assuming that such patches convey some kind of advantage. However, unless shadows are a good indicator of prey, a rudimentary light-detector – a primitive eye-spot – won't be much use. In such cases the animal needs a more structured eye, such as an array of light-sensitive spots. But then it must also have the right kind of nervous circuitry to respond only to patterns of stimulation that are associated with the presence of prey – or of predators that it would be wise to avoid. This observation provides advance warning that we shouldn't really think about senses without also thinking about their associated neural circuitry and what there is to sense.

Biology provides a broad range of 'components' for the manufacture of versatile neural circuitry: there are many types of nerve cell. Some increase their firing rate when they are stimulated by an incoming signal, some decrease it; there are even specialised nerve cells that turn on (and others that turn off) when stimulated, but these are rarer. Among the remarkable properties of nerve cells are their shape – they are typically extremely long, linking distant parts of the body – and their ability to transmit electrical impulses. They do this less like an electrical cable and more like a flame travelling along a fuse: instead of electrons flowing along them, ions move across the local cell membrane, triggered by similar movements just up the line. This ability allowed nerve cells to set themselves up as the 'communications network' for the senses. Nerve cells commonly produce whole trains of electrical pulses, a bit like the 'digital' signal stored on a CD, and they can vary the timing of the pulses or even suppress them altogether. Their activity is usually said to be 'binary' – meaning that they are either 'on' or 'off', like a switch. This is strictly true for each part of each nerve cell membrane, but usually a complete nerve cell will *vary* its rate of firing according to what other nerve cells, or cells of sense organs, are doing to it. So the output is much richer than just a binary on/off, just as the musical output produced from the binary code on a CD is.

The amount and quality of the neural circuitry attached to sensors and interpreting their input is usually limited, so it is important not to 'overload' the circuits with surplus information. Even if that's not the case, efficient neural computation carries its own advantages, and hence will be favoured by natural selection. So more sophisticated sense organs, like vertebrate eyes, ears, and even skin-temperature sensors, signal only the important *changes* to the environment. The brain really prefers not to be bothered with messages like 'That rock is *still* sitting where it always does' or 'That bird is still singing'. Sudden movement near the rock, or the bird stopping its song, are far more significant. By the same token 'Skin the right temperature' need not be signalled, but 'Too hot' should latch immediately into a reflex withdrawal program. For skin temperature this effect is achieved via a general property called 'habituation', which has of course been selected because of its evolutionary advantages. Habituation is usually a physico-chemical property of the nerve cell itself: after the stimulus has stayed at the same level for a time, the firing rate goes back to what it would have been had there been no stimulus at all. In effect the nerve cell recalibrates itself so that this new level of stimulation is seen as zero. If the stimulus level now changes – even if it goes back to the *old* zero level – the nerve cell firing rate will change. So habituation leads to nerve cells that naturally respond to *change*, which is what is usually important for producing an effective response.

Several responses may be available, even to the simplest animals: there are often circumstances in which 'Freeze!' is a valid alternative to 'Get the hell out of here!' So the animal has to choose a response that is appropriate. The brain, or more generally the layer or network of nerve cells between sensors (like eye-spots) and effectors (like muscles), has ways of connecting special arrays of receptors to special arrays of response. It is helpful to think of all the senses, together, as a 'sensorium', and all of the effector muscles and glands as constituting a 'motorium'. In between is a more or less complex process-and-control structure which accepts incoming signals from the sensorium, works out what they signify, and passes appropriate instructions to the motorium.

Reflex actions bypass most of this with direct connections – if you have the faintest suspicion that your hand is in the fire, best to pull it out quick and ask questions later. Nick Humphrey♪ has suggested that this kind of control structure, linking sensorium and motorium *directly*, evolves as a kind of 'internalisation' of a feedback loop that originally exists in the environment. We can explain this by describing a tiny robot, built by an acquaintance of ours. It is powered by 'solar cells' that receive sunlight and convert it into energy. When it is in a patch of sunlight, it moves forwards until it emerges from the edge of the patch. When it detects the absence of light, it spins at random and moves a short distance. It repeats this until it finds itself in sunlight again, and then it reverts to straight line motion. In practice what it appears to do is prowl around inside a patch of sunlight, moving away from edges whenever it encounters them. It can even follow a moving patch across the floor, say as the sun moves round in the sky. Actually it is employing a feedback loop out in the environment: spin randomly, look for sunlight, if not, spin again ... Now, some of its solar cells experience the drop in incident light first – the first ones to fall outside the patch. Suppose the robot accidentally developed an internal circuit that could use this information to 'predict' the direction of movement needed to stay in the bright areas. It wouldn't have to be a very good prediction to offer an advantage. If such robots could evolve, pretty soon they'd be able to see the edge coming and stay clear of it.

Organisms detecting food by chemical concentrations could learn to play the same kind of game, too. In effect they would learn to ride up the chemical gradient, towards its source. Evolution will 'build in' connections of this kind if they are possible at all. It is easy to observe such an 'internalised' feedback system in operation: you have plenty of your own, though they have evolved considerably further and have become extremely sophisticated. For instance, you can decide that you are going to reach for a nearby object, shut your eyes, and then immediately pick it up. (Try it now.) You will find that you can do so successfully without the 'hand-eye coordination' that we tend to imagine is responsible for our ability to manipulate objects. Not so (although it provides a useful check): we have a kind of inbuilt model of our surrounding geography, and we can plan *and carry out* actions within that model.

Where does such sophistication come from? How can simple circuits of nerve cells carry out complicated tasks? Working models of systems like this – usually programmed on computers rather than embodied in hardware – have shown that extremely simple neural nets can exhibit surprisingly complex behaviour. There is a rapidly growing area of research, involving mathematics, computation, biology, and psychology, which is investigating these possibilities. Their common topic is 'neural nets', in which layers of units that act like simplified model nerve cells are linked up in various ways, and 'trained' to carry out particular tasks such as recognising a particular input signal.

Even more interesting is a variety of experimental evidence for the self-organising power of these systems. For example a so-called 'hierarchical' network, obeying rules like 'cells that fire together grow more contacts with each other' can be 'taught' to recognise particular patterns among the signals to which its sensory cells are exposed. For example, a cell on the end of such a network, connected to a rectangular array of light-receptive patches, can be taught to fire on response to a pattern of inputs forming the shape of the letter T after only a few hundred trials. Such cells, which the system has trained to respond only to some specific pattern of inputs, are known in the field as 'grandmother cells', because they are supposed to be like that *one* cell in your brain that fires when you see your grandmother – and only her. Whether real brains possess grandmother cells is controversial (and we, JC & IS, suspect that they do not, as we argue in Chapter 7) but artificial neural nets certainly can have cells with such a property.

Instead of a simple hierarchy, several workers have produced what Victor Serebriakoff called a 'polyhierarchical' system, with about the same number of cells in each layer, but with several sense cells connecting to each cell of the next layer, and so on. Similarly, each motor element is (initially) served by several cells, each of which receives signals from several in the next layer. Systems of this kind 'learn' to associate certain behaviours with specific inputs, and if several such systems are connected to each other so that the input 'sees' some of the output, striking circuits are formed, each with its own individual character. However, the final structure of the circuit, to which it settles down after the learning process has modified it to carry out its task, can seldom be predicted from the initial set-up. Clearly the learning process passes through Ant Country: indeed the reason for this unpredictability is similar to that for winning strategies in games. In a game, a winning strategy is the result of a recursive pruning of moves that didn't work, and the complexity of the game tree results in a strategy that often bears no obvious relationship to the rules of the game. Training a neural network is very similar to playing a game: the moves are the training steps in which the network is shown various inputs, and it 'wins' if it produces the required outputs. The vast, complex, unclassifiable tree of possible networks is 'pruned' by the learning process, until it evolves a winning strategy and succeeds in recognising the inputs correctly. So again we do not expect the final outcome to bear any obvious relationship to the initial state – the opening position in the game. Or, for that matter, to the rules. What we see here is the phenomenon of *emergence*: the appearance of recognisable large-scale features in a system whose chains of small-scale causality are far too intricate to describe, let alone follow in detail.

The structure of the neural network connecting sensorium and motorium is usually described in terms of a computer analogy: some of them are viewed as 'hardware', the rest as 'software', with the hardware being wired in so that it cannot be changed, whereas the software is readily modified and changes during the learning process. Unfortunately this analogy is rather superficial, and it can be misleading if taken too literally. It is true that the initial stages of development resemble the construction of a hard-wired circuit, but thereafter the analogy breaks down. Let's see why. In a developing animal, nerve cells – like all types of cells – take up their positions in the growing, changing embryo; they grow special long structures, known as axons and dendrites, to make contact with other nerve cells, and with cells that will eventually develop into sensors. The rules that govern this growth process are those of embryology, in which cells move along gradients of chemical concentration, or the cells themselves change the distribution of chemicals so that different gradients are created. In the computer analogy this step is the 'manufacture of the hardware'. Then, when the embryo is born or hatched, the sense organs pick up signals from the outside world: this is held to be like feeding data into a computer to set up its programs.

For some animals – those that have almost completely pre-wired nervous systems – something like this may indeed happen. But even then, signals are being received by each cell *while the system is developing*. The 'computer' is already being programmed before its circuits are completed, and the structure of those circuits is affected by the programs. So the image of a finished computer into which programs are poured is somewhat inappropriate, and can easily lead us to make incorrect inferences about how the neural circuits of living creatures function. Animals do not program a built-in nervous system: instead their neural connections change in response to all of the stimuli that come in, before or after hatching or birth. They grow new cells, make more connections or reduce them, and a major part of the process involves killing off cells that have become surplus to requirements. The procedure is much closer to how Niven and Pournelle's Moties would have worked with computers – not by leaving them in their case and typing on the keys, as we do, but by tinkering with their circuits, throwing components out or rebuilding the circuitry in whatever manner they felt like, while the partially constructed machinery was still running. A Motie computer would not *have* a case, and its hardware would change from one minute to the next.

Brains, then, are like Motie computers, not IBM PCs, and are constantly rebuilding themselves. As a result of such running modifications, the entire nervous system evolves as a whole, 'hardware' and 'software' alike, homing in on a structure that performs the required functions. The hardware/software distinction is thus irrelevant, if not downright misleading, and best forgotten. This means that there is no sensible way to distinguish systematically between building a nervous system, programming it, or providing it with data. A batch of sensory data comes in, changes the program, and the circuitry is rebuilt accordingly. When the next batch of data comes in, it is received and processed by a *different* hardware system.

This distinction cannot be emphasised too strongly. You are a different person now from what you were before you read this sentence, and each of you readers out there is reacting differently from the rest. Philosophers get very excited about Heraclitus's point that 'you cannot step into the same river twice'. The usual justification of this remark is that the water is different next time round. We (JC & IS) are more excited by the less obvious fact that you cannot step into the same pond twice: the water may be the same, but it's a different *you* the second time round.♪ We gather that Heraclitus was aware of this too, but it seems to have been forgotten by most of his successors.

This organic co-evolution of the nervous system and its sensors – let us simply say 'brain' – means that they are very unlike our own designed-and-built artefacts like cars, washing machines, or CD players. The parts of the brain cannot be given neat labels stating what function they are 'for'. Yes, different parts of the brain do different things: we know quite a bit about the functional geometry of animal brains and to a smaller extent our own; we know that certain areas are associated with vision, or language, or fear. But parts of the brain get taken over, during evolution, by new functions; and any single part of the brain is likely to be involved in many different functions.

It is tempting to imagine that we would be in a far better position to analyse brain function if we could map out all of the brain's 'circuitry'. However, brains are far too complicated – the human brain is the most complex of all, with about 10^{12} connections between 10^{10} nerve cells – and most of the cells and connections are buried deep within a three-dimensional structure that starts to fall apart if we dig inside to find out what's there. But even if we could map out the complexity of a brain, we'd be lost deep in Ant Country. We wouldn't be able to deduce the brain's behaviour from the map. In fact, even when we can map part of the circuit of some simple creature, we usually don't know enough about how a single nerve cell works to deduce what that part of the brain is doing. For example, the first part of the lobster gut is controlled by a simple system of six interconnected neurons – but it has still proved too complex to model. So how can we possibly hope to model what systems of millions of nerve cells do? Let alone billions.

Modelling systems in detail, however, is not the only way to understand them. Science has a pretty good understanding of the flow of liquids, but it doesn't achieve this by tracking the detailed motions of individual molecules as they bounce off each other. Instead it works with the overall structure of a liquid, focusing on features such as conservation of volume, stickiness, and so on. In the same kind of way we can make general statements, usually mathematical ones, about the capabilities of neural networks and their high-level organisation. We can analyse the effects of symmetry, for example: symmetric circuits, such as those involved in locomotion, produce a characteristic range of output patterns. We have a rather good idea of how the 'central pattern generator' of the lamprey, a network of nerves along its spinal cord, causes it to move in its characteristic sinuous manner. We can describe the properties that such networks must have *in addition to* their token one ('token' because although we have discovered one function, we know that in different circumstances the same network may have other functions too). All complex systems are like this. We repeat a story from *Collapse* because it exemplifies this feature so beautifully. Schoolchildren of the 1940s, JC among them, knew that if you fed a ticket machine on the London Underground a tiny silver threepenny piece, instead of the slightly larger sixpenny piece that the designers had in mind, then the machine did not disgorge its usual sixpenny ticket. It disgorged the entire ticket roll, each ticket being stamped with the date but otherwise blank. This was not a *malfunction* of the machine, which was working perfectly; it was an unintended side-effect resulting from its design. The design could have been reworked to avoid that problem, but the machine would still be able to carry out unintended functions – what might it do if fed a lump of chewing-gum, for example? Or maybe if you give it just the right thump on the side ...

Examples abound in biology. The parasite in the ant's brain, described in the story at the head of Chapter 3, stumbled into its own silver threepenny piece. Another possible example occurs in birds. It is well established that in some bird species females prefer to mate with males that possess symmetrical tails. There are two competing theories to explain this, which may well both be right, and both of them are 'threepenny piece' explanations. One is that you need a properly functioning developmental system to ensure that your tail is symmetrical, so females that happen to have a genetically induced preference for mating with symmetrical males will tend to produce better offspring, thereby perpetuating the preference for symmetry. Over a fairly short period in evolutionary terms this tendency will be reinforced. Its function is that of an evolutionary short cut: it acts as a simple test for 'good fathers' that replaces the traditional trial-and-error method of letting them father babies and seeing which ones best survive. The other – the one that is relevant to the present discussion – is that the preference for symmetry is an incidental by-product of a visual system that has been trained to *recognise* tails. Instead of a grandmother cell, the visual system has a 'tail cell' that responds strongly when the eyes see a tail. Now, tails come in a variety of shapes, and for every tail that is lopsided on the left there is probably one that is similarly lopsided on the right. A neural network that responds strongly to both of these is likely to respond even more strongly to a symmetrical tail, because that resembles *both* lopsided versions and 'counts twice' when generating the final 'this is a tail' recognition response. So the preference for symmetrical tails results from how neural networks function – sixpence produces a ticket, but threepence a whole roll; lopsided tail produces 'oh, there's a tail', but symmetrical tail produces 'gosh wow what a super tail'. Incidentally, such behaviour is by no means confined to birdbrains: humans have a similar preference for symmetry or near symmetry. Recent research♪ indicates that women have more or deeper orgasms if they mate with males whose faces are nearly symmetric, and this kind of mechanism plays very much the same role as the tail preference in birds, so that facial symmetry becomes reinforced as the outward sign of a good mate. Our aesthetic preference for symmetry may have come about via this route. However, our brains seem to get bored if symmetry is too perfect, as artists are well aware: slight asymmetry is more interesting than perfect symmetry. Presumably slight imperfections 'tickle' our innate sensibilities by being close enough to symmetry to evoke a response but just sufficiently offbeat to excite some other kind of higher-level interest too.

By now we've told you a lot about brain function and structure, and we've argued that brains evolve because they're jolly good gadgets to run sensory and locomotory systems. Our brains, however, do much more: they are intelligent. Many biologists see intelligence as simply a further elaboration of the brain's inherent powers – a 'runaway' process leading to cleverer and cleverer neural networks just because that's what neural networks do.♪ The problem with this theory is that although artificial neural nets readily and naturally mimic sensory control and detection systems, nobody has yet constructed one with intelligence. So it looks as if some magic ingredient may be missing. Descartes thought it was a different kind of stuff, which turned mindless organs into minds, but we reckon it is the contextual element of co-evolving culture. If you think of phase space as 'stuff', then Descartes may have had a point, because culture-space is very different from neural-network-space. Somehow we doubt that's what he had in mind.

What we're looking for is an alternative story of human evolution, one that does not contradict the 'clever nerve cell' story but runs parallel to it. This story, which tells of organisms relating to their environment rather than cells relating to each other, will be important for later arguments about minds, too. The story could start almost anywhere, but to keep it reasonably short, let's begin with eggs.

Presumably the earliest egg cells differed in how much material and energy was stored in them, just as other body cells do today; those females whose eggs provided a more extensive 'packed lunch' for their offspring – perhaps those females that could afford to – loaded the dice in favour of their genes. It is not surprising that this stratagem, known as 'yolk', spread through many reproductive lines. Another trick – a real cheat, this one – was very close to the yolk trick in the phase space of possible lines of development, hence accessible to evolution and available for exploitation. This was keeping the egg inside the female. It is likely that very yolky eggs *had* to be fertilised inside the female, for two reasons: they were 'costly' to produce, so leaving them lying around unfertilised was a waste; and worse still, the investment in yolk might well be exploited by a predator. Moreover, sperms have problems locating the nucleus in a large egg with a lot of yolk. The yolky eggs of octopuses, sharks, snakes and birds are all fertilised inside the female today, and it's reasonable to expect the same of ancient ones. All of these eggs *begin* development inside the female, too. When a chicken egg is laid it already has some hundreds of cells in a flattish plate on the surface of the yolk. On the other hand most sharks keep the egg in mother's oviduct until it develops into a functional fish; effectively sharks are 'viviparous', or live-bearing. So are a few snakes, but no birds. This is rather odd since they start their development inside the mother.

Eggs that start their development inside mother have many advantages: one is the existence of an already well developed nervous system to look after them. This is a good example of a snooker-break reproductive loop: in order to have the stable environment needed to develop a complex nervous system, a complex nervous system must already be in existence. Eggs inside mother do not experience large variations in temperature, because mother can control temperature by moving into or out of the sun, flapping her wings to cool off, whatever. They don't have to be able to escape predators, because mother's nervous system senses them coming and takes action to defend against them – flight or fight, depending on the predator. Eggs that develop inside mother can even let her do much of the developing herself – they don't need to be equipped with contingency plans for unexpected circumstances, so the same quantity of 'building instructions' can lead to a more complex result. Mammals have *shorter* DNA sequences – genomes – than many amphibians, in part for this reason. With the eggs developing inside her, the mother can avoid the problem of producing all that yolk in one go: instead she can feed the embryo slowly as it develops. Many sharks and other fishes provide a placenta for this purpose: the quantity of material in their embryos increases enormously during development, like mammals. In other fishes, whose eggs are very yolky to start with, the eggs lose weight while the embryos develop. Mammals, of course, are the archetypal live-bearing organism. Human eggs are tiny, only slightly larger than normal body cells, and mothers provide all the nutrient for development in a wonderful thermostatically controlled uterus, protected by excellent sense organs and in a body brilliantly designed for running away.

Babies are incompetent organisms, unlikely to survive on their own. To deal with this problem, most viviparous mammals – giraffes, gnus, guinea-pigs – keep their babies inside for as long as they can, until the offspring can fend for themselves. However, this imposes quite a strain on the mother, so many mammalian groups have invented nests. Inside these nests, babies born while still premature can be looked after without mother being hampered by their weight. The prior evolution of milk and mammary glands to deliver it meant that apparatus for feeding babies outside the uterus was already in place. Adult animals can react to the behaviour of babies in the nest – for example, feeding the one with the biggest open mouth or making the most noise – so the babies are dependent on their own behaviour, as well as that of the parent, in order to be fed. This opens up a whole new set of possibilities – an evolutionary 'explosion' rather than a small-scale exploration – among them trial-and-error learning. When left to its own devices, any mistake on the part of a baby animal is usually fatal – an error-prone baby gnu quickly becomes hyena lunch. But within the protection of the nest, babies can fail without getting themselves killed. Instead they just make themselves uncomfortable. Mother–baby interaction therefore fosters learning and teaching. This trick soon extends to take in father too, who begins to participate in the care of the young; other members of the family can also be brought in and the relationships become very complex.

This is equally the case on Zarathustra.

Pursuer-of-sicknesses Ringmaster, I am becoming concerned about Creator-of-creations. I think he is going broody.

Ringmaster [*Worried.*] I hope he is not building a nest? Surely he knows that on a voyage of this kind there is no place for tadpo–

PoS I regret that he is. He has obtained a large quantity of shavings from Hewer-of-wood and when I last saw him he was arranging them around a pile of– of assorted fruit from the galley.

R Sad. A phantom clutch.

Destroyer-of-facts True, but at least the fruit will not hatch. [*In sudden panic.*] Unless we still have some of that bat-fruit that Performer-of-amusements traded his false proboscis for on Argyris III –

PoS No, we ate the last of those last night, with the pickled syrup sauce. You surely remember? One of them was so ripe it was on the verge of becoming airbor– but I digress. What shall we do about Creator? When his eggs do not hatch he will become distressed.

DoF [*After some deliberation.*] Steal his fruit.

R But that will cause him even more distress!

DoF Not if we steal them *gradually*.

R I do not understand.

DoF Let me explain. You recall how our evolutionary ancestors cared for their young? [*Without waiting for a reply, continues ...*] Then I shall explain it to you. No, no, do not interrupt, it is no trouble, no trouble at all – no, really ... The modern Zarathustran lays eggs, a habit that goes back to the seven protosymbionts that our vampire ancestor gradually replaced. Also like us, they made nests – as do some terrestrial species, oddly enough, such as the cuckoo, which lays its eggs in other birds' –

R Yes, but the terrestrial species build the nests *first*, and then lay the eggs in them. Whereas we lay the *eggs* first, and then construct the nests around them. Which is obviously superior.

PoS Why?

DoF Because it avoids wasted nest-building effort by Zarathustrans who turn out to be infertile, of course.

PoS Oh, right. Pardon my stupidity.

DoF Now, the tadpoles of the protosymbionts were very alike – more so than the adults.

PoS This is a very common phenomenon in biology, of course. As the adult form develops, so inherent differences become more apparent.

DoF [*With a degree of irritation.*] Indeed. The protosymbiont tadpoles were versatile parasites on *lots* of other animals, including adult protosymbionts of the other six species. They had inject-and-suck mouthparts. They injected psychotropic chemicals that made the parasitee – if there is such a word – feel good. And they sucked its blood. So each of the seven species of protosymbiont could exploit the other six as food for babies. The system was in balance, but it was highly unstable.

R What do you mean by that?

DoF I shall explain shortly. Our vampire ancestor went a stage further and exploited the entire system. It too sucked blood as a tadpole, but over a long period of time it evolved by dropping the adult stage from its development, becoming in effect an overgrown larva. The textbook term is 'neoteny'. The neotenous tadpole thereby retained its blood-sucking abilities throughout its adult life, enabling it to outbreed all seven protosymbiont species. That is the sense in which the system was unstable: it could easily be subverted by an invader. And so our ancestral vampire placed itself in the ideal position to become a reverse cuckoo / cuckold (delete neither since both are applicable).

PoS I do not understand.

DoF A reverse cuckoo builds its nest around other creatures' eggs. And a reverse cuckold – look, do I have to explain *everything*?

PoS [*Aside.*] He usually does.

DoF A reverse cuckoo with parasitic tadpoles gains the advantage of a regular food supply for its young. The advantages of being a reverse *cuckold* are manifest. And so our vampire ancestor prospered.

R But–

DoF You are, of course, going to say – most perceptively – that this advantage is minimal all the while there is an adequate food supply for all adults. Correct! But when the climate change came, and there were widespread droughts, our ancestor's strategy came into its own. Not only did our ancestor take over all the roles of the proto-symbionts: it *had* to, to keep the system working, as one by one they dropped out of contention.

R That makes sense. But why do you want to steal Creator's fruit?

DoF Well, you know that there are three Zarathustran sexes–

PoS [*Sarcastically.*] No!

DoF [*Who is immune to sarcasm.*] Oh, yes. Mother, father, and carer. [*Any adult Zarathustran can take on any sex, just as it can take on any role: the physical differences are mainly ones of inclination. For this reason we have employed the pronoun 'he' for all Zarathustrans throughout, instead of attempting to translate the convoluted alien terminology that captures the intricacies of social interaction among three* changeable *sexes.*]

Liar-to-adults [*Who has been half asleep in one corner.*] Three sexes? But that is totally unoctimist! I really cannot– oh. [*Now fully awake.*] Your pardon,

Destroyer-of-facts. I was pre-conscious and it had temporarily escaped my recollection that the typical brood contains four eggs – making a total of seven, to which we must add the Great Carer who looks after each family. [*The others nod in agreement, save for Destroyer-of-facts, who discovered as a recently metamorphosed ex-tadpole that the typical brood contains 2.3 eggs, and whose respect for religion never really recovered.*]

R Ah, now I see where Destroyer's strategy leads ... Suppose Creator's nest is reduced from four 'eggs' to three. Then it will lose its octimality, and he will realise that something is amiss. The intellectual jolt will no doubt return him to his senses.

DoF A strange new use of the word 'sense', but you have the gist of my proposal. It is risky, but that is of no concern to me. Now, one of you must distract Creator while I sneak up and steal an 'egg' from his nes– Oh! Creator-of-creations, I did not sniff your approach! I was, er, discussing the need to seal the leg of my, um, vest ... [*Stares hopelessly at the others, for once at a loss for words.*]

Creator-of-creations Have any of you guys got a spare kipper?

PoS [*Whispers to Ringmaster.*] Phantom cravings too! Definitely broody ... [*Delicately.*] Um – Creator, do you really think that a person in your condition should ... er ...

CoC Condition? What condition? I have been studying a terrestrial art-form known as 'painting'. I was merely looking for the final article for my projected still life: 'bunch of fruit and kipper with woodshavings'. [*Belatedly notices their expressions.*] Why are you all staring at me like that?

On Zarathustra, eggs created a demand for nests. On Earth, nests created a demand for learning, leading to a rudimentary form of culture. In wild dog packs, for example, only the dominant female produces pups that thrive, but they are looked after by all the members of the pack, their uncles and aunts as well as their brothers and sisters. It then becomes important to recognise your own pack, and each dog pack has its own calls and ritualised behaviours, which the pups learn from their relations as they grow. This trick is a doggy form of culture: a new kind of heredity. It operates in an interesting tension with sibling rivalry – which is not *always* evolutionarily valuable – and instead of being transmitted in genes and stored in organisms, it is transmitted by sense organs and stored in brains. Some birds learn and teach in the nest, but even parrots – among the most intelligent of birds – don't pass on many behavioural items in comparison to carnivorous mammals like cats and dogs, or even rodents like rats and squirrels. Mammals like gnus that are born 'grown up' don't have this kind of cultural inheritance either. But our group, the primates, has turned it into a way of life; for instance in baboon troups babies receive a lot of attention from all the other animals. Our closest relatives, the two species of chimpanzee, continue learning right into adulthood, as do we. This entire repertoire of non-genetic tricks for giving offspring a flying start is a very sophisticated form of privilege.

We can view the evolution of privilege as something that takes place within, and is driven by, environmental context. Many animals grow through several stages and in several environments. A moth develops as an egg, a caterpillar, a pupa, and a – well, a moth. A bird develops inside an egg, in a nest under the watchful eye of its parents, and on its own as an adult. A human develops inside the uterus and outside it. The environment in which a particular developmental stage takes place is a constraint that can be evaded only by developing in a different environment, and that's tricky because the animal is already developing successfully in that environment, so why change? So the environmental constraint is a 'given', a background rule that is simply obeyed – just as the rules 'gravity pulls you downwards' or 'fats do not stick to water' are given. The development at one stage must therefore prepare the animal for the next one – indeed for all the future ones, to some extent. Animals whose privilege does not include a nest nevertheless have inbuilt environmental expectations – the newly hatched maggot is ready to deal with a carcass, the newborn greenfly is ready to make its mouthparts drill into a leaf for food. These expectations affect both the animal's physiology and its nervous system. But in addition, the previous generation has to have made the right preparations: mother fly has found a carcass to lay her eggs on, mother greenfly has laid her daughters on the right kind of leaf. (Greenfly *can* be sexual, though they often clone.) Both offspring are privileged in their different ways, and the privilege must match their developmental programmes. A greenfly faced with a carcass can drill all the holes it wants, but it will never get the food it needs.

This is where brains – remember them? – come into the alternative picture that we have been piecing together over the last few pages. Brains make privilege a real winner. Versatile nervous systems can match more environmental possibilities, and can encompass a wider range of behaviours, all of which makes the generation-to-generation privilege trick much easier. Passing on privilege successfully requires a recursive process of 'snooker-break' type: mother fly must pass on a carcass (a feature of her own behaviour) and the ability to exploit it (a feature of her offspring's behaviour). This means that fly genetics, development, and behaviour must 'fit together' from one generation to the next in a stable, repeatable manner. The more you think about this the more amazing it seems – but like any recursive process, the evolving creatures eased into it by way of scaffolding that has now been discarded because it is no longer required. Passing on privilege is just like maintaining a snooker break: the current move (generation) must not only pot the next ball (produce offspring and provide privilege to help them develop), but it must achieve good position for the next shot (the offspring must be able to keep the process going).

Brains make it easier to keep the break going: they let you adjust your next move to cope with errors in the previous one. More accurately, they provide flexibility and enlarge the region of phase space from which the break can be kept going. Imagine a fly whose maggots instinctively move their jaws but have no idea what they are chewing. Then they can only eat a carcass if they happen to be lying face down. If the maggot develops so that it points the wrong way, the snooker break stops. But if the young fly can sense the position of the carcass, and take action accordingly, then the break can continue. Good snooker players have a whole armoury of special shots, involving spin and swerve, to help them recover from a bad positional shot and continue the break. Similarly, creatures with adaptable and versatile behaviour can keep a more complex break going.

This is just one of the reasons why brains are a good idea in evolutionary terms. Several animal groups have produced very good brains, opening up the possibility of versatile behaviour. The diversity of such creatures demonstrates that this strategy must be a highly successful one; this in turn suggests that braininess, though not necessarily using the specific neural architecture found on Earth, is an evolutionary universal. Even animals with fixed 'telephone exchange' nervous systems, such as simple insects, have more sophisticated responses than we usually credit them with. They must have the same kind of versatility as a snooker player, able to keep their break going when there are differences in terrain, wind speed, whatever. By 'break' here we mean their particular lifestyle. However we should not credit them with *too* well developed conceptual powers: on a meta-level their strategies may be simpler than they appear. For example a stick insect that has lost a leg, or several, can still walk. It does so not because it has complex and versatile neural circuitry that incorporates special contingency plans for such an eventuality – it is more like an old-fashioned telephone exchange with a fixed system of relays – but because the simple rule 'keep the horizon horizontal' plus eye–leg coordination stops the insect from falling over.

The first faint glimmerings of intelligence appear when animals go beyond such simple trickery to become genuinely versatile. You may not have thought of arthropods, crustaceans, insects, or spiders as being intelligent, but they can display behaviour that is far more versatile than that seen in animals with 'telephone exchange' nervous systems like the stick insect. For example, even the tiny brain of the honey-bee has the ability to take conceptual short-cuts as well as geographical ones. Imagine a hive of bees just outside a wood, with a flower patch on the far side. The bees find the patch by flying round the edge of the wood, and soon all bees take that route, it being the best available. One day someone cuts a wide track through the wood, on the other side from the bees' dog-legged path – not directly in line from hive to flowers, but off to one side. Then *immediately* all of the bees take the new short-cut to the flowers instead of their old route.

The most impressively intelligent arthropods, however, are probably the mantis shrimps, while the most intelligent of all invertebrates are the cephalopod molluscs – octopuses and squids. We shall say more about all of these animals in Chapter 8, when we come to consider consciousness. Their brainpower is so impressive that it is surprising that on the whole the invertebrates have not exploited their individual intelligence to get the next generation off to a good start. One reason may be that most cephalopods breed only once, when they are one year old, and die after breeding, so that all adults of the species come from the same generation. This makes it difficult to pass on behaviour to the young like intelligent mammals do. There are a few small exceptions: mantis shrimps guard their eggs and young, but no more cleverly than earwigs do, and although the worker bee's intelligence is indeed directed towards the feeding of the next generation, it doesn't pass on newly invented tactics.

As we've already explained, mammals have come up with a whole range of methods for passing on privilege: now we explore their effect on the development of brains. One method is to keep the babies inside the uterus, away from all the problems of the outside world, protected by mother's homeostatic mechanisms from changes in temperature or water availability, fed and oxygenated by her very effective blood system, and – especially – kept safe from predators and supplied with food and water by her intelligent pattern-recognition systems. This trick has implications, for the brain circuitry of such creatures must be able to respond to a great variety of features of their environments as soon as they are born. By 'features' here we mean something more than just 'aspects'. We mean identifiable structures and patterns that are associated with important events or things (an idea that we shall take further in the next chapter). For example, a baby gnu must be able to distinguish other baby gnus from wild dogs, without ever having seen either of them. In order to do so its visual system must be able to pick out, from the bombardment of incoming light signals, those particular patterns that tell it 'baby gnu over there' or 'wild dog over by those bushes'. This is not a simple ability, and computer manufacturers would give their eye teeth to know how it is done, because for all their amazing speed and memory the best computers cannot scan a visual image and pick out what objects are present in it. Tarantula wasps also know a few things that computer manufacturers don't, because they know how to distinguish spiders (in which they lay their eggs) from tarantula wasps of the opposite sex (with which they mate). A mistake here is likely to prove fatal, so they've become very good at it. Presumably the tarantula wasp's relatively simple nervous system (but still too complex for us to model) homes in on a few simple cues associated with spiders or other wasps, rather than recognising them in detail. Experiments have shown that far more complex animals, such as birds, take similar conceptual short-cuts.

The baby gnu, however, needs a really good image-analysis kit in its brain even to get started on the problem of analysing features of its surroundings. It certainly seems very unlikely that this system develops by a 'learning' process: the visual circuitry in the retina and brain matures into its functional state while the baby is inside its mother, and whatever signals it may have detected during that time must have borne little relationship to images of wild dogs. Yet its cells have interacted in the way that nerve cells do, resulting in a baby gnu with a reliable set of responses – 'avoid!', 'suck here', 'get this between you and danger!' – and a pre-programmed lick/suck/listen recognition of its mother.

The other main mammalian privilege technique that we identified is for babies to be born when they are still very undeveloped, into a nest. Here they are protected from nearly all dangers, and they will normally get food and warmth – though there is little they can do about it if they don't. This strategy is just as privileged as the alternative, but with very different rewards. For the first time in terrestrial evolution, trial-and-*error* can take place with very little risk. In an environment that lacks the privilege of the nest, any error carries a substantial penalty – commonly death, or at least the failure to be represented in the next generation. In the nest, however, learning becomes possible. Suddenly a new method for transmitting information across the generations makes an entrance on the evolutionary stage: information as privilege. This change is both universal and public, in our previous terminology: it is to be expected, in the fullness of time, and it affects *everything*. It is the main cause of one of the biggest evolutionary explosions of all time: the ability to *learn* from others of your own species. It means that mammalian babies that can remember which teat gives the most milk do better than their siblings. It means that baby leopards who get to practise their hunting techniques on last-gasp prey brought to the nest by a parent will be more adroit with real prey later. Among domestic animals, mother cats often bring their kittens a half-dead mouse to play with, and in this manner the kittens learn to hunt real live mice of their own when they grow up.

Learning is a way of passing things between the generations that is far more flexible than genetics and has a much shorter response time. The nest is where learning began, so nests are just as important for the evolution of intelligence as nerve cells are. However, once parent-mediated learning has started to evolve, the nest as a physical object can sometimes be dispensed with, especially in social animals. The immediate territory of a group of feeding baboons, or the suburban playgroup, protect youngsters more effectively than do the nests of rats or otters, and *much* better than the herding tactics of wildebeest or zebras can manage.

Offspring that learn more quickly, or more effectively, have a better chance of breeding, so evolution will favour the ability to learn; it will also select parents (especially mothers) for better teaching. Again we see a complicit process that naturally drives itself towards greater complexity, as the overall system modifies itself at each recursive step. We see many stages of this process in mammals today, from the weaning tactics of pet gerbils to the overt house-training of kittens and puppies. We see the instruction of young apes and monkeys by their elders. Above all we see our own elaborate cultural and educational systems. There is a distinct irony here. We have a habit of thinking of animals that employ nests as being more primitive than those that keep their young inside the mother. But the keep-'em-inside trick is a private one, with no interesting evolutionary potential except for the animals that do it, whereas the invention of parental care in the nest is a public one, opening up a whole new facet of evolution – leading, among other things, to this book.

Let us summarise.

The standard theory of the evolution of intelligence is reductionist: it focuses upon the data processing abilities of networks of nerve cells, and in effect appeals to Murphic resonance: if intelligence *can* arise, then it *will* arise, because of the obvious evolutionary advantage that it confers. Ever ready to discard – or at least question – our own principle, we nevertheless think that such a purely internal route towards intelligence is unlikely. It is not at all obvious how big a selective 'plus' intelligence *on its own* really is. As we said, intelligence has to have something to be intelligent *about*. Curiosity is closely tied to intelligence, and we all know what curiosity did to the proverbial cat. This is not just a liar-to-children teaching myth: cats really do get into trouble by being incorrigibly curious. IS had a cat that climbed unnoticed under the floor when his father took a board up. It was trapped underneath when the board was nailed back, and was rescued only because (a) it had gone missing, (b) under the floor was an obvious place to look *because cats are curious*, and (c) faint miaows could be heard from beneath the sitting-room carpet. At any rate, the standard reductionist view of the evolution of intelligence emphasises the role of ramifying nerve cells: as organisms get bigger brains their behaviour becomes more complex – at least potentially. Bigger brains, like bigger computers, can support more complex algorithms. This explanation of the human mind therefore starts with nervous circuitry and tells a story something like the development of the computer: more memory, faster processing, subroutines to handle common processes, 'shells' to organise those subroutines, and so on. 'Once upon a time there was a nerve cell and then we got Einstein.' This is the 'clever nerve cell' theory, and as you've seen, we don't believe that it is the best way to tell the story of intelligence. What it misses out is the crucial role of context. Intelligence needs a context in which to offer substantial selective advantages, and the most likely context is a social one. The main advantage offered by intelligence is that it helps you deal with your *own* species – your main competition, but also your main potential source of help.

The contextual story of the evolution of intelligence goes by way of the mammals with nests, and starts not with a nerve cell but with yolk. The road to Einstein is a road mapped out by privilege complicit with 'wetware', not by wetware alone. (By 'wetware' we mean neural circuits made up of squishy stuff, the way biology does everything, instead of nice clean silicon 'hardware'.) The complicit evolution of intelligence builds upon baby mammals in the nest, where security opens up the possibility of trial-and-error learning under mother's instruction. The passage of information by learning lays the groundwork for the evolution of social structure, making nests redundant; societies create privileged corners of the world in which babies can grow up more safely.

From there to human society – butchers, bakers and candlestick-makers, policemen and poets and pimps and priests – is a straight path. The key to our intelligence is neural nests – not neural nets.

7 Features Great and Small

JC's daughter Beth, at about the age of eight, was out with her parents in the car and noticed a line of birds sitting on telephone wires – black blobs spaced along a set of parallel lines.

'Oh, look,' she said. 'Music!'

Human minds do more than just recognising various bits and pieces of the universe. They look for patterns in what they recognise, and do their best to understand how the universe works. The universe, however, is very complex: in order to understand it we must also simplify it. Indeed the whole point of *understanding* something is that you can grasp it as a whole, and that necessitates some kind of simplification or data-compression. An explanation of the universe that was just as complex as the universe itself would merely replace one puzzle by another. In this chapter we shall argue that the brain organises its perceptions of the world into significant chunks, which we shall call 'features'. As usual we shall take an evolutionary and contextual view of this ability, as well as asking about the internal structure of the brain. Not just 'how does it work?' but 'how did it arise?' And to get started, we shall take a look at two simpler creatures: the mantis shrimp and the octopus.

Both the octopus and the mantis shrimp are effective organisms, even if they never meet another of their own kind to learn from. They learn to catch food – usually crabs, which are very good at defending themselves – by a variety of clever, individual tricks. They discover and exploit holes in their territory. They flee from bigger animals and hide until the threat has passed. They both seem to play with their prey – and with odd pieces of rock, coral, and seaweed. How do they recognise those bits and pieces of the universe that are significant, to them, for these activities? You can't flee a predator or play with a piece of coral unless you have some way to recognise it. Let's see how they develop their abilities, by observing how they behave as they grow.

We identified yolk as a key source of privilege. Both kinds of animal hatch from rather yolky eggs. Most mantis shrimp have a brief larval life among the plankton, during which they adopt no less than three distinct forms and lifestyles (figure 22). The first form consists mostly of (what will eventually develop into) a head, and is a plankton-eating filter feeder. So is the second, which has by now acquired abdominal limbs. The third form has large pincer-like claws and is carnivorous; it regards *anything* less than half its size, which moves, as food, which it attacks, snipping or tearing off any appendages. It then eats what's left – which is often appendage rather than prey, and could be a piece of floating seaweed, or even a lump of wood or plastic in the hands of an experimental zoologist. When the third form metamorphoses into the (small) adult shape, taking up life in the coral crevices, it continues this rather indiscriminate butchery. At this stage, however, it never attacks tiny stones, pieces of coral, or seaweed. It has now learned to discriminate between prey, predators, and the 'neutral environment'. Movement – especially movement that changes direction – is its cue for attack. You can see mantis shrimps doing this in an aquarium. If you present them with food held in a forceps, they will look very carefully at it, their eyes following it, and then they will attack. They will attack the forceps three or four times, as well, but after that they learn to take the food from it. If presented with an empty forceps they will eye it carefully, sometimes approach it, very rarely attack. But sometimes they will duck down into a crevice and then quickly come up again, as if they are 'hoping' that the forceps has now got some food in it ...

Surprisingly, little coral octopuses show just the same behaviour, learning to take food from forceps, and 'hoping' that empty forceps will fill with food if they turn their eyes away for a few moments. These similarities of behaviour in organisms whose brains are organised in totally different ways tells us that there

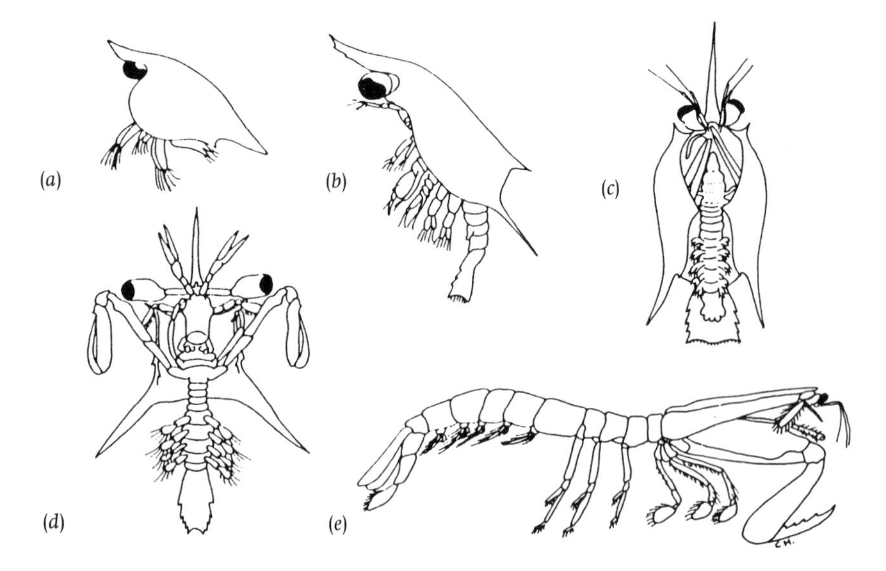

Figure 22 *Life cycle of the mantis shrimp* Squilla mantis. *(a) Metanauplius. (b) Pre-zoea. (c) Protozoea. (d) Antizoea. (e) Adult.*

are 'universal' properties of brains, and offers hope that we can understand the intelligent behaviour of other species.

We can tell what other species *don't* do more easily than we can find out what they *do* do. They do not carefully observe every property of an approaching crab – its shape, colour, the way it moves – and compare their observations with some internal picture, before deciding whether to attack. This would take too long even if it were possible, and it would be difficult for such a laborious method to evolve. Instead, they react swiftly to some feature of the prey – for example, that its motion is different from that of the nearby seaweed. The world of the octopus is neatly organised into 'food' (small things that move or are carried in forceps) and 'non-food' (everything else). However, the 'everything else' category is subdivided into 'useful crevices' (big enough to squeeze into, too small for a predator to follow), 'home crevice' (I live here), and 'the rest of geography' (not distinguished).

The behaviour of these two organisms supports the view that brains do not represent the outside world *as it is*, but in terms of a 'chunked' model in which certain types of stimulus are lumped together and interpreted as being 'the same'. These chunks stand out from the rest because the brain perceiving them has evolved detectors for such stimuli. We shall call them *features*. Features are not just convenient classifications of related sensory inputs: they are highly visible 'peaks' in the geography of the mind. When we look out of the window and see a bird in the sky, we pay a lot of attention to the bird and very little to the uniform blue of the background against which it is moving. The bird is a feature, 'blue' is a feature, and the concept 'sky' is a feature; but the patch of blue sky ten degrees to the left of where the bird is right now is not a feature – it's just a 'default' bit of sky. This kind of undemocratic discrimination between different kinds of sensory input is built into the physics/chemistry of phase space as a natural possibility, ready to be exploited by evolution.

There are two mutually complicit reasons why brains should work as feature-detectors. The internal, reductionist reason is the structure of neural networks. In Chapter 6 we described how the neural networks that are currently in vogue among cognitive psychologists can be taught to detect particular patterns. When they see granny, their 'grandmother cell' fires. The same is true if they see anything sufficiently *like* granny, such as a photograph or another person with a similar face. Granny is a feature because she elicits a yes/no response. The second, contextual reason is that feature-recognition is a good evolutionary survival trick. If you can pick out the lurking leopard from among the bushes, you won't get pounced upon. Any signal that is strongly associated with leopards, and not with bushes, will suffice to discriminate: the simpler the signal, the quicker the discrimination can be made. If a human being's peripheral vision catches sight of a dark spot on their shoulder that *might* be an insect, their first reaction is to swat at it with a hand. Only after that do they take a closer look to see whether it really was an insect. For quick reactions, you need feature-detectors.

The development of language in humans, which we discuss in Chapter 9, represents the current ultimate in structuring the world by its features, for each word represents a feature. The real world is continuous, but our inner world of features is discrete because features elicit a binary yes/no response. This is why we have so much trouble with variables like alive/dead which, despite appearances, are continuous – see below for justification of this outrageous claim. Our minds have polarised such variables into two opposite features, but when we look closely we start to see the continuity, and we find it hard to 'draw the line' between them. The reason that we find it so hard is that there is no clear-cut line; but our propensity to structure the world into discrete features makes us think there ought to be. If we can find a new feature that represents a neat boundary, we feel much happier. Many of our assumptions about how the world should be are based upon tacit comparisons with binary pairs such as male/female, hard/soft, conscious/unconscious, which aren't even reasonable distinctions to start with. In particular today's medical and legal professions, along with the rest of us, are grappling ineffectually with the fact that the alive/not-alive distinction often breaks down when we try to make it precise. The great biologist Peter Medawar complained that he was sometimes asked what happened to the body after burial, by people who had a morbid fear of being buried alive. Medawar would explain that if someone was shot through the head then the heart, kidneys, and other organs would retain their integrity, meaning that they were still alive. This is one reason why organs can be transplanted from accident victims into living people. Organs remain alive for several hours, about a day if cooled; cells within the tissues remain alive for several days. There is a persistent story at Birmingham University, where Medawar was Professor of Zoology, that he would culture living pig cells from pork sausages to make this point to his students. The answer he gave to those worried questioners, afraid of an Edgar Allen-Poe-style premature burial, was: 'Don't worry, you *will* be buried alive.' We are increasingly faced with the recognition that the alive/dead distinction for people is dubious: it affects issues such as 'brain death' (when should we switch off the resuscitator?) and abortion (is the fetus a part of its mother, or a separate entity?). The cartoon character makes the point as he is being wheeled into the heart transplant operating theatre: 'Excuse me, but am I the recipient or the donor?'

Even our brains find it easier to cope with simple features. Simpler brains *have* to work with simple features. The apparently simple polyp *Hydra* is famous among biochemists because (and only because) it is sensitive to glutathione, a chemical which is apparently released by many freshwater crustacea. As we've said, it uses this sensitivity to track down food. If it reacted *only* to glutathione, then its mental picture of the world would contain only two features: 'food' (I sense glutathione) and 'non-food' (I don't). Actually, it also responds to touch, so perhaps its mental picture is twice as rich. Many animals have more complex worlds. The honey-bee's world includes many features of flowers, its hive, general surrounding geography mapped by trees, grass, and garden fences. It can even communicate some of these features to other workers by way of a ritualised 'dance'. The language of the dance – the manner whereby it encodes information about nectar and where to find it – is almost certainly built into the brain circuits, because in a given bee species it does not differ from one hive to the next; however, it does differ between one species or variety of bee and another.

More complex still is the mental world of patas monkeys, which have tens of kinds of alarm calls; we know that each call affects the other monkeys in different ways, because their *behaviour* is different: a 'snake' alarm call sends them all up trees, whereas a 'hawk' alarm call has them all scuttling to shelter on the ground. These monkeys have a rich life, responding to many features of food, social interaction, and kinds of danger.

As human babies grow up they seem to respond to more and more features of their environment. The adults and children around them label many of these features using words – indeed a word is basically a label for a feature or a process, and a process is really just a different kind of feature, so words are codes for features. Some words come to represent complex features, 'ideas' that incorporate several proto-features. For example 'weather' is a label for a complex of features – rain, sunshine, fog, hail, snow. It takes a very specific and actually rather weird view of the universe to see sunshine and snow as two different aspects of the *same thing*: a view that focuses upon 'what do these things mean for me?' rather than their physical origins. Our brains lump them all together as 'the same kind of thing' because they are to some extent alternatives: if it isn't raining then it must be snowy, or cloudy, or sunny, or *something* like that, so all of those somethings must presumably be the same *kind* of thing. In this manner a baby builds up a geography of the space around it, but not so much a geography of the physical space in which it sits as that of the *space of features*. It ignores a cigarette burn in the carpet, for example, whereas an adult – especially a house-proud one

– might find that the most important feature around. The developing baby builds up a catalogue of features, and it learns to put them together, or to accept alternatives, such as babysitter in place of mother – another case of 'same kind of thing' once the shock of being abandoned to a stranger has worn off and the advantages start to sink in. What it does not do is classify its environment into finer and finer levels, taking in more and more detail: 'oh, look, that patch of wallpaper isn't quite the same as the patch next to it – I need two new words for those patches'. There is a good reason not to proceed in that manner: it would be useless. It would provide a mental map that is of the same size and complexity as the territory it represents.

What we want are maps that help us navigate, and those must represent the territory in a simple and accessible manner if they are to be useful. Simplified representation necessarily implies loss of detail, just as the word 'Coventry' on a map fails to include the statue of Lady Godiva on her traditional horse in the central square, Coventry cathedral, and the Coventry City football ground. Features can be large or small, and what we recognise as a feature changes over time. Features like 'continent' and 'polar ice-cap' became apparent to human beings only in the last few centuries. 'AIDS virus' and 'hole in the ozone layer' are features that we made explicit only very recently – as soon as we realised that they affected *us*.

How do we produce such maps? We classify the world in a very different manner, one that relies much more than we consciously realise upon the mathematical concept of symmetry. Before proceeding, we shall examine this concept in an informal manner. The word 'symmetry' is used rather loosely in ordinary speech, to mean some kind of repetitive pattern or even just 'elegance of form'. Mathematicians use the term in a much more specific way: a symmetry of an object is a transformation that leaves it looking exactly the same. The most familiar kinds of transformation are motions, which the mathematician classifies into various kinds – rotations, reflections, translations (a word that means 'slide rigidly' and has nothing to do with linguistics) and so forth. There are thus many kinds of symmetry. The most familiar is the bilateral symmetry of many animals, including humans: their left halves look much the same as their right halves, but reflected in a mirror. Indeed, if you do reflect the entire animal in a mirror (a transformation) it looks much the same as it did to begin with (it is symmetric under that transformation). Less superficial features, however, are not so symmetric – the position of the heart and other organs, the folding of the intestine, the function of the brain's two hemispheres. And even on the surface most animals are not quite as perfectly symmetric as we first imagine.

A less familiar kind of symmetry – well, it's familiar enough, but we don't normally think of it as a symmetry – is bland uniformity. A flat white-painted wall, for example, has a *huge* amount of symmetry – provided we ignore its edges, or adopt the mathematician's favourite trick of imagining the wall continuing to infinity in all directions. Such a wall looks exactly the same *whatever* transformation you apply to it. Rotate it through sixty-seven degrees, reflect it in any mirror-line you wish, translate it sideways by a millimetre, a mile, or a light-year – all you get is an infinite flat white wall with no distinguishing features. Every bit of it looks exactly like every other bit. Our brains are unimpressed by symmetry that is so extensive, and usually fail to notice its significance. One reason why it is so easy to ignore is that it represents a simple 'default' option – make everywhere the same as everywhere else. Our brains can simplify the innumerable sensory inputs from the entire wall into a single feature: *the same as usual*. Wherever you look, nothing changes. This is a case of habituation, which we first encountered in Chapter 1 and was explained in Chapter 6: our brains work by perceiving *differences* in space and time, not by perceiving everything that is 'there'. This type of simplification offers distinct advantages for an evolving brain, because it leaves more of its precious nerve cells free to perform other tasks. So our brains have evolved a liking for symmetry – perhaps for other reasons too, see below – and they make considerable use of it, as we shall see.

There are many analogues of the blank wall in nature: the uniform black of a starless sky, the desert that goes on forever, the pond as still and flat as a mirror. The BZ experiment of Chapter 1, before it starts to form patterns, is a good example – a flat dish full of orange-brown liquid, no region any different from any other. However, bland uniformity can be a major source of more interesting patterns, through a process known as 'symmetry-breaking'. This occurs when something disturbs the uniform serenity of the bland wall, the sandy desert, the flat pond. This disturbance can be 'spontaneous', as it is in the BZ reaction: the uniform state is unstable, and tiny changes in chemical concentrations – always present because of the random motion of individual molecules, the presence of dust particles, tiny bubbles, whatever – reinforce themselves recursively to create first blue spots, and then expanding concentric patterns of red and blue rings.

Where did all that pattern come from?

Better to ask 'where did the symmetry go?' A system of concentric rings possesses only rotational symmetry about its centre (and reflectional symmetry in mirror lines that pass through the centre if you want to be pedantic), but the original orange-brown liquid possessed a lot *more* symmetry – all possible motions of the (idealised and infinite) dish. It is not so much a case of gaining a pattern as losing some symmetry. However, since we start with bland uniformity, it is the *pattern* that attracts our attention.

The disturbance that breaks the symmetry can also be deliberate, as when a stone is thrown into a still pond. Again we find a system of concentric rings, which now take the physical form of waves on the water. The source of the pattern is the same: an extensive (but unnoticed) symmetry is broken to reveal a more striking (but lesser) degree of symmetry. In the same way, a spider breaks the symmetry of a blank wall. It thereby spoils the nice simple default description 'the same everywhere', which becomes 'the same everywhere except at that funny blotchy black bit'. A brain that is responsive to symmetries – as ours is – can hardly fail to focus upon the blotchy bit that breaks the symmetry. And so the spider becomes, to our perceptions, a *feature* – a coherent thing in its own right. The same goes for a lion in the otherwise uniform desert, a swan on the otherwise uniform pond, a hawk in the otherwise uniform sky.

Our visual systems rely upon and make use of symmetry in another way. They have to be able to recognise an object in different positions, different orientations, and at different distances. So they must cope with objects that have been translated, rotated, reflected, or dilated (changed in apparent size) and still recognise them as the *same* object. These motions are symmetry transformations, so there is an inherent reason why our visual senses should have some kind of inbuilt symmetry and/or respect for symmetry. In fact the eye has circular symmetry, and also an approximate dilational symmetry: its sensory rods and cones are more tightly packed towards the central region, the fovea. The top few layers of the visual cortex in the brain, which receives the signals from the eye, have the same symmetry as a tiled wall, a regular array of vast numbers of neurons. The circular symmetry of the eye is converted into the rectangular symmetry of the visual cortex by means of an elegant mathematical transformation, the 'complex logarithm' (figure 23).

This fact has been established by showing animals' eyes various images and observing the resulting patterns of activity in their visual cortexes. The logarithmic eye/brain transformation is also responsible for certain visual hallucinations, where waves of electrical activity travelling across the cortex are 'back-projected' and seen by the visual system as if they were signals from the eye that would produce that particular wave pattern. The wave patterns in the cortex are very simple: parallel rows of electrical activity, like waves rolling up to a long, straight beach. The 'detransformed' patterns that the eye would have to experience in order to create these regular trains of waves in the cortex are more unusual: concentric rings and spirals. And those are exactly the shapes that these particular hallucinations produce.

Clearly this beautiful mathematical structure has not come about by accident, and we can make some educated guesses about how it has arisen. Two scales are involved: the long timescale of evolution, in which the visual system of an organism slowly changes as it is passed on to descendants, and the shorter and more immediate timescale of biological development of a single organism. The evolving architecture of the eye and visual cortex seem to have been influenced by a common kind of pattern in the outside world. Namely, the outside world contains a lot of coherent individual objects whose images move across the eye by way of symmetry transformations. Recognition of those objects carries considerable evolutionary advantages – some of them are leopards, some are bananas. So the eye/brain system has evolved to incorporate those symmetries that the outside world presents to it: to represent visual signals in a manner that respects the natural transformations (translation, reflection, rotation, and dilation) that images are subjected to when the object generating them moves. The symmetry-based structure of the visual system is also fine-tuned during biological development, when the sense organs are still changing: the brain learns how to extract meaning from sensory inputs and influences the growth of physiological structures that refine and develop that ability.

Notice that structures that strongly influence mental activity are here constrained by external physical realities – the rules obeyed by light rays and the fact that leopards stay roughly the same shape as they move. Notice also that a brain that has evolved to exploit those symmetries will be especially good at picking up coherent patches of any incoming image that seem to move as a whole, relative to everything else. So it will naturally tend to decompose the image into patches of that type – visual features. Evolution will reinforce this

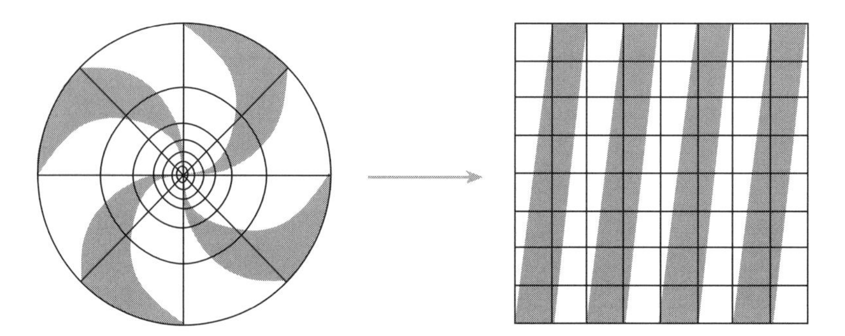

Figure 23 *Complex logarithm transformation between retina (left) and visual cortex (right). Parallel waves of electrical activity in the cortex 'detransform' to give spiral hallucinations.*

tendency because those mental features aren't just accidents: they are, to invoke the title of this book, 'figments of reality' – traces in the mind of real things that are actually out there in the universe in which that mind evolved.

It is probably no coincidence that we use the word 'features' to mean both 'prominent trait' and 'characteristic part of a face'. Human faces loom large in the early training of the neural nets in our visual system, and judging by the language the general concept of a feature seems to have been abstracted, long ago, from facial instances. A remarkable recent discovery shows just how strongly our perceptions are influenced by features, and also provides some fascinating hints about how our brains' real neural nets might carry out the chunking process required when the incoming data range over a continuous phase space. Or in this case 'face space', since the features being detected really are facial ones. In 1996 Alice O'Toole and Thomas Vetter carried out a computer analysis of black-and-white images of faces from 65 men and 65 women.[♪] They first found the 'average face', the centre of gravity of face space, so to speak. By 'average' we really do mean average, not 'typical'. In a computer, a grey-scale image is in effect a huge list of numbers, with the number corresponding to each pixel of the screen representing the level of grey. We can carry out all sorts of arithmetic and algebraic operations on the array of grey-scale pixel values, so we can do arithmetic and algebra with images. In particular, images can be averaged just by averaging the numbers that occur at each pixel. For things like faces it is necessary to standardise the position and size of each image first, if the average is to have any useful meaning. Having done that, any given face can then be compared with the average by subtracting the average face away and seeing what patterns result. These 'correction terms' to the average face won't look like *faces* any more – they will be more like 'a thinning of the lips combined with a lengthening of the nose, a difference in spacing of the eyes, and a softening of the chin'. Not a *face* with thin lips, long nose, and so on – just all the *differences* between a face and the average one, all rolled into one image.

There is a mathematical technique for finding the most important patterns among such correction terms, which works in terms of things known as 'eigenvectors' – so we shall call the resulting images 'eigenfaces'. An ugly word, perhaps: part German, part English – but you'll surely remember it when you see it again, which is why we decided to use it. An eigenface is an unusually prevalent pattern in the correction terms to the average face. If this description sounds woolly it is because it *is* woolly: the actual definition is rather technical.[♪] However, here's one possible way to think about it, using a simpler example: landscapes.

The 'average landscape' is pretty close to a flat plane, because hills and valleys tend to cancel out when you add them together. What is the commonest pattern in the difference between a real landscape and this average 'flat' landscape? If the images being averaged feature a lot of landscapes with only one peak or valley, then the commonest 'correction term' will be a single peak, like this: ⁀ . If you add it to the flat landscape you get a hill; if you subtract it (turn it upside down) you get a valley ‿ .

Faces are like complicated landscapes. You might expect to find several eigenfaces – just as in a real landscape you might get two hills, or a hill and a valley, or whatever. However, the two scientists found that for human faces there is only *one* eigenface, and it represents the difference between 'typical' male and female faces. If the eigenface is added to the average face, the result is visibly male; if it is subtracted, the result is female (figure 24). It is no surprise that 'male' and 'female' is an important distinction in human faces, of course: the surprise is that these differences can be captured by a 'one-dimensional' set of images – a single variable, the amount of eigenface added or subtracted. This fact suggests that a fairly simple mechanism in our visual system is used to distinguish male faces from female ones. It also looks as if one of the functions of the appropriate neural net is to compute eigenfaces – a reasonable idea, since the concept is mathematically simple to describe and to carry out, and it is a natural for exactly the kind of 'parallel processing' task that neural nets are very good at.

Taking the idea a little further, we can speculate that this strong male/female facial axis in vision has not come about by accident, but by sexual selection. Recall that female birds have a preference for males with symmetric

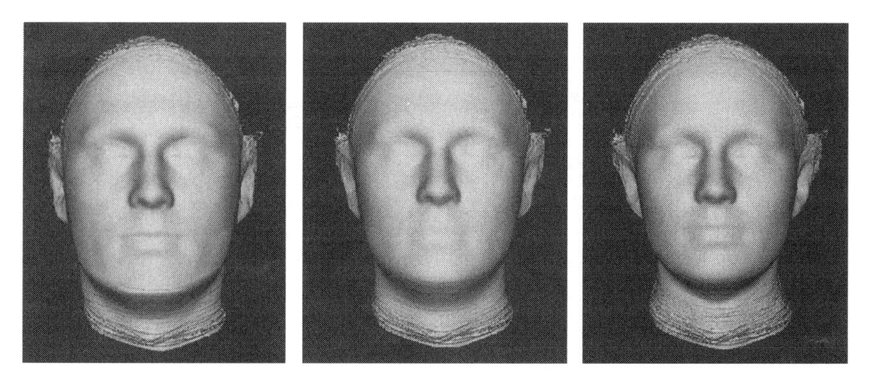

Figure 24 *Eigenfaces. A single variable captures the difference between a male face (left) and a female one (right). Central neutral face is the overall average.*

tails, and that one reason may be a quirk of the neural network for tail recognition. Whatever the reason, anything that the females prefer will be selected over many generations, so the image in their heads actually gets projected back out into the anatomy of the male birds. Similarly, in human vision there could well have arisen a preference for eigenfaces, because those are what naturally emerge from the simplest mathematical computations on images of faces. If so, the females will prefer mates whose faces resemble eigenfaces, and the same goes for males. Sexual selection will then rapidly drive the population towards the brain's image of the 'ideal' face. This is wild speculation, but it looks very plausible. Why else should there now be such a strong principal axis along the male/female direction in face space?

We can also speculate that what a 'grandmother cell' detects is not granny, but eigengranny. Or maybe – we actually think this rather likely, and we are sceptical about actual grandmother cells in human brains – a big enough component along the eigengranny direction in face space elicits a response from the relevant neural net as a whole. Something in the brain then has to attach the label 'granny' to what we are seeing, but the message can be transferred to the linguistic areas of the brain without using a dedicated cell whose only task is to recognise granny. Instead of grandmother cells, the brain may employ eigengranny networks.

This kind of feature-detecting structure has gradually evolved over the last 600 million years, so that by now it has become very sophisticated indeed. As we have said and will keep saying, evolution is recursive, building upon its own results, erecting scaffolding and discarding it when the resulting process becomes able to continue of its own accord. Brains did not just acquire feature-detecting abilities in visual perception: they acquired them in hearing, touch, smell, all the senses, and for the same reason. It is reasonable to suppose that the brain would eventually evolve a kind of generalised 'feature detection system', and so it did. We know that this feature-detecting ability is especially well developed in humans because it shows in their behaviour; we assume that it is less well developed in other animals because those behavioural clues seem to be reduced or absent. Among these clues are language, which arises when a brain starts to associate symbolic labels with features, and consciousness of self, which occurs when the feature-detecting system becomes recursive, detecting some of its *own* features. We will discuss both of these topics at length in later chapters, so we skim over them here in favour of the third clue, the way we form 'conceptual maps' of intellectual territory.

Animals often form some kind of mental representation of real territory. We have already mentioned the honey-bee's ability to recall and pass on some kind of coded description of where nectar can be found, and to take short-cuts. There is evidence that squirrels, who bury their food, know where to find it later because they have created some kind of mental map. They seem not to rely upon a sense of smell – which makes sense, because if the owner of the nuts and acorns could find them by smell alone, then so could any other passing squirrel, which rather removes the point of squirreling them away in the first place. Humans have taken this technique *much* further, extending it to abstract maps of spaces of ideas. This is an extremely important feature of the human brain, and it underlies our ability to reason, talk, and write – the most obvious things that distinguish us from other terrestrial animals, the ones of which we are most proud, and the ones that are most difficult for science to explain and understand.

An evolutionary biologist's mental map, for example, is built up from simple pieces of information about animals – living or extinct – and fossils found in various places at various times. However, this 'raw information' is not present as just a random list: it is structured in the biologist's mind. There are obvious forms of structuring, by space and time: where and when did *Triceratops* live? There are less direct structures too: what other creatures resemble *Triceratops*? What does its 'family tree' look like? Then the map starts to become recursive, mapping structures generated by the map itself: the *concept* of a family tree, the curious regularities and progressions that seem to occur in many such family trees, the idea of *evolution* as a way to explain those regularities ... The map is multilayered and richly textured, and the structures and patterns on one level become objects in the landscape at the next level down. We say 'down' here because the most attractive image is one of digging *deeper* into the *under*lying meaning of what we observe; however, you can imagine the more abstract, general concepts *building up*on the raw materials if you prefer. We hasten to add that by talking of 'layers' and 'levels' we don't mean to imply that the map is neatly stratified like a brick wall, so that we can sensibly talk about 'level seven' of the map. It is more like a hand-built drystone wall, where some rocks are so big that they occupy several levels in comparison to their neighbours; despite the overall irregularity, however, some rocks are definitely lower down than others and there is a general layered effect.

There is more, much more, to the evolutionary biologist's mental map. There is the chemistry of DNA, the understanding that genes are involved in animal development, the recognition of radiation as a source of mutations ... There is the wonderful, illuminating, dangerous idea of natural selection. But the whole map has a structure, a geography: some ideas are closely connected with others, some are isolated facts; some have an enormous range of implications, some just don't seem to tie in properly ... We are using standard images from everyday language here, but look at the words: 'close', 'connected', 'isolated', 'range', 'tie in'. There is a deep wisdom in our language, and it is often revealing to examine the metaphorical images contained within it. The idea that intellectual concepts are somehow arranged in a 'conceptual space' was implicit in our language long before cognitive psychologists made it explicit.

Mathematics is a very different subject from biology, with a different worldview and different criteria for what is important. JC, a biologist, and IS, a mathematician, know this full well – when writing *Collapse* we spent about four years arguing before we evolved a common viewpoint. But slowly we realised that although our mental maps differed in detail, they had many features in common – what we eventually came to call meta-features, features not so much *in* the maps as common to the way in which we both went about constructing our maps. We both looked for common patterns in diverse examples, we both sought unifying principles, we were both aware that there are logical constraints that worthwhile ideas must respect, and so on.

Science is a general, powerful, and fruitful method of constructing mental maps of complex territory. It is in many respects the *only* really successful method, because it alone builds in a degree of protection against natural, but dangerous tendencies of human psychology – the worst being, as we said earlier, to start with the answer and then choose evidence that favours it. This is not to say that science possesses an inherent purity that no other system of thought can approach: there are many features of science as practised today, most of them stemming from its social and political organisation, that tend to subvert its own built-in protections. The system of 'peer review' of grant proposals, for example, embodies an entirely sensible wish to avoid wasting public money on trash. What better way than to appoint a panel of experts to review new proposals, and let them decide what is trash and what is worth funding? Except that unfortunately this approach encourages the dominance of particular schools of thought, so that wonderful new ideas get thrown out along with the obvious trash. And as the school ossifies, *everything* it funds turns to trash.

A central aim of science is – or certainly should be – to *explain* the world around us. In the terms we are now using, this means that we use it to build a conceptual map that can guide us through the complexities of the natural world. As we said in Chapter 2, there are many senses to the word 'explain' – ranging from immediate, simple causality ('it is raining because water is condensing from steam in the clouds') to lengthy chains of implication ('water is a liquid because the interatomic forces between hydrogen and oxygen atoms obey so-and-so's equation ...'). Explanations of this kind, whether simple or complicated, are reductionist. Reductionism, you will recall, explains phenomena by looking at internal structure, digging down layer by layer in an ever-branching tree of explanation, and it provides useful understanding precisely when it does not pursue that ever-branching tree into nightmare territory. Newton's laws of motion and his law of gravity are a wonderful example of the power of reductionist methods: instead of millions of different kinds of motion, from thistle seeds wafting in the wind to the stately revolution of the Andromeda galaxy, we have a few general principles which, fed the appropriate data, generate the resulting movement. (For the thistle seed we need aerodynamic laws too, but those actually come fairly directly from Newton's laws of motion plus a few other simple physical rules – oh, and by the way, we can't *actually* work out how either a seed or a galaxy will move in the detail we'd like. But we can understand many features of their motion this way.) In such cases, reductionism helps us to generate a better cognitive map, wiping away vast swathes of detail in return for constructing a prominent meta-feature a layer or two down.

We also pointed out in Chapter 2 that there is an alternative approach to explanation: 'contextualism'. Instead of looking (conceptually) inside things to work out how they tick, a contextualist view looks at the outsides of things. What external constraints moulded this thing as it developed? It must surely be clear by now that we don't think that either reductionism or contextualism *alone* is the 'right' way to think about the world. Who says there's only one right way? For that matter, who says there's a *right* way at all? We think that they both have the potential to offer useful insights, and that they are most effective in combination. But most of science emphasises the reductionist viewpoint, and contextual considerations tend to be neglected because that kind of thinking is not encouraged by scientific training – which is a pity, because even very simple contextual considerations can often suggest radically different ways of tackling scientific problems, as we saw with the sense of smell.

There is a famous argument that claims to demolish any prospect of a reductionist theory of intelligence – and by extension, of understanding and mind. It was proposed by John Searle and is known as the 'Chinese room'. He asks what it means to *understand* Chinese, and imagines a thought experiment in which a person who does not understand Chinese manipulates huge stacks of paper according to rigid, pre-prepared instructions. Questions in Chinese come in from outside the room, the pieces of paper get moved around, and eventually an answer in Chinese goes back out. Searle asks us to assume that although the person does not understand Chinese, those instructions have been set up so effectively that they permit him to respond to the incoming questions just like a native Chinese speaker. (Leave aside whether this is really *feasible* – just suppose, OK?) Then, says Searle, we *know* that the person does not understand Chinese. Any reductionist theory of intelligent understanding is analogous to the room and its rules for manipulating paper, so *real* intelligence cannot be reduced to a set of underlying physical rules. There are many things to say about this argument – they can be found in books like Douglas Hofstadter and Daniel Dennett's *The Mind's I*, Dennett's *Consciousness Explained*, and even *Collapse*. We make two brief points here. The first is that Searle's claim that the Chinese room evidently does not understand 'Chinese' rests on a false analogy. The proper analogy is between a person who, by the operation of their mind, understands Chinese, and the Chinese room which, by its manipulation of pieces of paper, appears to 'understand' Chinese. Agreed, we know that the person pushing the pieces of paper around in the Chinese room does not understand Chinese, but his role is not analogous to that of the person who understands Chinese. It is the entire *room*, rules, paper-pusher and all, that is analogous to such a person. The person in Searle's room is analogous to one nerve cell in the Chinese-speaker's brain. So what Searle's argument establishes is that when someone understands Chinese, their individual nerve cells do not. Agreed, but that's not really the point, is it? After all, nobody thinks *our* nerve cells understand what our brain is doing. Since the system can carry out Chinese conversations flawlessly, then *the whole system* can reasonably be held to 'understand' Chinese – by which word we mean the analogue, for that system, of understanding in an individual human.

This objection to Searle's argument takes the Chinese room at face value and examines the flawed analogy that surrounds it. Experience shows that it carries considerable weight for anyone who seriously wants to imagine how machines might be intelligent. It is meaningless nonsense to anyone who already knows they can't be, and is seeking a form of words to limit their imagination before it comes up with something that might shake that belief. The whole idea is a test of prejudices, not a serious argument. If you can't imagine how a system could 'understand' things, then you've ruled out machine intelligence before you start to think about how it might go – and you've also ruled out any explanation of your own brain, other than dualist mysticism. But there is a more basic objection to the Chinese room, which does not require taking it at face value. Namely, it is very easy to invent 'thought experiments' whose ingredients are subtly self-contradictory, in which case you can 'deduce' anything you like from them. As Dennett says, the fact that something is 'possible in principle' – in the sense that we can imagine it happening if we relax our critical instincts far enough – is much less informative than the fact that it is wildly impossible in practice. How could the Chinese room possibly work? The instructions on the paper would have to include contingency plans for every possible Chinese question. They could not be merely a vast catalogue of questions and answers – no conceivable room would be big enough to hold them, no such scheme could ever anticipate the full range of what might be asked. The rules for moving the paper would have to be prepared by somebody with an *exquisite* understanding of Chinese. In which case it is not such a leap of imagination to say that any system that carries out those rules is, in a sense, imbued with the same understanding.

We therefore disagree totally with Searle: we see no reason *in principle* to exclude the possibility of a reductionist theory of intelligence, and given that, the possibility of an intelligent, thoughtful machine. But as we've just said, possibility in principle is not as interesting as what is possible in practice, and in practice there are enormous obstacles to making an intelligent machine. Indeed if the central thesis of *Figments* is correct – that minds cannot be made except by complicity with cultures – then you would have to build an entire machine culture, *and* let it evolve for millions of years, not just one fancy mechanical brain, to get anything intelligent. We seriously doubt if really intelligent machines could interface successfully with our culture, anyway.

The Zarathustrans have their own reservations about reductionist approaches to intelligence, different from anything we've yet discussed.

Liar-to-children It is becoming increasingly apparent to me that these single-minded humans simply do not think like normal intelligent beings.

Destroyer-of-facts You should be wary of cultural absolutism, Liar-to-children. What you mean is that they do not think like *us*.

LtC That is precisely my point.

DoF And mine.

Ringmaster You are in agreement, then.

Performer-of-amusements Yes. They have no reason to fight because they both know they are right.

R [*Aware that something is wrong with the logic but unsure what.*] Let it be so graffited.

Liar-to-adults [*To Liar-to-children.*] You are saying that the humans are our intellectual inferiors, of course?

LtC Not necessarily. Their thought patterns intrigue me. I have been trying to make sense of their way of thinking. They appear to – well, I was going to say 'place the universe in a different phase space', but that is not how they think. The closest image I can find in humanspeak is that they *carve up* the universe differently from us. A curious way to smell things.

Hewer-of-wood You ought to ask me about that, I'm an expert in carving. I've just been carving this snoozo–

R You represent an untapped source of talent, Hewer.

LtC [*Excitedly.*] What I mean is, they seem to think of the world in terms of fixed kinds of *thing* instead of fluxy processes. And the way they understand *things* is to carve them up – uh, sometimes literally. [*Recalls a human TV programme about Christmas turkeys, looks down at his own feather-covered body, and shudders.*]

DoF And despite that you have come to the conclusion that their approach might offer some advantages?

LtC Possibly. It appears complementary to our way of thinking, rather than opposed.

LtA Let me tell you something, Liar-to-children. There can be *no* complement to the Zarathustran way of thinking, for the simple reason that it includes its own phase space, so that there is nowhere for a complement to poke out into. You might as well ask what is west of the West Pole. Moreover, this 'carving up' philosophy is uncomfortably close to the heretical beliefs of the Fragmentalists, an obscure sect that flourished briefly during the latter part of the third octillennium.

PoA What is a 'sect'?

LtA Like an oct, but smaller. It means 'a piece cut off'. Which is precisely what made them heretical: they 'carved the universe up' in a totally unorthodoct manner.

DoF [*Remembering.*] They invented the Law of Conservation of Anarchy, did they not?

LtA Yes. They believed that anarchy ruled the universe, that at root everything is primal chaos. The greater the degree of chaos, the higher the anarchy level.

LtC Ah, yes, we did this in the philosophistry subsidiary. There is kinetic anarchy, produced by moving about very fast, and potential anarchy, caused by resisting an outside force passively. The Law of Conservation of Anarchy concerns the conversion of potential anarchy into kinetic, and conversely.

LtA Exactly. In fact it states that anarchy can be so converted, but never lost.

DoF Absolute nonsense. For instance, anarchy can be turned into heat by setting fire to things. But the Fragmentalists just redefined heat as a hitherto unrecognised form of anarchy, and so it went. They managed to hold the theory together for a time, but eventually they were brought down by the new trino scandal.

PoA [*Waggles hindfeathers in nudge-nudge mode.*] Tell me more.

DoF The Fragmentalists carved the universe up into smaller and smaller pieces until they reached such a fine level of subdivision that they could no longer observe the pieces directly. They had to infer their existence from various assumptions, the most prominent being the Law of Conservation of Anarchy. They would carve something up one way, and calculate the total anarchy; then carve it up another, and do the same. If the two totals agreed, they were happy. If not, they assumed that some unobserved pieces must exist to supply the missing anarchy. They called the process 'recarvery'.

R [*Horrified.*] That is a highly unoctimistic approach, and clearly fallacious. [*Even more horrified at having expressed an opinion without getting it from somebody else, he looks round to see if anybody has noticed. Of course, they have not.*] The correct deduction is that you have observed all the pieces, and anarchy is not conserved. We agree on that, of course?

DoF Of course. But the Fragmentalists seemed to feel better if they had made a good recarvery. At any rate, they carved the universe up into so-called 'anarchicles' with outlandish names like carryons, partyons, rabbitons, trions, and so forth. They claimed that the universe was built from anarchicles, but of course the evidence did not entirely agree, so they were forced to invent a dual system of archicles, one for each anarchicle – carrynos, partynos, rabbitnos, and trinos. That fixed things up for a time, and the Standard Model of Fragmental Archicles was looking pretty good. But then they discovered an anomalous trino recarvery in which a tiny quantity of anarchy seemed to have gone missing.

LtA [*In disdain.*] Obviously an instance of the spontaneous creation of order from chaos.

DoF Quite. But of course they had to go and invent a new archicle instead, to carry the missing anarchy. Known, inevitably, as the new trino. Unfortunately, try as they might, they could not detect new trinos. Not even using planet-sized crystals of octium. They claimed that new trinos just passed right through without interacting, but of course it was all nonsense and everybody knew it. Within half an oct the whole enterprise had collapsed of its own inanity.

LtA But only after infecting innumerable innocent minds with a heresy.

DoF And ruining the futures market for octium.

LtA [*Gravely.*] Liar-to-children, do you see the moral of this sad story?

LtC I certainly do, Liar-to-adults, and I renounce the Fragmentalist heresy forthwith – indeed eighthwith, if that will make you feel more octimal.

LtA One can be octimal, but I am unhappy about 'more'. It is like being very unique. Nonetheless, I accept your clumsy apology with my customary grace and charm.

R [*Wearily.*] Let it be so graffited.

LtC Wonderful, I feel octually cleansed. [*Takes deep breath.*] Now, as I was saying, I think that this human way of thinking may have merit. Take this novel concept they have of 'mass', for example, and its amazing equivalence to ener–

LtA [*Exasperated.*] Liar-to-children, have you not listened to a word I have said? Destroyer, enlighten him.

DoF The humans have made precisely the same mistake as the Fragmentalists. What is this 'mass', for example? It is a quantity assigned to pieces of matter so small that they cannot be detected. The mass of a macroscopic object is then computed as the total masses of all these invisible components.

PoA Ah. It is totals all the way down.

DoF [*Unamused by the frivolity.*] If the observed mass does not agree with observations, then the humans assume that some components must be missing, or else they adjust the masses of the components until the numbers do agree. It is quite mad – empirical numbo-gumbo. It has no octual significance whatsover.

LtC [*Disappointed.*] Oh. But what about this Albert-of-einsteins person that the humans revere so much?

DoF A genius. [*They look puzzled and he explains this obscure term.*] It is human for 'eccentric but plausible lunatic'. Albert-of-einstein is the inventor of a formula that any human would be proud to wear on a shirt-of-T: they believe it to be a step towards the Theory of Everything.

LtA [*With menace.*] What is this vile, unoctimistic formula?

LtC Oh, I know this from my psychics subsidiary! [*Writes.*] $E = mc^2$.

LtA [*Sneers down his proboscis at the alien scrawl.*] I take it that the symbol E stands for Everything?

LtC No, energy. And *m* is mass, while *c* is the speed of light.

LtA The *speed* of light? They think light has a *speed*?

LtC Apparently.

DoF Regrettably they seem unaware that the pulsiform vorticles considered on their own produce a fractimal distribution of illuminance whose integral not only diverges, but does so in an oscillatory manner. In those circumstances even *position* makes no sense, let alone spee–

LtC Absolutely. Anyway, they have this formula with E, m, and c –

PoA What does the little duck thing mean?

DoF Pardon?

PoA The other symbol. [*Draws a '2'.*]

DoF Oh, that. It is not a duck. It means 'squared'.

PoA Square light?

DoF Square speed.

PoA That is *better*?

DoF 'Square' means to multiply a thing by itself.

LtC [*Helpfully.*] The terrestrial organism *Amoeba* multiplies by itself.

DoF [*Irritated.*] Yes, but it does that by dividing.

PoA [*Confused.*] I thought *Amoeba* was a kind of blobby thing, not a square.

LtC [*In exasperation.*] Look, this is human mathematics. Nobody is expected to understand it. Not even humans.

PoA True.

DoF What is worse, a genuine Theory of Everything is not exactly difficult to find. But it is not $E = mc^2$.

PoA No? What is it then?

DoF [*Taken aback.*] You do not know? What else could it be? It is $E = 8$.

As the tale of the Fragmentalists demonstrates, reductionist science usually looks for a mathematical equation, formula, or process that describes general features of the universe. Quite *why* the universe is so mathematical is a mystery: the physicist Eugene Wigner marvelled at 'the unreasonable effectiveness of mathematics', and many people since have tried to explain it. The explanations include various forms of 'God is a mathematician' – that is, 'the universe is just like that', a type of explanation that we have already declared a willingness to accept for such 'Deep Thought' questions. A different point of view holds that our minds operate along mathematical lines, and *select* mathematical features of the universe while failing to notice other forms of behaviour. We suspect that both positions have some validity, and we note that the first is reductionist, the second contextual. What do we get if we try to put them together?

One way round we get the philosopher's perennial worry about reality. If our minds are merely generating mathematical patterns of their own accord and imposing them upon a non-mathematical universe, then our perceptions of reality are very different from reality itself. This leads us to doubt the evidence of our senses, which is the ultimate aim of this particular 'reality is a figment of imagination' philosophical game.

However, we can put them together the other way round. *Why* do human minds operate along mathematical lines? Because they evolved in a world that is full of patterns that can be captured in quasi-mathematical terms, and exploited to improve survival prospects. The sensory images in our heads are not the *same* as the real world: in that sense, those images are figments. But the matter that makes up our brains, the electrical impulses that course along our neural pathways – those are just as real as anything else. Our minds are 'figments of reality', developed in order to represent significant features of the objective world outside us, but representing them imperfectly and in coded form. The 'figmentation' process arises because those electrical impulses have two distinct interpretations. They *are* physical processes in the real world; but to the owner of that particular brain they carry an interpretation as *models* of the real world. The models are imperfect but the physics that runs them obeys all the usual rules.

We have already seen how our sense organs exploit mathematical features of the universe, such as symmetries; and we have also seen that the physical structure of our brains is constrained by those features. The dual interpretation of nerve impulses, either as features of reality or models of it, permits a remarkable kind of feedback between the patterning of the physical world and the patterning of the brain that perceives it. Indeed it is this feedback that may well be responsible not only for the unreasonable effectiveness of mathematics, but for its very existence within human culture. We offer two examples before developing this point further.

First, we have seen that the geometry of the eye, and that of the visual cortex, are closely tied to rigid motions and changes of scale in the outside world. Moreover, the standard raw materials of geometry, the notion of a point and a line, match the physiology of our visual system closely. A point image is one that triggers a response in a single retinal cone, or perhaps in a close cluster. A line becomes highly significant only a few layers down in the visual cortex: the upper layer detects the presence of a signal at a given position in the image, the next layer down computes the orientation of locally aligned image-points, and the layer below that looks for edges – boundaries of significant features: *lines*. So the basic ingredients of geometry, as understood by mathematicians, are presented to us fully fledged in our visual systems. Moreover, the more sophisticated modern view of geometry, which is based upon symmetries and rigid motions, is also present in the processes that our visual system is required to carry out when it tries to detect moving features of an image. Geometry, then, may well be a figment of our visual system – but it is a figment that corresponds to significant features of the outside world, hence its unreasonable effectiveness.♪

A similar case can be made for Newton's law of motion. For simplicity we shall tell the story in liar-to-children mode, without claiming historical accuracy. Newton observed moving bodies, and came to the conclusion that bodies left to their own devices will continue to move at constant speed along a straight line. More interesting was what happened to make a body deviate from such a course: a force had to be applied. For example a frictional force directed along the line of travel causes the body to slow down; a force directed at an angle to its path causes the path to curve *and* probably affects the speed. Newton deduced that the crucial 'positional' physical quantity is not speed but *acceleration* – rate of change of speed. The crucial variable that affects it is force. Acceleration, said Newton, is proportional to force; moreover the constant of proportionality is the body's mass. Now let us review the story, remembering that Newton's mind is a figment of reality. Observation of a moving body is an essentially visual act, and constant speed along a straight line is a very simple kind of visual geometry. Human minds detect forces using a different organ, the ear. Tiny hairs in the inner ear, surrounded by fluid, bend when a force disturbs the body to which the ear is attached. How does the ear detect this force? Because the force applies an *acceleration* to the head, which moves the fluid, which bends the hairs. So Newton's ear is already applying his law of motion, before his mind makes that law explicit. The same goes for mass. Newton's mind was aware of mass because his muscles had to work harder to pick up heavy objects. That is, it required more force. Now when you pick up an object, you change both its position (at ground level) and its speed (zero). That is, you have to impart an acceleration. The heavier the object, the more force you have to apply to impart a given acceleration: now Newton's physiology has provided him with the mathematical intuition required to formulate his law of motion in its full glory. The whole law, and associated concepts like rates of change, is a mathematical pattern that existed in the external world, became built into Newton's sensory system as a result of millions of years of evolution on the part of his ancestors, and was finally made explicit because Newton was trying to build a cognitive map of the appropriate intellectual territory.

An earlier attempt at the cognitive map of physics, made by Aristotle, failed – but for instructive reasons. In Newtonian physics, moving objects continue moving unless some force slows them down: it is the physics of objects moving in a vacuum. In Aristotelian physics, moving objects eventually 'run out of push' and stop. Newton would explain this as the action of a frictional force, but Aristotle took it as given. Mathematically the distinction is that Newton's laws work with accelerations, whereas Aristotle's ideas reduce to similar equations but using velocities instead. You do find something very like Aristotelian physics in the outside world, because there's a lot of friction out there, but it doesn't match experiments terribly well if the effects of friction are reduced. You also find something very like Aristotelian physics in the human mind. That is one reason why cartoon physics ♪ is so weird: only when Wile E. Coyote *notices* he has walked off the edge of the cliff does he suddenly fall, and only after Road-Runner has got his legs up to speed does he start to move. For a less frivolous example, we mentioned earlier that our minds make maps of our local geography, so that we can reach for an object with our eyes closed. Psychologists think that we plan the motion in advance using some kind of mental model of space and movement, so that after shutting our eyes we still have the model to guide us. Experiments on the speeds and directions in which people move their hands when grasping for objects suggest that this mental model is set up along Aristotelian principles rather than Newtonian ones ♪ – for example we plan velocities, not accelerations. The Aristotelian model is simpler, and it fits quite a lot of the real world, where friction is high. Another advantage is that with an Aristotelian model our grasping hand can arrange to 'run out of push' just as it reaches the object. If our mental model was a frictionless Newtonian one then our hands would oscillate around the object, but not stop, and probably we would knock it over.

Physical scientists have long remarked upon the deep mathematical structure of reality – even declaring, as just observed, that 'God is a mathematician' – in an effort to capture this feeling. However, the human mind is an inveterate pattern-seeker, and the evidence is that it likes *mathematical* patterns. This opens up a possible counter-argument: it is not God who is the mathematician, but we. Perhaps the mathematical universe is a case of selective reporting: when we find a mathematical pattern we notice it, and when we don't – we don't. Because we can't. This is a seductive argument, but it loses a lot of its force if we remember that mind is a figment of reality. There definitely *are* things in the real world that function along mathematical principles – the most obvious being the mind of a mathematician. But a mathematical mind can function along mathematical principles only because the germs of those principles are present in the real world from which the brain housing that mind was constructed and in which it evolved and developed. The same goes for sense organs in general – as we have seen for vision, hearing, and smell. Mathematics, a human mental construct, is unreasonably effective because, being a figment of reality, it captures significant patterns of the real world – the world from which mathematics itself ultimately derives.

If there is a core message to *Figments*, then this is it. Our minds lead a dual existence. Descartes' mistake was to view this as a duality of *materials*, which it is not: it is a duality of interpretations, just as a map can *be* a sheet of paper but *represent* a world. Features of the outside world are converted, via our senses, into 'figments' in our brains. On one level (brain) these are ordinary real-world processes involving chemicals, electrons, whatever; but simultaneously on another level (mind) they are mental maps of a very different order of reality, tigers and cows and people's faces.

This kind of two-level feedback – what in *Collapse* we called a 'strange loop' – provides a key to the curious 'dual' nature of brain/mind. For example, why does the real world seem so vivid? Why does red look so utterly different from green – and yet why do we find it impossible to imagine a colour that is different from the standard repertoire? Why is touch so sensual, why is pain so immediate, impossible to ignore, and just plain *nasty*? This is the philosophical problem of 'qualia' in conscious perception, and we shall have much more to say about it in Chapter 8, but we can offer a brief and partial answer now. On the 'figment' level our brains do not perceive the universe in a passive manner; instead, they project the inner world of figments back on to (our conception of) the outer world of reality, so that our private inner world appears to us – but not to anybody else – to be 'out there'. (What others perceive 'out there' is their *own* back-projection of their mental figments. However, on the whole different observers agree on what is projected, because it all stems from that common external reality, and is produced by similar brains, trained by similar Make-a-Human Kits.) Our brains, in this sense, create their own realities – and this enables them to attach vivid labels to prosaic reality, labels that are vivid because they are inside our minds where our personal identities also reside; but also labels that have evolved to be vivid because we survive much better if they are. This is very like Nick Humphrey's feedback loop story, which we described in Chapter 6 in terms of a light-seeking robot. Labels and associations that originally exist in the external world can, over time, be replaced by internal feedback loops in the mind which mimic the external loop sufficiently closely to have survival value. So our inner world of vivid figments *must* match the external realities well; for if it did not then we might easily imagine a tiger to be a rock, and try to sit on it, an action that would *not* be conducive to survival. It is evolution that binds the brain/mind strange loop together so that it evolves as a whole, ensuring that what mind chooses to perceive is usefully related to what is really there. And mind 'decorates' the important sensory messages with qualia like 'red', 'bang!' and 'ouch!'

This leads to a delightful paradox. Perceived reality (as opposed to real reality) seems vivid to our perceptions, *not* because it is real, but because it is virtual. 'Red' is a vivid construct of our minds, which we plaster over our perceptions by projecting them back into the outside world. There is an objective sense in which the outside world is red too – it reflects light of an appropriate wavelength. But that is a different kind of 'redness' altogether, with none of the vividness that our minds use for 'red' decoration of London buses and blood. It's just light bouncing around. Indeed 'wavelength red' does not correlate perfectly with 'sensual red': our colour vision is buffered against severe variations in observing conditions, such as changes in light intensity created by shadows or bright sunlight. Edwin Land, the main inventor of colour photography, discovered that in some circumstances our senses can perceive 'red' just as vividly, even though all red wavelengths of light have been filtered out. If you don't like this line of thought, bear in mind that many animals – bees in particular – see light at ultraviolet wavelengths, and hence pick up vivid 'colours' that we do not see at all. The bee's virtual world is different from *our* virtual world, and while they both are rooted in the same objective reality, they are different interpretations of it, a point that we take up in the next chapter.

Smell, which we have described at length, and taste, which we have not, are perhaps more obvious cases where our vivid sensual impression has no direct external match: we smell 'bacon' but the real world just produces molecules; the response they excite has much more to do with our sensory apparatus than with any natural feature of the molecules. (The receptors in the nose *respond* to vibrational frequencies, if Turin's theory is anywhere near correct, but we don't smell things *as* frequencies. 'Darling, that bacon has a mouthwatering spectral line at 25.3 kilohertz.' No.) As we said earlier, most adult humans are 'smell-blind' along at least one dimension of smell-space. So our personal experiences of smell, and yours, are very probably different – an interesting case where we *can* do experiments on 'what it is like to be' somebody else.

If you really want proof that the world of our senses is a figment of reality, go to the nearest amusement arcade and put on a Virtual Reality headset. The crude, blocky computer-generated images that these gadgets present to the eye ‘possess’ – that is give our minds a vivid impression of – the same solidity as the more refined images of reality that our eyes present to our brains. Yet here the actual external reality is quite different: a pair of tiny TV screens carrying images that have been tailored specifically to create the illusion of depth. The three-dimensional world that they appear to depict exists only as a mathematical map in the computer’s memory. Despite this, they have depth, presence ... *they look real*. This is because ‘red’ *is* the ‘decorated’ picture that the brain cooks up when the eye is stimulated by light of certain wavelengths: our decorated version of reality *is* virtual.

8 What is it Like to be a Human?

When JC's children David and Rebecca were about seven and eight years old, the family had many pets including cats, a tokay gecko, a corn snake, hooded rats, and several tanks of tropical fish. JC fed mice, baby rats, and cockroaches to the gecko and the snake, and large wriggly worms to the larger fish. The children invented a rationale for this, a tiny morality: some animals (worms, cockroaches, most fish) 'don't have minds'; some (geckoes, snakes, mature rats and mice, cichlid fish) have 'minds for themselves'; and a few (cats and people) have 'minds for others'. Rebecca was very worried when she was about thirteen, because she felt that most of the time she didn't have a mind for others, and was therefore not a real person. She stopped worrying only when she was told that she was not, as she thought, the only person with that problem. Indeed, most of the time people have no minds, sometimes they have minds for themselves, and only rarely does *anybody* have a mind for others.

In *The Philosophical Review* for October 1974 Thomas Nagel wrote a celebrated essay: 'What is it like to be a bat?' In it he examined the difference between an external observer's understanding of the physical processes that occur in a bat's brain, and the bat's own mental perceptions. He argued that no amount of external observation can tell us what being a bat feels like to the bat. His essay is often interpreted as ruling out an objective theory of consciousness, but that's not exactly what he was saying. Indeed at the end he argued that we should try to find objective ways to express what subjective experiences 'feel like' to those experiencing them. He didn't expect anyone to succeed in doing so, but he felt it would be well worth trying, because it would focus attention on the distinction between subjective and objective experience.

This distinction lies at the heart of one of the big current scientific issues, a problem that has puzzled philosophers since the dawn of philosophy. What is it like to 'be' a mind, and why is it like that? We are, by the way, thinking mainly of the modern mind, the kind of mind that *you*'ve got. Perhaps ancient minds were much the same as ours; perhaps a Neolithic goat-herder would find a modern Bedouin to be an interesting guest, and vice versa. Perhaps ancient Abraham, or Priam of Troy, could swap stories with a modern rabbi or a Coventry city councillor.

Perhaps.

Julian Jaynes' book *The Origin of Consciousness in the Breakdown of the Bicameral Mind* says not. It asserts that Abraham and Priam were more different from us than, perhaps, chimpanzees are. They were not conscious, instead they had bicameral ('two-chambered') minds. The left hemisphere of the brain drove the body, unthinkingly, just as we drive a car along a familiar road; the right brain spotted problems, pitfalls, and potentialities, and informed the left brain by talking to it. The gods of the ancient Greeks and Trojans, in this view, were right-brain advisors. Jaynes' view is radical, but it is taken very seriously by most students of consciousness. We're not sure about it either way. Our intention is to explain today's kind of consciousness, and even if its provenance was so recent, so sudden, and so dramatic, the story we are about to tell would still stand.

The mind has many attributes, and in particular we must be careful to distinguish between intelligence, awareness, consciousness, and free will. Previous generations of scientists adopted the view that only humans are consciously aware of their own wishes and decisions: animals obeyed 'drives', but *we* knew what we were doing. So when a cat stares at its empty food bowl, walks over to the cupboard where the food is kept, looks up at you, and miaows, all this shows is that it has a food drive, not that it feels hungry or wants to be fed. Within the last five years or so it has started to become respectable to blur the dividing line between the mental processes of animals and humans. The reasons include behaviour (animals, especially the higher mammals, give much too strong an impression that they have some idea of what they want) and evolution (our ability to experience subjective feelings presumably evolved along with the brain that handles the physical processes involved, and our brains evolved from similar structures in animals, just as the rest of us did). In this chapter we take a look at a number of theories of consciousness, in both animals and humans, and attempt not only to answer Nagel's question, but the question that forms the title of this chapter.

The problem of consciousness is a big one, but we at least *know* what it's like to be conscious. We have no idea at all 'what it's like to be' *sub*conscious. Indeed, it is pointless to wonder what 'being' a subconscious process feels like, because it doesn't – that's why it's subconscious. Silly Question, then. Except that it points to another neat paradox: the problem of subconsciousness is in some respects deeper than, and harder than, that of consciousness. Ironically, most scientists consider investigation of the subconscious mind to be an entirely acceptable scientific activity, because we cannot 'observe' our own subconsciouses. This is *really* weird. The part of the brain that we *can* gain access to easily is ruled out of order. Why? Because that access is 'subjective' – experienced by a human. On the other hand, when a scientist looks at a pointer on a voltmeter and writes down the corresponding numerical value in a notebook, that is an objective measurement – because it is experienced by a human. Strange.

Before discussing mind as such, we must first distinguish it clearly from reality, which is tricky since the only way we experience reality is through our minds. The objective/subjective distinction is often misunderstood as a result of forgetting this point: the question that gets forgotten is 'objective or subjective *relative to whom*?' If I watch an image of your brain while you are experiencing the beauty of a sunset, then the usual assumption is that the image of your brain's activity is objective, but your view of the sunset is subjective. What is often missed is the fact that my view of the allegedly objective image is itself a subjective experience – but for me, not for you. For this reason we shall mostly avoid the objective/subjective terminology from now on. The important distinction is actually between the physical processes that occur in a brain, and the experiences of the associated mind that accompany those processes: between cold reality and the warm mind that it engenders.

'Reality' is a very tricky concept indeed. We shall take a rather commonsense view of what it means: whatever it is that is actually out there in the physical world, *doing* everything. When you experience a sunset, you see orange clouds, but the reality behind that perception is a matter of light rays, light-sensitive molecules, neural pathways, and electrochemical brain activity. Nagel's point is that no amount of analysis of the latter can actually tell you what the former 'feels like'; but we think he overstates his case, because as human beings we already *know* what a lot of experience feels like, and the main question is to work out what physical processes in the brain are associated with the production of those feelings. Anyway, let's start by taking a closer look at the concept of reality.

Throughout the ages philosophers have made great play of the possibility that reality – the world around us that we perceive through our senses – may be a figment of our imagination. For instance, Ludwig Wittgenstein recognised that the sensations that we humans describe as 'reality' are actually produced in our own minds, so that our observations of the outside world are necessarily subjective. However, he also recognised that usually different people seem to have very *similar* subjective opinions about what they think the outside reality is. The commonsense position must surely be that there is therefore some kind of reality that is independent of human observation, and that our senses are picking up some of its features – but because different individuals have their own viewpoints and their senses are all imperfect, those sensory perceptions do not fully agree with each other. Instead, in his early work, Wittgenstein convinced himself that 'objective' reality is *only* a matter of agreement between a lot of subjectives.♪ So if we all thought that the Earth was flat, then it *really* would be, for example. Wittgenstein could be right (though we doubt it). Commonsense realism could be right (and we're sure that in some sense it is: the problem is to make that sense precise). But they surely can't *both* be right. Unfortunately a lot of Wittgenstein's followers *say* that he's right, but *act* as if he isn't. They don't just decide to agree that they are outside a room, for instance: they get up and go out through the door like everybody else.

An extreme denial of the commonsense position is the solipsist worldview: I know that *I* really exist (or I wouldn't be experiencing what I experience or thinking these thoughts); but the only evidence I have for *your* existence is what my mind constructs when I think that I perceive you. The only reality that I *know* is the one inside my own head. So it is entirely conceivable that I am the only real thing in the universe, and all the rest is just my imagination at work. There seems to be no logical way to disprove this position, but it does require a highly inventive and oddly selective imagination. We, of course – JC & IS – don't really exist: you, the reader, made it all up yourself, for *only* you really exist. So here you are, cleverly giving yourself the impression that you have no idea which words are coming next, making it all up as you go along. There's no point in us asking why you bothered to buy the book, because of course you just imagined yourself buying it. There's no way we can convince you that you are *not* making everything up as you go – but it's hard to see why you would take the trouble. For this reason we find the solipsist view to be philosophically incoherent and just plain silly. If you don't like that conclusion, don't blame us – *you're* the one imagining it. (You can be as rude as you like to a solipsist.)

Another similar line of thought is the one summed up in the question 'If a tree falls in the forest and no one hears it, does it make a sound?' For that matter, if no one sees it fall, does it really fall at all? These questions lead to something much more interesting than solipsism, because they impinge upon the role of a human observer in an objective reality. The 'sound' question is largely a matter of definition. If by 'sound' you mean vibrations of a certain kind travelling through the air, then the falling tree does indeed make a sound – unless you believe that the answer to the second question is 'no', and that without a human observer nothing can happen at all. Let's leave that kind of argument on the back burner for a moment, and consider the alternative definition: sound is a particular kind of sensation occurring in a human mind. Clearly with *that* definition, there is no sound if a tree falls unobserved. If you think this position is unreasonable, try replacing 'sound' by an attribute that depends more obviously on human judgement. How about 'Van Gogh's "Road with Cypresses" is a beautiful painting'? Is that statement true even when nobody is looking at the painting, or is beauty truly only in the eye (actually mind) of the beholder? ♪ And what about 'hydrogen cyanide smells like almonds'? What does a molecule 'smell like' if there's nobody there to smell it?

Back to the tree. Does it really 'fall' if no human observes it? Does it produce vibrations of the air if no human detects them?

If you walk through a forest you will find many logs lying on the ground. They look exactly like logs that people have observed falling, they behave exactly like logs that people have observed falling, they act as habitats for the same creatures as logs that people have observed falling. Sometimes they even occupy positions right next to what, two days ago, you observed as an upright tree. Either the universe is playing a very silly trick, or those trees fell even though nobody saw them doing it. Similarly, we know from zoological experiments that animals and birds can hear sounds: make a loud noise and they jump or take flight. Of course we can't conclude that when an animal hears a noise it experiences the same kind of sensation that we do when we hear, but that's not important in this context, and is probably meaningless anyway. Suppose you are watching the forest from a distance and a tree falls. You don't hear anything, but you do see a large flock of birds take wing. The natural conclusion is that the birds heard the noise, and in particular that the tree made a sound. Or you could leave a tape recorder in the forest, wait until the tree has fallen, and listen to the tape. Funny, the record on the tape is just what you would have expected if there had been a human being around to listen to the sound when it happened.

All such indirect evidence indicates that a tree falls in the same manner, with the same side-effect on air molecules, whether or not a human being is present. We don't *know* this conclusion is correct, but it seems a fair bet.

Philosophers worry about these problems because they underline the extent to which our individual perceptions of reality are distorted by the fact that we are the things that do the perceiving. To some extent they worry too much: the way you gain brownie points as an academic philosopher is to defend, successfully, a view of the world that everybody else finds absurd. Similarly, lawyers gain the admiration of their peers by successfully defending a mass murderer found covered in blood with a knife in his hand and his fingerprints all over the corpses, mountaineers achieve recognition by climbing the North face of the Eiger in winter, and mathematicians achieve recognition by finding some vast and intricate proof of a notoriously difficult but totally useless theorem. It is a sad comment on the human condition that these peer group proficiency tests focus solely on the technical *difficulty* of the feat, and often ignore the common-sense question of whether it was worth performing to begin with. On the other hand, there are times when the difficult feat does have value outside the peer group, and that's the case here. Philosophers who worry about how an observer's view of reality depends upon the observer aren't just playing silly intellectual games. They're circling around some central issues in science, which first became unavoidable in Relativity and Quantum Mechanics. Those are Big Ideas that need Big Capital Letters to show how Important they are, but we can illustrate the principle with a small idea, and become more ambitious once we've understood that.

Imagine yourself standing at one end of a field. A friend is standing a few yards to your right. At the far end is a horse; to its left (with reference to figure 25) is a large tree; to its right a windmill. A sheep grazes calmly in the middle of the field. Now, what you, as observer, see is the sheep standing in front of the horse. Your friend sees the sheep standing in front of the tree. Does this difference in perception imply that there is no such thing as an objective reality, independent of human observation?

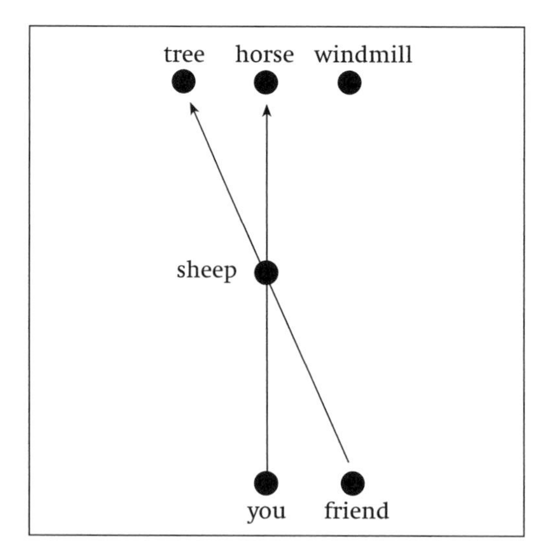

Figure 25 *Different perceptions of a sheep in a field.*

Some philosophers seem to think so. But notice: neither you nor your friend sees the sheep in front of the *windmill*. The observer does not get a free choice of what appears where; the different scenes viewed by different observers are not independent and they are not arbitrary. They depend – as we all know – on where the sheep is and where the observer is. What you see, and what your friend sees, are mathematically related by the geometry of the field. The difference is *not* evidence that there is no objective reality. On the contrary, there *is* an objective reality, which is observed from two different places by you and your friend, so that each sees different things. Reality determines rules that relate what one of you observes to what the other observes; conversely the existence of such rules carries a strong implication that your apparently contradictory observations can be reconciled by interpreting them within a broader, objective reality.

This kind of difficulty came to a head in Einstein's theory of Relativity. People had got used to the idea that observers in different frames of reference may perceive the same events in different ways. However, it turned out that the way such observers perceived light didn't fit the expected rule for 'translating' events from one frame to another. Einstein, along with a few other assorted mathematicians and physicists, found a way to fix up the translation rules so that they agreed with observations. The resulting picture of reality was unusual: for example the order in which events 'happened' was no longer sacred. One observer may see a train pass a station before the light in the waiting room switches on, whereas the other sees them occurring in the reverse order. This does not mean that space and time do not exist; neither does it render the concept of a train philosophically incoherent: it means that they are intertwined in a higher reality and that what we see is a projection of that reality into our own consciousnesses.

The role of the observer in quantum mechanics – as you'd expect – is even more elusive. According to most physicists, human observations of a quantum universe cause a whole collection of potential realities to 'collapse' into a single actual reality. (*We* – JC & IS – think that this interpretation is just as philosophically incoherent as solipsism, and therefore wrong; indeed it appears remarkably like the belief that a tree only falls if a human observer sees it. That doesn't alter the fact that most physicists think it is right.)

Both kinds of observer-dependence are used by some schools of philosophy to deny the existence of an objective reality. But in both cases there are constraints on what different observers can perceive; indeed it is the formulation of those constraints that constitutes the corresponding mathematical theory of the physical world. The message is not that reality doesn't exist, nor that it can be anything you want it to be. The message is that reality is not the *same* as your observation of it: that is just one shadow out of many. Reality is whatever *casts* all those interrelated shadows, and you can infer some aspects of its structure by looking at how different shadows are interrelated. Plato knew that: in fact the image of shadows is his.♪ Many have forgotten it; Wittgenstein denied it. Unfortunately Plato also had a rather mystical view of what it all meant, and that bit *ought* to be forgotten by all but philosophers.

Anyway, so much for reality as a figment of imagination. Much more interesting, and absolutely central to this book, is the contention that imagination is a figment of reality.

We mean at least two things by this. One is that what our minds observe, through our senses, is determined by the reality outside ourselves, subject to imperfections in what is doing the observing, and bearing in mind that we observe only a shadow. The second is that 'mind' does not require any ingredients that do not already exist in the material universe within which it evolved. It follows the same rules of behaviour as ordinary matter, space, and time, and it forms a part of the reality that it is attempting to observe. Mind is a *process* carried out inside an ordinary material brain made from ordinary brain material.

Our language is poor at describing processes, and especially so when it comes to describing the mind. Part of the problem arises from confusion between what we see, touch, and hear 'out there', and the actual external world. In fact any discussion of how minds experience their surroundings has to grapple with the relationship between three distinct things: the external physical reality, the physical processes going on in the brain, and what the mind engendered by those processes experiences. For example, the relationship between

- A London bus reflecting light at the wavelength we call 'red',
- The electronic computations that our brain carries out when it responds to that light,
- The feeling of vivid redness that we experience when we see the bus.

If you bear these distinctions in mind, you'll find it easier to understand what we're trying to say. We have to relate them to each other without getting them mixed up.

The most difficult one, from the point of view of physics, is our own subjective experience. Our senses provide us with a vivid impression that the external world contains things that are red, furry, hot, or hard – impressions known as *qualia* (singular *quale*, pronounced 'quah-lay'). They also produce the illusion that those impressions take place 'out there' – as if we are sitting inside our own heads looking out through our eyes. But in point of fact all those vivid sensations are produced *inside* our heads. They originate in physical features of the outside world – wavelengths of light, arrangement of components, vibrations of molecules, chemical composition – but the physical processes that we *interpret* as qualia are not the same as those outside physical features. They are representations, coded versions, and they are constrained, incomplete, and imperfect – but decorated very prettily with qualia, *exactly* to our taste. Imperfect or not, they are what mind-stuff is made of. Each of us has nailed our own little art gallery to the walls of our mind. When our senses communicate signals to our brain that resemble one of the pictures on the art gallery wall closely enough, we conceptually nail a copy of that picture, and the entry about it in the gallery's catalogue, to the outside world. We drape our perceptions of external reality with our internal images and associations, and by so doing, we also drape ourselves.

Although we can observe reality only by way of sense impressions, we can be sure that our impressions of reality are *not* the same as reality itself. One reason is that there are many ways to fool the observer-in-our-heads into 'seeing' things that aren't there. As mentioned in the previous chapter, the popular computer technique of 'virtual reality' operates in exactly this manner, sending slightly different images to each eye to create an impression of three-dimensionality. The 'objects' that appear to exist are vivid – though usually crude – but neither they nor the space they appear to inhabit has any real existence. Virtual reality works *because* the representation in our minds is not the same as the reality that it represents: our minds have evolved the facility to project this internal representation back 'out there' so that it is easy to work out which sensations are associated with what. We even do this when presented with a photograph, which we *know* is an arrangement of pigment on a sheet of paper. We close the strange loop between reality and its mental representation by pretending to ourselves that they are one and the same.

The internal representation has its own reality, too: it is constructed from processes involving ordinary matter: molecules, electrons, whatever. Such matter leads a dual existence, or more accurately has a dual interpretation: electron in one, part of a representation of a moving sheep in another. This is not especially mysterious: the word 'octopus' has a similarly dual interpretation as a collection of atoms in the ink on a page, and as a representation of an eight-tentacled marine animal. An interpretation depends upon a point of view, a context: the same physical thing can have many contexts, hence many interpretations. Such are the games that minds play, and it is important to remember them when thinking about any questions concerning mind – especially the more slippery philosophical ones. Descartes' mistake was to confuse a duality of interpretation with a duality of material substance.

Back to the central questions of intelligence, awareness, consciousness (we leave free will to Chapter 9). Where did they come from, how do they work? Despite the ongoing controversies about Artificial Intelligence, the easiest of the four is intelligence. It's a lot easier to imagine the possibility of an intelligent computer than it is to imagine the possibility of a conscious computer or a computer with free will. While our ability to imagine something is not necessarily a reliable indicator of its feasibility, most of us feel more comfortable about machine intelligence than machine consciousness, probably for good reasons. As the previous chapter shows, intelligence seems to be mainly a matter of having enough brain power and making use of it effectively.

Awareness has a different quality: it is the feeling that there is a world 'out there'. Awareness is not just a matter of receiving sensory inputs: a video camera does that, but few would consider such a machine to be aware. Animals are a different matter: while we can't yet *prove* that other animals possess awareness, there seems to be no particular reason to doubt that the higher mammals, for example, do. They have sense organs very like ours, they have brains very like ours, and if they're chimpanzees they have an awful lot of genetics in common with us. They *behave* as if they possess awareness.

Consciousness is another matter entirely: it is not just awareness, but a kind of introspective awareness in which the possessor has a definite feeling of individuality. A conscious being has an 'I' in its mind, *and it knows that it has*. The precise meaning of 'conscious' varies according to who is using the term: the fact that we are 'unconscious' when asleep seems to be very important to some philosophers, but others recognise that our consciousness remains active during dreams and may actually be turned off when we are awake doing complex but routine tasks like driving in to work.

One reason why the study of consciousness has not been a terribly respectable part of science is that it is difficult to imagine how any experiment could begin to demonstrate its existence – or not. On the other hand, it is hard to imagine an experiment that would demonstrate the existence of the Big Bang in any direct manner: instead the Big Bang is inferred from observations, interpreted within a particular paradigm. And it is not as big a step as many people imagine to go from 'I detected traces of the background microwave radiation' to 'I feel conscious'. All observations are reported by people. At any rate, by asking how consciousness might come about, and by thinking about how a conscious being is likely to behave, we can make some educated guesses about its existence and its nature – possibly even its mechanisms.

What we cannot do is put ourselves inside another animal's mind. Indeed it is very difficult even to imagine being a mind that is *very* different from our own. One aim of this chapter is to make out the case that we don't know, and can't imagine, what it would be like to *be* an octopus or a mantis shrimp. (However, despite what Nagel says, we can and will take a serious stab at his question about bats – because bats are a lot closer to us, their sonar sense notwithstanding.) This will help establish limits on what kinds of question about consciousness make sense. The octopus and mantis shrimp are definitely intelligent – but tempted as we may be to suppose that there is 'somebody inside', there is not the slightest evidence that these animals are conscious – by which we mean they are aware of themselves as a concept, as a feature that their own brains can detect. What does it feel like to 'be' an animal that has no consciousness, no 'self'? As well to ask what it is like to be a rock. There is no *you* to do the being, there is no you to do the feeling.

In contrast, mammals such as dogs, probably cats, and perhaps horses do show behaviour that allows us to impute to them something very like a self-aware mind. Everyone who keeps such animals can tell stories of occasions when they seemed to have a rather clear idea of what was going on – not just the guilty cat that has just been caught stealing food, but the cat that sits on the floor, eyes the food, glances at its owner, visibly makes the wise decision, and walks out of the kitchen. And sneaks back later once its owner has gone. Yes, there is no way we can *prove* that there is anybody in there, but science is about finding the simplest explanation consistent with the evidence, not about proof. We have a strong feeling that there might well be somebody in there, because it explains the behaviour so much more reasonably than mechanical 'drives' do. Of course we must also recognise that people tend to anthropomorphise, so that Gerry the pet gerbil is seen as a veritable rodent Einstein because it knows which end of its treadmill is the way in; however, that's no reason to go overboard the other way. The apparently 'conscious' behaviour of the higher mammals is nearly all about relationships with creatures of the same kind – or with people, in which case they act as if they've decided *we* are their kind too, despite the funny shape and smell.

So, we reckon that Fido definitely thinks he's a somebody, but an intelligent octopus doesn't. In order to explain this view, we will have to analyse – and to some extent explain – consciousness. That's a tall order – it is a task which has stumped philosophers from Aristotle to Jurgen Habermas, a guru public philosopher of the 70s. Plato and Goethe, Descartes and Hume all thought that they had explained it – or explained it away. One pitfall that it is inordinately difficult not to fall into is to think of consciousness as a person, a 'self' sitting in our head and looking out of our eyes. Philosophers call this little person in our heads a 'homunculus'. Human beings all have a strong feeling that consciousness is 'homuncular' in nature, but there is a wealth of evidence that such a simple picture cannot be correct. In a musical analogy, the patterns of brain activity do not indicate any kind of 'conductor' that acts as a recipient for sensory perception and takes the decisions: the brain seems more like a chamber orchestra in which each player is reacting to the others, but not under central direction. Another problem with the homunculus theory is: how does the *homunculus* know what it sees? Is there an even tinier minihomunculus inside its head, and a microhomunculus inside that, and so on? Is it homunculi all the way down? And if this chain stops somewhere, why not at the very first step? A third difficulty is that although we have a very strong sense of being a unified person, we are terribly confused about *which* of our feelings, sense of identity, emotions and behaviour contribute to this person-ness; so much so that there is now a good case to be made that what Sigmund Freud called the 'ego' and the 'id' are now interpreted the other way round by his intellectual successors. Freud saw the id (Latin for 'it') as the raw stuff of human emotion and the ego (Latin for 'I') as a much purer sense of identity; now it seems that it is the id that generates the sense of 'I-ness'.

How does this 'I-ness' come about? We have already said that for Descartes the mind was simply made (if that was the word) of different *stuff* from the rest of creation, the theory of 'Cartesian dualism'. Mind was an immaterial spirit that observed the pictures thrown on to the walls of the brain as if they were sprayed out of the optic nerves into an empty room. Descartes' image is that of a disembodied personality (yes, a homunculus) sitting in an otherwise empty theatre while images of reality are played out on a stage – what the cognitive scientist Daniel Dennett calls the 'Cartesian Theatre'. But if mind is made from different stuff altogether, why can't we isolate any of that stuff from brains, and why do you need a brain at all? Which you surely do, because damage to the brain often results in damage to the mind. Indeed, damage to specific regions of brain affect mind in equally specific ways, so that a nasty bang on *this* area of the brain causes problems with language, and a chunk taken out of *that* area means you can't recognise faces any more.

For a thousand similar reasons we are led to reject the Cartesian

Theatre in favour of something much more dynamic and material. Mind is not a thing but a process, one carried out by ordinary matter. If there is a 'dualist' aspect it is that – as just mentioned – that process can be interpreted in two ways. In one interpretation it is a vast complex of scurrying electrons and chemical molecules (reality), in the other it is what those movements *mean* to the mind that they constitute (figments). The first description emphasises reductionist rules, the second emphasises the emergent phenomena that they generate. The two seem very different because emergence is like that. For a homely analogy, a bridge is both a lot of sand grains and other types of matter stuck together in a particular way, and a span that lets us cross a river. This type of 'duality' must be remembered whenever questions relating the experiences of minds to real physical phenomena are under discussion.

When discussing the mind, it is extraordinarily difficult to select metaphors that are not tainted by a hint of the Cartesian Theatre. The reason is that we are trying to explain how the mind creates an illusion of an internal observer, even though that's not really what's there – but we are trying to get that point over to a person who has the strongest feeling that they *are* an internal observer, sitting inside their own head, *experiencing* that self-same illusion. It *feels* like being in a Cartesian Theatre, and our language for describing minds has built itself around that model. So our metaphors and images will almost certainly carry overtones of Cartesian Theatre, and of naive Cartesian duality, because they have to be expressed in ordinary language. Moreover, we have to explain why it *feels* like a Cartesian Theatre. We don't know any way round these problems, except to say 'if it sounds as if we're taking a Cartesian Theatre viewpoint, then that's not what we mean.'

Back to the vexed question of the intensity of the experiential features of the mind – the problem of 'qualia', sometimes referred to in the literature as the 'Hard Problem of Consciousness'. (The 'easy' problem, by the way, is to understand the physical organisation of the brain that *corresponds* to conscious experience, a curious new use of the word 'easy'.) Our belief is that on one level the Hard Problem of Consciousness is not hard at all, but what in Chapter 2 we called Deep Thought – any attempt to provide an answer sooner or later ends up by saying 'well, the universe is just like that'. If you feel this is a cop-out, you haven't understood that the same difficulty arises in many other areas of enquiry. For example, while we can describe an electromagnetic wave with mathematical equations, we have no idea *why* we live in a universe rich enough to include electromagnetic waves. We can't say *why* the Big Bang happened when it did, we can't even say, on a genuinely deep level, *why* water is wet. So it's no big deal that we can't say why 'red' looks like it does to a human mind.

We can, however, explain exactly why, in a universe rich enough to permit qualia, qualia will come into being. The reason is evolutionary: the more vivid your qualia, the more effectively you will react to your sensory impressions, and the more likely you are to survive. The rabbit's fox-detecting qualia improve because that way it's a lot easier to detect the fox; ditto for the fox detecting the rabbit; so an evolutionary arms race sets in driving both towards ever more vivid qualia. Here 'vivid' of course means 'exciting a clear and definite response'. Just as sexual selection mechanisms led to peacocks with striking, brightly coloured tails, so it also led to peahens whose *perception* of those tails was striking and brightly coloured. Different senses evolved at different times, and therefore acquired different qualia. That is one reason why hearing is so different, subjectively, from seeing, and both differ completely from smelling, sensing temperature, feeling texture ... There is also an advantage in not letting distinct qualia become confused: the brain receives the greatest information when it is obvious which sense is reporting it. So evolution encourages qualia to be unique to each sense – so much so that taste and smell, while biologically very similar, have related but different qualia. You may feel unhappy that none of this explains what qualia feel like (true: the universe is just like that). But you shouldn't be upset about the 'unreal' nature of qualia: they are virtual, things experienced by minds that decorate their important sensory perceptions, things whose physical description does not and cannot convey 'what they feel like' to the mind experiencing them. The problem, as indicated earlier, is a difference of contexts: what I observe happening in your brain with my PET (Proton Emission Tomography) scanner may be the same *thing* that causes you to see red, but it is not 'red' itself. In your context it's 'red', in mine it's a pattern on a computer screen. But this is true of all observations: we don't ever experience an electron 'as itself' – just various derivative phenomena which, in the appropriate context, we identify with the concept 'electron'. So before getting too carried away about how weird and mystical conscious phenomena are, compared to physical reality, it's worth getting carried away about how weird and mystical physical reality is.

Qualia are perfectly real: they 'are' processes that go on in brains in response to various inputs. But they are not 'just' processes that go on in brains in response to various inputs, which is the standard mistake made by everyone who forgets that mental events have more than one interpretation, depending upon context. A process can be a burst of nerve impulses from one point of view and 'red' from another. The difficulty of thinking about such questions is exacerbated because sometimes we *can* find relatively simple physical correlates for qualia: for example 'brightness' in vision qualia is clearly linked to intensity of light in physical reality, and the red–purple spectrum of colour qualia is similar to a spectrum of wavelengths of light waves. Unfortunately these simple correlations are probably unrepresentative of most qualia, hence misleading. Qualia of smell, for example, seem to be processes mediated by neural nets of rather arbitrary structure – to some extent correlated with vibrating chemical bonds, but maybe not very closely or very repeatably. Since nearly all adults are 'smell-blind' to some categories of smell, your smell qualia and ours are probably *different*. There is no reason to expect qualia to match simple physical quantities of a kind that a human scientist might measure. On the contrary, between physics and qualia lies Ant Country, which the universe can traverse in a blink, but which is inaccessible to human reason. This does not prevent reason leaping over the logical gap – but it does create a gap, at least on a reductionist level.

Nagel's question about 'being' a bat can be interpreted in several ways, but what he was really getting at was bat qualia. The use of the word 'be' indicates the intended viewpoint: that of the bat. If you were a bat, what would your qualia be like?

Let's try a warm-up question: what is it like to be a Zarathustran?

Hewer-of-wood [*Picks up oddly shaped piece of rough wood.*] Oh, *yes*! Fantastic! Just nurphlerise *this*!

Liar-to-adults There are times when I get the feeling that you and I are not actually part of the same species, Hewer. I have not nurphlerised anything for *ages*. To be honest [*smiles ingratiating smile of a used groundbug salesperson*] I can hardly recall what it is like to nurphlerise.

Performer-of-amusements You can hardly recall what it is like to be honest, either, Liar-to-adults.

LtA OK, Performer, when did *you* last nurphlerise? For that matter, when did you last exercise your sense of gullibi–

Ringmaster You two, stop it! Your lack of consensus is glaring!

Liar-to-children Is it? It is not at all obvious to me, Ringmaster. But then, I have always been consensus-blind myself. Especially to the difference between a medium strong consensus and a weak but strengthening one ...

R But that is *so* obvious. You really do not – no, you do not, I can tell. That is even more obvious, and quite ungraffitable. Which is distressing, since I rather enjoy the experience of graffiting. Makes my feet go all tingly.

Destroyer-of-facts As with us all, Ringmaster. Graffit-thrill is one of the few qualia that we all express in our present configuration.

Pursuer-of-sicknesses I do not smell your path.

PoA You should know that there is no point in trying to read the scent-trails.

LtA Oh, shut up, both of you! Destroyer, what did you mean by that strange remark of yours?

DoF I was referring to our suite of senses and the associated qualia.

LtC Qualia?

DoF Vivid sensory experiences. They are hypermassively parallel-processed by our neural nets, and become causally dislocated because the logic paths must traverse Ant Country, and therefore cannot be mapped.

PoA You mean there is no point in trying to read the ant-trails.

DoF Most amusing. Where was I? Oh, yes: you will of course recall that we evolved from a vampire-like parasite. Originally our main sense was taste, centred around blood. Over time we evolved an exquisite awareness of subtle chemicals in our host's bloodstream, and as we began to displace our hosts from their roles in the primal burrow we developed an entire suite of senses, analogous to those of our various hosts. We all *possess* the same suite of senses, but when a group of eight Zarathustrans segregates spontaneously into a standard octuplet of roles, each of us suppresses most senses and expresses only a few preferentially. [*Looks at Hewer.*] Excuse me, but I wish to perform a small experiment, and I shall borrow your lump of wood. [*Places it on the table.*] Now, each of you: what do you sense most strongly?

LtC It made quite a bang when you put it down. Rather like a lump of wood hitting a table. [*Long pause.*] Oh, right.

PoS Yes, and it has quite an interesting smell, a bit like roasted quux. But I am sure that must be obvious to you all.

Creator-of-creations I cannot smell a thing. [*Picks the wood up and licks it.*] Yuk. It has a definite modicum of croquonut about it – maybe two modicums, I am not very good at quantifying qualia.

DoF My turn, now. It is a sickly groonish-brungey colour with a high degree of contrast, not unlike a Skreenish sunset.

LtA I confess I can scarcely perceive it at all. Its gullibility is so low that it is off the bottom of my scale.

HoW Nonsense, it is as plain as the proboscis on your face! In fact, I nurphlerise it *as* a proboscis. Where is my knife? [*Rummages ineffectually under the table.*]

DoF Hewer-of-wood means that when he looks at it he experiences the immediate and indelibly vivid impression that inside it is a proboscis, waiting to be carved. He has no choice: that is the main thing he perceives.

HoW [*With heat.*] There *is* a proboscis inside it, waiting to be carved! Can you not nurphlerise something so utterly self-evident?

PoA Seems more like a rather rude part of the anatomy of a hypnoceros to me. [*Giggles.*]

R Whatever emotions this object stirs in us all, it creates no consensus. [*Stops, thinks.*] Mmmph. We *all agree* that there is no consensus ... Hpmmm. That is a form of meta-consensus that has not previously come to my attention. Perhaps I *will* graffit something after all. [*Does so.*] *Ooooh. Tingly.*

DoF Wonderful! Do you all see?

PoS See what? What is there to *see*? It looks like a silly bit of woody stuff. But it smells most strongly of roasted quux, possibly with a hint of marmojam or sparsely (delete whichever is–)

DoF Excuse my use of loaded qualia. I meant, do you sense my implication? [*Waits expectantly.*]

All the others [*In unison.*] No.

DoF Never mind, I shall explain it to you, as I anticipated would be necessary when I asked my rhetorical question. Let us start with our Ringmaster. It is highly significant, is it not, that the quale he sensed was not related to the wood at all, but to our varying reactions to it.

R Of course. I sense the qualia of consensus extremely strongly; that is why I am Ringmaster.

DoF Actually, it is *because* you are Ringmaster that you sense the qualia of consensus so strongly, but that is beside the point. You also sense other qualia? Vision, sound?

R Of course, but they are much less dominant. And I must confess that I too have not nurphlerised for a very long–

DoF When our attention was drawn to Hewer's lump of wood, each of us experienced qualia related to our role in our octuplet. The Ringmaster senses consensus, or its absence. Hewer's primary quale is a vivid sense of what lies inside a piece of wood, waiting for the rest to be carved away to reveal it. [*Hewer leans over and regains his piece of wood before someone spoils its elegant nurphle.*] Liar-to-adults is attuned primarily – no *solely* – to gullibility-qualia. Liar-to-children's forte is the auditory sense, mine is vision. Creator's is taste, because originally his role evolved from that of the burrow bartender. Pursuer, whose role descended from that of corpse-tender, is – not surprisingly – receptive to smell.

R What about Performer-of-amusements?

DoF He has a sense of humour.

That's put the Hard Problem in its place – but what of consciousness itself? Since Descartes, many different theories of consciousness have been proposed, and we shall focus on some of the most modern contributions, from people like Patricia Churchland, Daniel Dennett, Stuart Hameroff, and Roger Penrose. Over the next few pages we will survey some of the ideas currently under discussion. We begin with a topic that has greatly excited many philosophers: along with Dennett, we think it is a complete red herring. We mention it because you may well run into it, and deduce from the awe in which it seems to be held in some quarters that it is a serious contribution to the subject – which it is not. This is the concept of a 'zombie', which is *exactly like a human in every respect* except that it is not conscious. Zombies are thought experiments, often used to focus attention on the (alleged) ineffable quality of consciousness that is so different from the physico-chemical activities of material brains. Imagine a zombie, the argument goes. It functions just like a real human being, right down to the electrical activity of its brain cells – but you *know* it's not conscious, because that's how you imagine it. Conclusion: consciousness is different from the physical activity of a material brain.

There are so many things wrong with this suggestion that it is hard to know where to start. The main one is that it's a cheat. It's a Silly Question in disguise, one of the 'but what if you *saw* a ghost?' variety, which we dismissed in Chapter 2. Suppose that mind is an emergent property of brains, as seems likely. Then mind is inaccessible to detailed reductionist analysis – not because brains contain some ineffable non-material quality, but because reductionist explanations of *all* emergent properties have to traverse Ant Country. Now in Ant Country, the chain of logic that links the emergent property to its reductionist causes *exists*; it is just far too intricate for anyone to comprehend. Nature, running trillions of causes-and-effects in parallel, traverses Ant Country with ease – it is *we* who get bogged down. So if mind is an emergent property of the brain (in the context of a human being in a human culture, we must add) then it would not be possible to have a creature identical in every physical respect to a human being, without consciousness emerging from its structure and activity – just as it does in a human being. Emergence is not the *absence* of causality, something that *disconnects* effects from causes: it is causality too intricate to follow in detail, but which still achieves its effect. You might as well encourage people to imagine a zombike, which is *exactly like a bicycle in every way* except that it does not move when the pedals are pushed. Oh, mystic miracle of ineffable immateriality, the source of motion in a bicycle is not anything physical! Or a plant that has exactly the same molecular structure as grass, but is pink with blue spots: wow, colour is not a physical property at all! Or as Dennett suggests, imagine a creature that is exactly like a human being right down to its DNA structure, which walks and talks like a human being, and is not alive. In any case, if you really do want to allow that game, how about imagining a being that is exactly like a pocket calculator in every respect, except that it is conscious?

In order for a thought experiment to illuminate a problem, the experiment must bear some relation to reality. Being able to *imagine* a zombie, a ghost, or a conscious pocket calculator does not imply that such a thing can actually exist, so any 'ah, but what if one *did* exist?' argument falls flat on its face at step one. But such arguments are insidious: as Dennett rightly points out, they 'work' because they invite people *not* to use their imaginations. The question 'is consciousness an emergent property of material brains?' does focus our attention, and it is an excellent mental exercise to try to imagine how such a thing could possibly be. Notice that this question is not posed in a manner that rules out either of the answers 'yes' or 'no' in advance. In contrast, the zombie concept actually invites us to try very hard *not* to imagine how such a thing could possibly be; and if we accept the possibility of zombies then we immediately concede that consciousness is *not* a property of material brains. Thought experiments like this rely upon us limiting our imaginations, and in this respect the zombie resembles Searle's Chinese room, which plays a similar distracting role with regard to intelligence.

So much for zombies. Another common misconception in the literature concerns the *location* of mind, or indeed consciousness. Where is it? In the brain? Sure, but *where* in the brain? All over? No, that's not a location. But surely if mind exists in a material sense, then it must be *somewhere*? Sounds sensible – until you put your own mind in gear for a moment. Mind is a process, not a thing. Processes take place within things, but they do not have a specific location. Where in a car is its ability to move located? In the wheels? No, because you need an engine too. In the engine, then? No, because you need the wheels. In *both* (along with transmission and so on)? But that's not a location! In part this way of thinking is a flashback to Descartes, who thought that the mind came together in the pineal gland – a curious organ in the brain whose function is still rather mysterious. It also comes to light in discussions of the location of thoughts. Does a thought have a location in physical space? Obviously not, otherwise there would be a distance between thoughts. 'How far away is a thought of a purple cow from that of a rosebush, ho ho ho?' Those of us brought up on phase spaces know better. Thoughts live in thought-space, and while that space may not possess a metric (a well defined concept of distance) it clearly has a topology (a way to tell when two things are close together or not). A thought about a purple cow wearing green Wellington boots is fairly close, in the topology of thought-space, to one about a purple cow wearing green slippers. A thought about a purple cow wearing blue Wellington boots is a bit closer to the first than the second. All three are a long way from a thought about a rosebush, and all of those are even further away from an unwarrantedly sarcastic thought about philosophers who don't understand the first thing about phase spaces ...

Enough of this nonsense: now for some sensible suggestions about consciousness. We begin with a reductionist theory, originated in part by Stuart Hameroff, which has been promoted by the mathematical physicist Roger Penrose in his books *The Emperor's New Mind* and *Shadows of the Mind*, and in the *Journal of Consciousness Studies* (now four years old). We discuss it now, to show you some of the new kinds of thinking that are entering the field, and to convince you that this one is wrong. Penrose starts from the fact that humans make conscious choices between alternative courses of action – or, at least, they feel very strongly that they do. This is really the question of free will, not of consciousness, but the two are closely linked because in order to have a free will you must first have a will, that is, an 'I' inside.

A brain can be thought of as a dynamical system: its internal state changes, very rapidly, as time passes. We can ask what kinds of mathematical or physical rules the brain's dynamic obeys. Could it, for example, be a deterministic dynamical system of the type currently studied by mathematicians? Do some sort of 'Newton's Laws of Thought' exist? It looks as if they cannot, for such laws would contradict free will. A deterministic dynamical system has a *unique* future given its present state, so it cannot make a choice. The same goes for any algorithmic system – one that carries out a 'prescribed' process whose final result is implicit in its initial state. Therefore, Penrose argues, the brain is not algorithmic. To justify this contention further, he cites Gödel's theorem in mathematical logic as a proof that brains can think non-algorithmic thoughts, and he offers the phenomenon of quasicrystals as evidence that physics itself is non-algorithmic. We will say more about those arguments later, but to do so now would lead us off the main track, so we will accept, for the sake of argument, that non-algorithmic processes are required for the conscious exercise of free will. This implies that we must introduce an element of indeterminacy.

The obvious candidate is quantum mechanics, which – according to conventional physics – has an inherent and irreducible random element. Quantum effects occur on extremely small scales, so there must be structures in the mind that 'amplify' quantum effects into human-scale choices. What are those structures? Penrose, following Hameroff, settles for tubulin – a remarkable hollow protein structure like a molecular drinking straw, built from a chequer-board assembly of two very similar units, alpha- and beta-tubulin. Tubulin forms structures known as microtubules, part of the cell's 'skeleton'; it can unzip rapidly like a peeled banana or grow more slowly like the brick wall around a well. These facilities allow cells to move by tearing bits of their skeletons down and rebuilding them elsewhere: Niven and Pournelles' Moties would have approved of tubulin. It is also implicated in cell division: the cell goes fishing for its own chromosomes using tubulin rods, and having hooked them it pulls them – and itself – apart into two. It is also involved in the transport of molecules around the cell, which can be pulled along by tiny 'stepper motor' molecules running along the tubulin rods like trains on a track. As we said in Chapter 1, molecules are tiny machines. Cells are more like miniature cities than diminutive lumps of jelly. Notice, in passing, how opportunist evolution is, and how silly scientists are who think that once they have found one function for a molecule, that's it. What is 'the' function of tubulin?

Tubulin is a major component of all cells, including nerve cells. Being tubular, it can act as a 'wave guide' for quantum wave phenomena – rather like optical fibres guiding pulses of light in telecommunications – and can thus act as an amplifier to turn quantum indeterminacy into macroscopic choices. This is the Hameroff–Penrose story, except that we have omitted to discuss the role of gravity, which provides an element of nonlinearity. It is an extremely ingenious and interesting theory, which suggests, temptingly, that the brain's true information storage capacity is many orders of magnitude larger than we have hitherto thought. Instead of each nerve cell being 'on' or 'off' and thus storing only one binary digit, it contains vast lengths of tubulin, with corresponding storage capacity. Imagine, for example, ringing the changes on tubulin structure by using alpha-tubulin for the digit 0 and beta-tubulin for the digit 1. And to clinch the argument, anaesthetics may have some effect on tubulin, and we all know that anaesthesia renders people unconscious.

Neat, but as we said, we don't believe it. Our objections are similar to those voiced by Patricia Churchland and Rick Grush, who refer to the whole idea as 'Penrose's toilings'. This is a pun on 'Penrose tilings', beautiful aperiodic tiling patterns related to quasicrystals, which we describe in a moment. There are two classes of objections to Penrose's theory. One is reductionist, the other structural. On a reductionist level the theory just has too many loose ends. For instance Penrose justifies the contention that the brain is non-algorithmic by appealing to Gödel's theorem, which states (in one interpretation) that any consistent logical system rich enough to incorporate arithmetic includes unprovable truths. That is, there exist non-algorithmic mathematical processes. Churchland and Grush argue that Gödel's theorem is irrelevant because we do not even know that the brain's operation is logically consistent; indeed people often hold mutually contradictory views without recognising the contradiction. The force of this criticism can be understood more easily by applying similar reasoning to another of Penrose's pieces of evidence for non-algorithmic processes in nature, the existence of quasicrystals. Quasicrystals are a recently discovered form of matter whose atoms are not arranged in periodic lattices – as happens for crystals – but instead form almost-periodic latticelike structures, one class being the aforementioned Penrose tiling. It is known that the question 'does a given set of tiles cover the entire plane non-periodically?' has no algorithmic solution. Penrose argues that because quasicrystals answer this question by physical means (and the answer is 'yes'), then they must be carrying out a non-algorithmic process. This argument is so full of holes that it really should never have been put forward. It confuses the general question of whether an arbitrary set of tiles covers the plane with a specific question for one particular set; it assumes that quasicrystal structure contains no imperfections, which is not even true of real crystal lattices; and it ignores the fact that real quasicrystals are finite, whereas all of the non-algorithmicity of tiling problems stems from covering the *infinite* plane. It really is nonsensical.

These are objections on a rather esoteric philosophical level. There are more concrete ones, mostly centred upon tubulin. 'Unconscious' is too glib a description of the anaesthetised brain: it is our awareness of the outside world that gets lost. We have already said that when asleep, the brain is often conscious – for instance during dreams, when we get a distinct impression of an 'I' watching the dream. And when we are awake, our consciousness sometimes goes on autopilot. Moreover, whether or not anaesthesia reduces consciousness, there is no direct evidence that changes in neural tubulin are responsible for the effect of anaesthetics. However, the drug colchicine, used in the treatment of gout, *is* known to affect tubulin: it disrupts the function of microtubules. If Hameroff and Penrose are right, then administering colchicine ought to disrupt consciousness too. There is no evidence that it does, and a lot that it doesn't. Finally, the core of microtubules is usually not empty, but almost certainly contains ions of calcium and sodium, not to mention molecules of water. These would prevent the tubulin acting as a waveguide for quantum effects, and stop it amplifying them.

Penrose has attempted to rebut the arguments of Churchland and Grush, but the rebuttal largely amounts to a restatement of his original position and a lot of ifs and buts. We remain unconvinced. However, our main personal cause for disbelief is not anything reductionist, but a structural incongruity: the explanation by quantum effects is simply the wrong *kind* of explanation. Quantum indeterminacy is random, so a quantum consciousness would make random choices, not the structured ones that we experience ourselves making. As we said in *Collapse*, it is as if the question is 'Chartres cathedral?' (meaning 'where does its beauty come from?') and Penrose's answer is 'bricks'. (Yes, we know they're actually stones, but that's what we said in *Collapse*, for reasons of euphony.) Each brick, he in effect says, has a tiny bit of cathedral in it, and a tiny bit of beauty, so you get a complete, breathtakingly beautiful cathedral by using a lot of bricks. Trouble is, you can also get *Coventry* cathedral, or a fast food emporium, by piling up bricks. The beauty of Chartres cathedral lies in its organisation, not in its components. It could have been built (ignoring practical engineering factors for the sake of dramatic irony) from blocks of compressed carrier bags, baked bean tins, or dried cow dung: it would still have been just as beautiful – at least until you got close up.

Indeed, if the key to conscious choice is just to introduce an element of randomness, we don't need quantum mechanics anyway. Instead, we can equip the mind with its own non-quantum dice. Any chaotic classical (that is, non-quantum) system will do: for all practical purposes chaos behaves like randomness. The real problem is how tiny elements of indeterminacy – real or apparent – become translated into *coherent* large-scale choices like 'let's go to the movies tonight'. That's a structural and organisational problem, and if we knew how to do it with quantum components we could almost certainly play the same game with classical ones.

In contrast to Penrose's reductionist approach, in which the subtle qualities of the mind such as consciousness are derived from analogous physical qualities in a reductionist way, there is Daniel Dennett's view of the mind as a conglomerate of loosely knit processes, each semi-independent of the others, which he refers to as 'pandemonium'. It is an image inspired as much by modern computer architecture as by visions of Hades. Today's computer operating systems involve large numbers of subprograms known as 'demons', which wait until they are called on, do their thing, report their results, and shut up shop again. There are demons that monitor memory usage, check that communication lines are functioning, set up the screen display, and so on. Pandemonium is not one big demon: it is a host of semi-autonomous demons, gabbling away to each other, arguing, interfering – and collectively producing something that 'works'.

It may sound odd, but that's how today's computers are 'organised' – and they wouldn't work as well if they were organised in a more obviously hierarchical manner, like older computers were. That doesn't imply that minds have to work the same way – but it shows that such a process can deliver coherent results. And it opens up a radical alternative to the homunculus in the Cartesian theatre. After all, we've just said that computer demons 'wait until they are called on' and 'report their results'. Called on by what, report to what?

To other demons.

The apparently organised behaviour of the computer *emerges* from the interactions between demons. Of course in this case everything was pre-orchestrated by a master programmer working for the computer company – not in detail, incidentally, but certainly in overall conception. But a computer operating system that *evolved* could blunder into a similarly pandemonic set-up without there being any kind of master programmer at all.

Dennett tells us that the human mind is like that. You – with your strong, overriding sense of you-ness, the feeling that what you experience is experienced by a single entity, and that this entity is very much in charge – may well feel that the idea that you are an emergent feature of pandemonium is ludicrous. However, there is a great deal of evidence that the brain/mind is organised in just that manner. Here are a few pieces of that evidence. firstly, a great deal of your mind is inaccessible to conscious introspection, and uncontrollable. You cannot – at least, without training and a lot of practice – control your own heartbeat. Your brain sends signals to your heart, telling it when to contract, but 'you' have no choice in when those signals are sent, or what they are. Even if your heart is going wrong, and you are consciously aware that it is, you still can't get inside your own head to control it. You *can* control your lung rate, to some extent, but you can't choose to suffocate yourself to death by holding your breath. You cannot 'decide' to stop your visual system detecting images, except perhaps in rare psychological disorders where undesirable images may be filtered out. The best you can do is decide to shut your eyes – and that's very different, because your visual system is still operating, as you can check by shutting your eyes and waving your hand just in front of them. You can't shut your ears at all, and you can't decide not to hear intrusive noises: you have to opt for a technological solution and stuff something in your ears to plug them. And all of these pretty much autonomous processes go on simultaneously, sometimes in overlapping areas of the brain.

Another piece of the jigsaw is the observation that because it takes time for your senses to send messages to your brain, and time for your brain to send instructions out to your muscles, then 'you' are constantly running about a third of a second behind reality. Indeed different senses are running *different amounts* behind. The problem is not just that all of this fits together into an apparently seamless whole, which is strange in its own right: it is that you react to events in the real world that are over before you could possibly register them in your brain. For instance, think about a cricketer or baseball player catching a ball. They see the ball up in the sky, follow its fall with their eyes, put out their hand, and watch the ball smack into it. Great – except that it actually smacked into their hand or glove a third of a second ago. So their hand must have been out there a third of a second ago too, otherwise the ball would have flashed past long before the hand got into position. Which means that the brain must in some manner have 'extrapolated' the ball's motion, and put the hand into the position where it estimated the ball would shortly reach – not where it actually was at the time. None of this internal processing and anticipation is accessible to conscious thought: the world that we (think we) observe is synchronised with what our conscious mind considers to be *now*. This, incidentally, is presumably why a fielder who fumbles the ball, so that it bounces into the air, is temporarily unaware of where it has gone: it takes a definite time to recognise that soemething has gone wrong and resynchronise the mental extrapolation mechanisms with reality. Usually the ball has hit the dirt long before. This is not just a question of 'reaction time', but of projecting some of your mental processes a third of a second ahead of what the rest of those processes perceive the real world to be doing, without ever noticing any discrepancy.♪ (Some people have seized upon this ability as further evidence for the quantum nature of the brain, because in quantum theory the timing of an event can be 'advanced' or 'retarded' compared to what is observed. This is the same error as bricks/Chartres cathedral: a false analogy that operates on the wrong structural level. It is the *organised* nature of the 'prediction' that matters, not just the raw ability to predict.) These are just a few odd bits of evidence: for chapter and verse on all the reasons why the brain must be composed of numerous quasi-independent demons, you should consult Dennett's *Consciousness Explained*.

We find Dennett's story the most convincing among those currently on offer. However, we wish to add a final gloss, the idea that the brain's independent units are brought together by a general feature-detecting system, which does not organise them, but instead rationalises their independent decisions. We call this unit the 'ringmaster', by analogy with circus usage. Its role is much like that of the Ringmaster in a Zarathustran octuplet – the octuplet itself, of course, being a metaphor for pandemonium – and it is closely implicated in the question of self-awareness, which is distinct from consciousness but part of the same package.♪ If you go to a circus, what you think you see is a carefully orchestrated sequence of events – trapeze artistes, lion tamer, clowns. There is an obvious and very important personage, who wears a top hat and wanders around telling everybody what to do: the ringmaster. Obviously he is the one in charge who is organising everything.

Not so.

All of those events go on with very little in the way of cues from the ringmaster. His job is not to control the events: it is to give the impression that they are under control by interpreting them to the audience. If a clown accidentally falls off the shetland pony, the ringmaster's job is to pretend that it was a deliberate part of the act. The clowns, indeed, are the bane of the ringmaster's life, so he spends a lot of time looking as if he's in control of them, when in fact they are largely in control of him.

Like the ringmaster in a circus, the ringmaster in our heads gives the impression of being in charge when in fact it is not. We emphasise that the ringmaster is *not* a homunculus sitting in a Cartesian Theatre, observing the play of sensory impressions on a screen. Neither is it the 'ghost in the machine', a phrase popularised by Arthur Koestler (which we think is actually the ice-cream seller in the Cartesian Theatre, and so is demolished along with the theatre by modern cognitive science). The ringmaster is just another demon in the pandemonium, and its role is to *appear* to the emergent phenomenon that is 'me' to be making sense of everything else that is going on.

We think that the ringmaster came about for contextual reasons, like all of the big puzzles of mind – a conclusion that is based upon experience with animals. We shall describe experiments and observations on many kinds of animals, and will conclude that recursive interactions between organisms of a similar kind engender consciousness of a self performing those interactions (so that ironically 'self' evolves because it has to be distinguished from other non-selves). And we shall delight in the paradox that there is no continuing self in these animals, while there is continuation in just those animals that don't have a self.

JC has kept a variety of animals – llamas, horses, donkeys, goats, rabbits, cats, dogs, tortoises, geckos, guinea-fowl, small tropical octopuses, mantis shrimps, *Hydra*, *Amoeba* ... He has interacted professionally with many others, from dolphins to birds of prey. He gave an 'animal handling' course in the continuing education department of Birmingham University for many years. IS's regular animal contacts have been more limited – cats, goldfish, and koi, mainly. The many opportunities for observing people in contact with very different life-forms (and IS insists that cats are *definitely* different life-forms) have influenced our views about the very different nature of intelligence as exhibited by people, by other social animals, and by asocial animals. The latter very often *look* as if they are communicating. If a mantis shrimp – a kind of shrimp that can grow to about twenty centimetres long – is given a series of tasks that must be completed in order to get food, then it takes a long time, a lengthy series of attempts, before it can solve the first puzzle. When it gets the food it will sometimes repeat some of the steps of the task, as if rehearsing them. Then it rapidly becomes adept. Dougal, a mantis shrimp, lived in the Birmingham Zoology Department. It took Dougal roughly twenty variously successful attempts to get pieces of dead fish from inside closed Petri dishes. One technique it used was breaking the glass, after which plastic dishes were substituted. Then it adopted one of two successful strategies, depending on which way up the dish was. It adapted these techniques instantly to 'nested' Chinese-box Petris and to very large ones. It took only five attempts to learn to unroll plastic scrolls, and everybody present had the strongest impression of a nasty look when the string was replaced with an elastic band – which Dougal succeeded in pulling off the end at the third attempt. It would play with any new item in its tank – including live ones: soon only two large predatory poisonous-skinned fishes remained as its companions – and if anything the amount of play increased after starting it on these tasks. Everyone, especially the laboratory staff who had most to do with the animal, agreed about its intelligence and its very positive reactions to these tasks. It obviously enjoyed them, that's the way JC thought about it then. *Now* he thinks that there wasn't an 'it' to do the enjoying – at least not at the start.

Zoology students on marine field courses sometimes caught little cuttlefishes and octopuses, and occasionally these could be kept alive on the 'wet bench' – a bench equipped with tanks, bowls, and dishes, brimming over with running seawater. Here the animals would convince students of their intelligence. They would change colour in a flash to match new environments, but they would also flash a pair of large black 'eye spots' at large animals, creep up behind shrimps to catch them, 'build' a den to hide in ... The larger tropical octopuses, kept in a tank in JC's room, greeted different graduate students appropriately – one often brought them little crabs, and they would come out to meet him; others just poked them, and were not greeted as enthusiastically, if at all. The octopuses behaved like we expected rats and mice to behave; not like animal Einsteins, but more intelligently than anticipated. Visitors were particularly impressed. At the time JC believed that these animals were, to some very small extent, partners in the interactions. Now he doesn't: the word 'partner' carries the wrong overtones.

It is as bad to underestimate the mental abilities and sophistication of animals as to overestimate them. The recent literature on animal behaviour is replete with examples showing that animals that we thought were stupid (like pigeons) or just organic automata (like worker bees) possess a cognitive map of their surroundings. They can take short-cuts in real space if a new path is opened, as we have described for bees in Chapter 6. Because they do this immediately, rather than learning it laboriously from explorations of the outside world, the change in geographical route must be triggered by events inside their brains. So the same behaviour shows that they can take short-cuts in mental space if shown new clues. This ability strongly suggests that the animal's 'self' is somehow represented in these maps, since it appears to know where *it* is. Similarly, the chess player may put himself in the position of a knight and map out alternative moves, and someone playing a computer game may 'become' the hero. But this is quite different from the knight, the software hero, or the mantis shrimp being conscious of its own existence as a continuing 'I'. This is a subtle point – well, it ought to be obvious but many people must find it subtle since they often get it wrong, just as JC did with Dougal – and it must be made clear.

Suppose you are playing a computer game, in which various 'characters' on the screen interact with your commands. *You* know you're playing a game, and that 'the hero' is responding – apparently – to the virtual situations that he is going through. But you also know that, rational and intelligent as his actions seem to be, he is both your puppet and the pawn of the computer's program. He doesn't know what he's doing – there's no 'he' to know it – whereas you do. Bees, mantis shrimps, and octopuses are more like the computer game hero than they are like you, and when you and they play games together it's mostly you playing the game and them following their inbuilt rules as usual. Except, perhaps, for *pet* octopuses and mantis shrimps, who have learned tasks: they are changed by their experience, they really do seem to be enjoying solving the puzzles you set. Sometimes when you set up a new task they get into a 'ready' position – so they might perhaps be a tiny bit conscious ...

There is a general problem with evolutionary terminology and viewpoint: we seem to need to find 'rudiments'. It is easy to fall into the trap of assuming that nothing can come into being without precedents. If so, everything that is around now must trace back almost indefinitely. However, emergent phenomena do not possess rudiments: that is what makes them emergent. And consciousness is emergent if anything is, so it may be difficult to find a convincing evolutionary story for the gradual evolution of consciousness. If it is a 'threshold' phenomenon, a sudden expansion of the phase space for brain function, then on an evolutionary timescale it would appear to happen virtually overnight. The underlying hardware – wetware – might evolve continuously, but an emergent feature like consciousness can appear only when that evolution reaches a critical 'trigger point'. So we should not expect to find any obvious evolutionary continuity of consciousness – but we should expect to find a general tendency towards more complex brains with more nerve cells, more memory, and faster transmission of nerve impulses. Which we do. For the same reason, we should not expect to find a series of organisms that fills the gap between non-conscious and conscious animals in a continuous manner, any more than we can find a series that fills the gap between two-legged and four-legged locomotion – we might just hope to plug the gap at three legs (old man with a stick?), but three-and-a-half legs remains glaringly open.

Whatever consciousness may be, our minds certainly have it: we are all well aware of this – we know the 'inside story'. Few people would deny the same of the great apes – gorilla, orang-utan, both species of chimpanzee. Most of us would accept the same of many other social primates; some would include dogs but be less sure about cats, and it could also be true of some parrots (see the title story for Chapter 10). This series of animals, by the way, does not contradict our assertion that no continuous series between conscious and non-conscious is likely to exist: if there is any continuity here it is in our assessment of how *likely* such animals are to be conscious, not in how conscious they are. It is also worth pointing out that a scale of animals that are alive today is not in any sense an evolutionary scale. If it illuminates the evolution of consciousness it can do so only indirectly.

We cannot *know* what animal consciousness is like. If only we could get inside their heads and find out– but we can't, and even if we could, it wouldn't work. In order to find out exactly what it 'really' *is* like to be a bat you have to *be* a bat, not a human mind somehow downloaded into a bat. (Similarly, if you could download a fish's mental outlook into a land animal, you wouldn't get a fish that appreciated the terrestrial mode of living – just a fish out of water.) Because of the logical impossibility of mental downloading, not just its practical impossibility right now, we have to infer what's in there from the animal's behaviour, and we're unsure of the accuracy of those inferences.

We know that *we* are intelligent. From behavioural inferences we can be pretty certain that so is the octopus. We know we are conscious: behavioural inferences strongly indicate that the octopus is not. Why is there this difference? Like all of the animals in the above list, we humans have interacted with other members of our species and changed ourselves: our sense of identity has changed recursively as each interaction builds on the results of previous ones. The octopus, in contrast, is a solitary creature which does not engage in that kind of social interaction. Each individual octopus learns, but it does not and cannot pass that learning on to its offspring: it has to reinvent the same wheels all over again. Parrots, cats, dogs, apes, and humans have to reinvent *some* wheels over again – the constant bane of parents watching their children repeat the same mistakes that were made a million years ago, and every year since – but they definitely get a head start. Kittens are taught hunting techniques by their mother, and humans are taught how to weed the flower-bed by their parents. This observation leads to a wonderful paradox: a creature that seems to itself and to others to have a continuing identity – an 'I' inside that remains the *same* 'I' as time passes – is constantly changing itself. One that does not have a 'person' inside can learn, but it cannot change: its identity is far more stable, and it really does remain the same 'it' as time passes.

This distinction between 'I' and 'it' is much the same as that made by Freud, except that he used the Latin words 'ego' and 'id'. He invented the idea of a vast subconscious part of our minds, with the ego only a tiny observer in a little boat awash on a turbulent sea of unconscious brain activity. Lurking in the depths like a giant monster is the id, ready to overwhelm the ego with primitive urges such as sexual libido. Hovering anxiously in the skies is the chaperone superego, ever ready to pour conscience on the little ego and save it from sin. It is not necessary to swallow Freud whole to recognise that he had an important idea about minds. Despite our strong feeling that there is a unique 'I' that is experiencing everything, in charge of everything, and taking the decisions, actually our minds carry out many activities of which we are totally unaware. As Dennett emphasises, there seem to be a host of semi-autonomous units, sub- or co-processes, all going on inside our brains *in parallel*. Our illusion of having 'a' mind may be necessary to bind our ideas together into one conception of our world. Both Dennett and Marvin Minsky have emphasised that our minds have many subroutines – largely autonomous sub-minds. A few paragraphs back we added one extra twist, which we called the 'ringmaster'. The ringmaster is a rationaliser, with an answer for everything. The things that it rationalises and reconciles are features, either reported by other demons or cooked up by the ringmaster itself. The visual system works in just the same manner whether it is directed at a cow or a meaningless set of random dots, but in the first case its deeper processing layers eventually end up causing a 'cow' cell to fire – or more likely there's a neural net that pulls an 'eigencow' out of the visual data, like facial recognition in Chapter 7. Whichever, at this point the ringmaster gets a tap on the shoulder, accompanied a split second later by a whisper in the ear from the language facility: 'that thing is called a COW, remember?' Now – and only now – does the ringmaster become aware that there's a cow in the field. But he rationalises the sequence of events into '*I* looked into the field and *I* saw a cow'. That's how the ringmaster works. It's not a homunculus who has wandered into the Cartesian Theatre to see a cow movie.

How did the ringmaster evolve? It is possible that its development was triggered in mammals when the dinosaurs died out, an idea that goes back to Jerison. Recall from Chapter 1 his suggestion that mammals were originally active during the day, but became nocturnal for their own protection when the dinosaurs took over. Originally their visual senses were dominant, although they could hear as well. So their brains evolved a very effective capacity to detect visual features, and a *separate*, less effective capacity to detect features in sounds. But when mammals were forced to become nocturnal, sounds were their main source of information about their surroundings; so the sense of hearing improved rapidly, along with its feature-detection system for extracting meaning from what was heard. Their visual senses faded into the background, and their visual-feature detector became 'available' for auditory patterns. Then, when the K/T meteorite wiped out the dinosaurs and the tiny mammals emerged blinking into the sunlight, they were in possession of effective vision *and* effective hearing. Now nerve cells have an interesting property, whose importance was promoted by Donald Hebb: *cells that fire together, grow together*. So the brain's cellular architecture automatically associates neural pathways that tend to function in synchrony. But why *should* two distinct neural pathways function synchronously? By chance? Not at all: because they are both responding to the same thing in the outside world. So the brain short-circuits that external loop, creating an internal loop to mimic its effects. Jerison argued that this same process, carried out repeatedly, can provide a method whereby the evolving brain can, and will, improve its ability to pick up common features from many different sources of sensory input. This, perhaps, is how the mind's ringmaster got put together.

At any rate, the ringmaster is a master-rationaliser. So what happens if (when!) it directs its rationalising propensities at itself?

It becomes aware of an *apparent* 'I' inside.

This is where self-awareness comes from: it is what you get when a generalised feature-detector makes a recursive attempt to detect itself. In short: the problem of self-awareness is a special case of awareness – feature-detection – in general. As soon as such a system recognises some aspect of 'self' as a feature, hence the kind of thing that it can detect, the recursive loop is closed.

We repeat, yet again: the ringmaster is not the 'self' itself. It is a mental demon involved in creating the illusion of there being a self. 'Self' is something that is distributed throughout all of the processes that make up the emergent mind, just as 'running the program' is distributed among all of the individual demons in a computer. Recall from Chapter 6 the quote of Heraclitus: 'you cannot step into the same river twice' (the water is different next time round), and our gloss on it: 'you cannot step into the same pond twice' (it's a different *you* next time round). 'Self' is not a thing, but a process, which preserves an apparent sense of identity even as it changes complicitly with everything around it, both inside and outside the mind. As time passes, what seems to be the same 'you' changes, but it does so with enough continuity that it still *seems* to be the same 'you' even though it isn't. Analogously, think of the fish in the river: they too stay 'the same', although they are learning, accumulating 'knowledge' of the river, and so on. And the fisherman on the bank looks the same, from one day to the next, to the fish, even though he too has changed. In a way, the presence of the fish maintains the apparent continuity of the fisherman, just as his presence maintains the apparent continuity of the fish. In a similar manner, environment and culture maintain the continuity of the human sense of self, and that, repeated across many individuals, in turn maintains the continuity of environment and culture.

That is what it is like to be a human, and we have answered our title question. Now, finally, we are ready to take up the bat again. What *is* it like to be a bat? Nagel said we don't know, can't be sure, but it would be worth trying to start; nearly all of his readers thought he said we could never find out no matter how hard we tried. To be sure, there is no way we can *know* what battiness feels like to the bat that is experiencing it, but as we've just said, there is no way we can *know* what an electromagnetic wave is *really* like, either. In both cases, however, we can approach the question by way of analogies with what we ourselves experience. It is how even the leading physicists approach the quantum world – they never really come to grips with a quantum universe, instead they use analogues like 'wave function' to capture odd bits and pieces, more or less accurately. And so it should be with bats. At least, let's try: it's really more informative than you might think. For simplicity we will concentrate on one aspect of bathood, the reason Nagel chose the bat: its use of sonar to sense the world around it.

The big mistake is to set up false analogies. For example, we might start with the idea that sonar is sound, so the bat's sonar qualia might be like music, or rapid-fire Morse code, or something else analogous to human sound qualia. However, there are two good reasons why this analogy is likely to be false. The first is that bats' sonar qualia evolved mainly in order to help the animal navigate in three dimensions (and detect prey). It is presumably the bat's most vivid sense, inasmuch as that makes sense, because it is the one that is in constant use. The analogous sense in humans is not hearing, but sight, which serves the same purposes. The second is that we know a lot about the architecture of both human and bat brains, and the part of the bat brain that processes its sonar signals is very similar to the human visual cortex.♪ In short, bats *see* with their ears, and their sonar qualia might well be like our visual ones. Intensity of sound might come over to the bat as a kind of 'brightness', and so on. Possibly the bat's sonar qualia 'see' the world in black and white and shades of grey, but they could also pick up and render vivid various more subtle features of sound reflections. The closest analogy in humans is *texture*, which we sense by touch, but the bat could sense by sound. Soft objects reflect sound less well than hard ones, for instance. So bats may well 'see' textured sound. If so – and here our analogy is intended only as a very rough way to convey the idea – the sonar quale for a soft surface might 'look' green to the bat's mind, that for hard ones might look red, that for liquid ones like a colour only bats can see, and so on ... We agree that this description makes no rational sense – but it's the best metaphor we can come up with.

Just as the equations used by mathematical physicists to describe electricity and magnetism are the best metaphor for an electromagnetic wave that we can come up with – and fall equally short of capturing its essence.

9 We Wanted to Have a Chapter on Free Will, but We Decided not to, so Here It Is

A senior Royal Air Force officer had organised an official reunion for World War II veterans, all in full dress uniform, covered in medals and ribbons, aged about seventy. The highlight of the event was a fly-past of restored aircraft – Spitfires, Lancaster bombers, and so on – and he stood in front of the veterans to watch them. Suddenly, sensing something odd, he turned round – to find that the veterans had disappeared. Then he realised that they were all lying flat on the grass. The explanation?

A Fokker (a WWII German fighter) had roared across the field, flying *low* ...

'It would be very singular,' wrote Voltaire, 'that all nature, all the planets, should obey eternal laws, and that there should be a little animal, five feet high, who, in contempt of these laws, could act as he pleased, solely according to his caprice.' It is an eloquent statement of the problem of free will, and it is the place where our figments run slap up against reality, like the proverbial irresistible force meeting the immovable object. We have a distinct, overwhelming impression that we have a free choice concerning the actions that we take: free, that is, subject to the evident constraints of physical law. We cannot choose to float into the air, for example. Yet there is absolutely nothing in the inorganic world that possesses that kind of freedom. Even a computer, however intricate, runs a program and carries out the instructions therein: it does not have a genuine choice about what to do next.

Indeed this is the classic 'clockwork world' image of the inorganic domain of physics and chemistry: bound to obey fixed, inexorable rules. To be sure, the quantum rules have a statistical interpretation, but the quantum wave function upon which the rules operate is a deterministic mathematical object. The indeterminacy comes only when you try to measure it, and the whole area is so full of fudges and unwarranted assumptions about the nature of measurement that a quantum solution to free will is no more plausible than a quantum solution to consciousness. Which, despite the brilliant arguments of Penrose, is not plausible at all. So how can a process that operates by obeying fixed physical rules make choices? On the other hand, if we do not really make choices, is it right to punish criminals – who, after all, are merely doing what the rules of the physical universe force them to do?

Do we really have free will? What do we mean by that? If we do, where does it come from? If not, why do we feel so strongly that we do, and why does our behaviour give everybody *else* the strongest impression that we do?

We'll warm up by looking at free will from the outside – embedded within human culture. Later we'll dig inside the mind and try to work out what's involved. In today's western culture, the question of free will is being obscured by an unfortunate side-effect of the simpleminded mental image of DNA as blueprint – a kind of genetic determinism. This adds genetic reinforcement to a growing cultural tendency for people not to accept responsibility for their own actions, which in a sense is an attempt to deny that they have free will.

In the United States a burglar who fell through a skylight successfully sued the owner of the house he was trying to rob.

In Britain a man who punched someone he caught stealing his wife's underwear from the clothesline was sent to jail for three months. The thief was set free with a small fine.

In Cyprus three British soldiers who raped a young woman and beat her to death with a spade pled in their defence that they were so drunk that they did not know what they were doing. The Cypriot legal system wasn't impressed, but the defence lawyer clearly thought it was worth a try. In the very near future – if it has not happened already – a man will get drunk, kill somebody, and be acquitted because while drunk he is not responsible for his actions.

In only the slightly more distant future, a man will kill somebody over a trivial argument, and be acquitted because he possesses a 'gene for aggression'. Those who *lack* a gene for aggression, yet fight back when attacked, will have no such excuse. If they are unfortunate enough also to possess a gene for rational decision-making, they will receive punitive sentences. The meek shall not inherit the Earth, because as soon as they do some mean-minded bastard will take it away from them and the law will condone the theft. He couldn't help it, *his* genes are those of a mean-minded bastard. It was *their* fault for being so provocatively meek.

Heaven help us all.

The law, as Mr Bumble said, is 'a ass'. And when it comes to genetics it is in serious danger of being not just a ass, but a accessory – to its own flouting. The simpleminded genetic determinism that is being more and more widely espoused is nonsense. (It is leading to the same mistake in education, writ large and in spades.) The underlying idea – that gross human characteristics such as aggression are somehow caused by a single segment of genetic code – stems from a grievous misunderstanding of human development. The genome is more a recipe than a blueprint, and the ingredients and the skill of the cook are at least as important. A true human being has free will – in the only sense that matters, the relation of an individual to their culture – and is in control of its own destiny. A man who knows he gets aggressive when drunk, and kills while drunk, can try to excuse the murder – but he has no excuse for the drunkenness that he himself claims caused him to kill, because when he chose to get drunk he was sober. People who cannot control their tempers when drunk should consider themselves as being under a greater social obligation not to drink than those who can. Far from being a defence, drunkenness should compound the crime.

Disagree? Well, think not of a drunk, but of someone who is such a bad driver that they have been disqualified from taking the wheel of a car. One day they get into a car, crash into a bus queue, and kill twenty people. 'It wasn't my fault, I'm *naturally* a dangerous driver.'

Or think of a drunk driver doing the same. *Two* excuses.

Apparently it's OK to disown responsibility for wielding a knife when drunk, but not for wielding a car. Moreover, it seems horribly plausible that if the law ever notices this discrepancy, it will feel obliged to condone drunk driving. At least then the meek may get fair treatment again, on account of their unfortunate gene for meekness. It *isn't* their fault after all.

We (JC & IS) see it all very differently. The message is one of social responsibility. Stay away from cars if you're drunk, stay away from knives if you're drunk. And if you're naturally a dangerous drinker, stay away from drink. It's up to you. Are you a real person or a cardboard cut-out?

Even if you think that all human behaviour is ultimately genetically determined, that still provides no reason for excusing murderers. If genes do actually correspond to characteristics, then as well as there being a gene for aggression♪ there must also be a gene for controlling one's aggression, a gene for avoiding getting into tense situations, a gene for considering the effects of one's actions on other people ... a gene for taking responsibility. And if you are unfortunate enough to lack all of these genes, and I put you behind bars, don't complain: it's just that I happen to have a gene for incarcerating killers, it's not my fault.

If this sounds nonsense, that's because it is. Genes don't work like that.

And neither does the alleged 'gene for aggression' upon which casual killers – aided, abetted, and encouraged by their lawyers – are starting to pin their hopes of freedom. What you are is not written irrevocably in your genes. To be fair, some of it is: in a sense genetics determines a baseline from which upbringing, education, and culture start. So any fair-minded person ought to have a bit of sympathy for someone with a seriously biased baseline, especially if their upbringing was – through no fault of their own – deficient. They genuinely may not know better.

Even so, they should still be able to learn. If their baseline is so biased that they can't, then they need protection from themselves as much as we need protection from them. They really can't have it both ways, retaining the freedom accorded to a responsible member of society precisely because their lawyer has established that they are *not* a responsible member of society. We all grow up in a culture that values certain kinds of behaviour, and encourages them, while discouraging other kinds of behaviour. Within reasonable limits, whatever our genetic makeup, we can learn what our own weaknesses are and do our best to compensate for them. If we do not, then we do not deserve to be treated as a rational, self-controlled person. We are merely a dangerous machine, and as such we have lost any right to complain if, when we go wrong, a real human being decides we should be turned off.

The same goes for government officials. In the UK, it has in recent years become routine for ministers to separate policy (for which they are responsible) from provision (for which somebody *else* is responsible). The minister then accepts responsibility only for the overt *objectives* of policy – not for the consequences of policies, and not even for having appointed those who have to carry the policies out. For example, suppose that the Home Secretary sets policy by increasing the severity of jail sentences for crimes of violence. Subsequently there are riots in prisons, which have become overcrowded as a direct result of that policy decision. He then argues that this is not his fault, but that of the prison service, who as providers have failed in their duty to provide adequate prison places.

Ah, but it isn't it the Home Secretary who sets the amount of money that the prison service can spend on prisons?

Yes, but it's *still* their fault. Nowhere in the policy does it say *explicitly* that they should build an inadequate number of prisons.

It's a neat trick for avoiding responsibility, but ultimately it suffers from the same flaw as the 'gene for aggression' defence. *That* sort of Home Secretary is a serious danger to himself and the public, and he should be removed from office without delay. And voters, faced with the sort of government that can stomach keeping such a person on as Home Secretary, had better remove that government from office, because it poses an even greater danger.

We are social creatures, and we cannot permit ourselves to behave as if the rest of the world does not exist. Without responsibility, one person's freedom becomes another's captivity – so freedom entails responsibility. Indeed, as the SF writer Robert Heinlein repeatedly made clear in his writing, freedom *is* responsibility; and so did Robert Persig in *Zen and the Art of Motorcycle Maintenance*. Real human beings accept responsibility for their own actions.

So do real Zarathustrans – whether they want to or not.

Creator-of-creations Ringmaster, there has been an unauthorised abstraction of communal property.

Ringmaster Pardon?

CoC Somebody has eaten *Watcher-of-Moons*' entire supply of groose-grease paté.

R Who has done this appallingly anti-octual thing?

CoC I regret to inform you that the evidence points overwhelmingly to Liar-to-adults.

R Has he confessed?

CoC Willingly. But he requests an audience to explain his position.

R The consensus must be that his position is inexplicable, but in the interests of justice he deserves a fair hearing / immediate incarceration (delete whichever is inapplicable). Bring him to me. [*There is a pause, and Liar-to-adults is dragged into the room, his crest-feathers symbolically and humiliatingly decorated with discs of thin metallic foil.*] Liar, what do you have to say for yourself?

Liar-to-adults [*Grovelling.*] Ringmaster, it was not my fault.

CoC But we have video evidence showing you gorging yourself on groose-grease paté until the juices ran down your toes!

LtA [*Offended.*] I did not say I did not *eat* the groose-grease. I said that it was not my fault.

CoC It was the groose's fault for being so unavoidably edible, perhaps?

R Hrrmf. Let me handle this, Creator. Why was it not your fault?

LtA I am zenetically predisposed towards gluttony. My zenome includes an overactive zene for eating.

CoC What absolute twad–

R I *said*, let me handle this. [*To Liar-to-adults.*] That is unfortunate. Terrible misfortune, I *do* sympathise most deeply. [*Suddenly remembers he must ask a standard legal question.*] Naturally you can prove your contention?

LtA I have brought my personal zenome-scan. It has been verity-stamped, see here on the side?

R [*Peruses it.*] Yes, it is true.

LtA Then I shall be set free?

R Sadly, no, though it grieves me desperately to see the injustice involved in incarcerating one who is zenetically challenged in such a manner.

LtA But why should I not be set free? My case is undeniable.

R Sadly so. You see, while it is of course in no way your fault, it has been found that there is a strong zenetic linkage between gluttony, especially when it is expressed as unauthorised consumption of common goods, and autoincarceration.

LtA Pardon? What is auto–

R An unavoidable zenetic tendency that causes others unjustly to place one in close confinement. I regret that my actions are just as irretrievably constrained by your zenetic idiosyncracies as yours are. [*Pauses.*] Take him away!

Having examined the cultural dimension, let us now start to dig inside a mind and try to discover whether it might possess genuine free will – and if so, how; and if not, what do we do about it?

The old idea of determinism, which gained irresistible momentum under Isaac Newton and was formalised by the Marquis de Laplace, boiled down to a simple property of any mathematical system whose dynamic did not involve randomness or choice. Given an initial state, chosen once and for all, and a fixed rule for the transition from the present state to the next, then the entire sequence of subsequent states must *also* be determined once and for all. Laplace pointed out one consequence of this idea on the grand scale: given the state of the universe *now*, its future is completely determined, forever. There is no choice, no free will, and no loophole. Voltaire's complaint is answered in brutal fashion: the five-foot high animal does not possess free will, he merely thinks he does. Indeed part of the brutality is that the working out of the universal laws requires him to think he does. We are clockwork automata in a clockwork universe – worse than zombies, we imagine ourselves to be conscious beings when we are not. We are that staple of science fiction, the android that thinks he is human.

We said 'no loopholes'. Actually there are at least two places where Laplace's reasoning falls down, and both are often offered as an explanation of free will. We think not, but to explain why, we must first set the scene. The first loophole is chaos, the second is quantum mechanics.

The concept of chaos arises by following Dennett's injunction to replace vast perfections 'in principle' with what is actually performable in practice by us imperfect beings. There are two difficulties in implementing Laplace's argument, as a way of actually *predicting* how the universe will behave in the future. One is fairly obvious: the calculations become so complicated that they can't possibly be carried out. Laplace is inviting us to traverse Ant Country, and we can't. The other difficulty is less obvious: it is that we cannot determine the *present* state of the universe accurately enough. The difficulty does not lie in passing from the present to the future, but in determining the present. The discovery of chaos shows that small errors in the present often generate huge errors in the future – so huge that the prediction is meaningless. This is the infamous 'butterfly effect' – one flap of a butterfly's wings can change the weather completely a month or so later. This is a metaphor, not a recipe for an experiment: it is intended to emphasise that if the universe could be run twice, with a single very tiny change the second time round, then it would behave *very* differently on the second occasion. No one ♪ is saying that in the real world there is a butterfly somewhere that changes the weather. The 'butterfly effect' is a statement about the phase space of a chaotic system, about potential behaviour rather than actual.

Does chaos offer a loophole in Laplacian determinism big enough to drive free will through? It certainly opens up a big hole – but that hole admits *unpredictability* into a deterministic system, not choice. Suppose you are a cat, faced with a choice between new Miaow with nourishing minnowbone jello, and Brand Y, which consists of pictures of fish on cardboard cut-outs. Given free will, you would feel that you actually have to make up your mind, that you really could eat Miaow *or* Brand Y if you so decided. (Then, as any real cat would do, you turn up your nose at both and exit via the catflap, or pig them both ... but that's going beyond the philosophical bounds of the discussion.) If your brain were chaotic, in the mathematical sense, that would explain why nobody else could predict which brand you were going to settle on – but it wouldn't offer you a genuine choice. On the occasion you chose Brand Y, that was predetermined – only nobody could actually perform the calculation to see *what* was predetermined. If you were faced with the identical choice again under identical circumstances – but 'you can never eat out of the same cat bowl twice'. The butterfly effect would scramble the possibilities, and whether you chose the same thing again or not would prove absolutely nothing. Worse, if the chain of causality happened to traverse Ant Country, then even in principle nobody could perform the calculations needed to work out, in advance, what choice you were going to make. So if chaos were the answer to free will, then choices would not actually *be* choices: they would be random tossings of God's deterministic but chaotic dice. The thing that we mean by 'free will', whether genuine or an illusion, isn't like that; so chaos is not the answer to free will.

The same goes for quantum effects, which (allegedly) render the physical rules indeterminate. Let's accept that they do (although this is by no means as rock solid as most physicists claim). The snag is that quantum indeterminacy bears little resemblance to choice. The standard quantum example is the feature of an electron known as 'spin'. The usual image is that of a little ball rotating about an axis, but that's a classical image, not a quantum one. An electron's spin, about any chosen axis, is always either ½ unit or −½ unit (clockwise or anticlockwise relative to an oriented axis). The typical spin state of an electron is a 'superposition' of spins about various axes – say spin ½ about the northern axis *plus* spin −½ about the eastern one. The spins do not cancel out: they are along distinct axes. Now suppose you decide to measure the electron's spin. Then, miraculously, the result of that measurement is just *one* of the component spins. On one occasion you get spin ½ about the northern axis, on another you get spin −½ about the eastern one. You get one or other of the so-called 'pure' states that are superposed in the 'real' electron.

Always.

And as far as anyone can tell, which pure state you get is totally random – 'God playing dice,' as Einstein complained. That doesn't sound much like *choice*. The electron doesn't choose which pure state to be measured in: on the contrary, the measuring technique wallops the superposed wave function and squashes it down to just one component pure state. And that causes difficulties for a Penrose-like approach to free will, parlaying quantum indeterminacy into an ability to choose. Quantum systems involve indeterminacy, the argument would go, and free will involves indeterminacy; so if you stack enough little bits of indeterminacy on top of each other you get free will. Yes? Sorry, no. You get an incoherent mess. Quantum measurement is a bit like having two wooden shutters that open on to a windowsill, upon which rest a number of flowerpots. You can pull the shutters inwards, or push them outwards – but when you try to open them you have no control over which direction they take. Before making the measurement, the shutters are closed, and flowerpots are scattered along the sill. Is there a flowerpot behind the left-hand shutter? You fling it open – and see nothing. Now, did it open inwards, meaning there never was a flowerpot there, or did it open outwards, and push it off the sill? In the real world you could look down to the ground and see if there was a shattered flowerpot; but in the quantum world you don't have that luxury – if you disturbed the flowerpot, all trace of its existence will be gone. The point of this image is that the measurement process interacts with the quantum state in a manner that is not at all well understood, and we really have very little idea what happens between the original superposed wave function and the final 'measured' pure state. Indeed some physicists deny that anything describable happens. At any rate, just as for chaos, a quantum-indeterminate system does not make coherent choices: it tosses dice.

So whatever the 'answer' to the indeterminacy of free will might be, it cannot be just a matter of chaos or quantum mechanics. Of course it could be such things in conjunction with something else, but then it is the something else that holds the key to the riddle, and we can probably get rid of the chaos and quantum mechanics and still make the thing work. The core problem is how to put genuine choice into a deterministic system.

Mathematically, that's impossible. But mathematics is one thing, and its interpretation may be quite another. So let's think a bit about how we humans make choices, and about how it feels. Maybe that will provide some useful clues.

Our chapter title exemplifies the dilemma that such introspection opens up. It is a bit like being asked to 'be natural' when someone points a TV camera at you. Like the caterpillar who 'inquired which leg goes after which' and lost the ability to run, you promptly sit there thinking 'Be natural. Be natural.' Since you have no conscious idea how to evoke all of the subconscious thoughts involved in being natural, you can't do it any more. So you stick a silly grin on your face and end up looking as if you were drunk. Similarly, the more closely you examine allegedly 'free' choices that you make, the more you begin to wonder if you really had a choice at all. For example, at one stage during the writing of *Figments*, quite near the end, this chapter did not exist at all. It wasn't in the outline plan, it wasn't in our heads – it just wasn't. Then, lunching in the pub and discussing the book, as was our custom, we suddenly came up with the central insight that this chapter is leading up to. At that point we *chose* to add an extra chapter to the book, this one.

We could have chosen not to.

Couldn't we?

Well ... even as I (IS at this point) write, I'm not so sure. I could wipe the draft of this chapter from the computer – but I'm not doing that, which is why you're now able to read it. I'm not wiping it because the context for my choice – both mental and cultural – makes it rather obvious what I'm going to 'choose'.

Jack (JC) and I agreed in the pub that I would add this chapter. That may have been a free choice, but if so, it constrained my subsequent choices. Being the sort of social animal that I am, I really couldn't face Jack's wrath if I didn't do what I'd promised. Of course I could *still* wipe it, just to prove how free my will is – but then I'd have wiped the chapter because I was constrained by the line of argument that the book was taking, driven by my sudden need to demonstrate my own autonomy.

And did Jack and I *really* decide, freely, to put the chapter in? In the context of our sudden realisation that we *knew* what free will was all about – right or wrong, that's not the point – it followed as night follows day that we had answered one of the main questions that this book *must* tackle. Up until that moment we hadn't noticed the omission – though we did have about seven pages of fragmentary writings about free will and genetics stashed away from an earlier bout of authorial *angst* – but suddenly it was obvious. And so was the remedy. We didn't choose to put this chapter in: if anything, it made the choice for us. Wild horses wouldn't have been able to *stop* us putting it in.

Yes, but ... we could leave it out, couldn't we, just to prove that we had free will? Not to prove it to you – if this chapter wasn't here, you wouldn't read the proof. Constrained by context again, bother. Yes, but surely we *could*, just for our own satisfaction ...

Not if you believe that Dennett is right about the mind being pandemonium. By the time the ringmaster registers the 'choice', one of the demons – or a whole caucus, democracy, or riot of them – *has already made it*. But a demon is not individually conscious, so the choice was made unconsciously before it seemed to be made consciously. This is a genuine problem, not just conjecture. Experiments have been done♪ in which a subject was wired up to a brain scanner or a set of electrodes that detect brain activity, and told to carry out some task 'whenever they felt like it'. They were also asked to push a button to mark the first instant that they decided to carry out the task. Universally, their brain activity changed a split second *before* they pushed the button to indicate the moment of 'free' choice. This was the demons at work, behind the scenes, rigging the ballot before a vote was cast.

How on Earth can pandemonium 'make a choice'? Think of a crowd of people in a building, forced to make a sudden evacuation, perhaps because of a fire alert. There are a number of exits, and the crowd must 'choose' how to divide itself among these exits. At first sight, you might think that the crowd's 'choice' is just the consequence of a mass of individual choices, as each person selects an exit. But in a sufficiently crowded building, that isn't what happens. firstly, the choices made by individuals are not genuine choices at all: for most of them there is an obvious exit, such as the nearest one, or the only one they can see, or the one that they can see an EXIT sign for, so they head that way. Secondly, many individual choices may have to be revised as the crowd surges to and fro: a person may head down a corridor only to be confronted with a mass of people heading towards them, cutting off their exit, and so on. In practice what happens gives an observer the distinct impression that the crowd develops a 'mind of its own', and takes a collective decision which is based upon the geometry of the building and the distribution of people within it. (In fact, computer simulations of crowd flow♪ show that these are the main factors influencing the overall patterns of crowds: it makes hardly any difference at all whether the individuals apply 'intelligent' rules or rough-and-ready ones, for instance.) Is this a *genuine* choice? Probably not – but we have not yet decided whether the apparent choice made by an individual is genuine, either. On the level of the people concerned the motion of the crowd looks like pandemonium in the usual sense of the word, but on the level of the *crowd* it looks like a decision between alternatives – a choice. Which is exactly the point. Pandemonium on the small scale can translate into concerted action in the large. There is a general principle about analogies: it is *roles* in the analogy that must be matched, not the nature of constituents. Here the analogy is between a mass of demons in the mind of a person, and a mass of people in a crowd. The appropriate 'dictionary' is

crowd → person
person → demon

and not

person → person

so the 'choices' made by individual people in a crowd are not analogous to the 'choice' made by the crowd itself. This is what we meant when we said that the collective movement of the crowd seems to have its own 'mind'. Similarly you often hear talk of 'mob psychology' in a riot. By this kind of language we do not mean some mystic melding of the minds of individuals – wrong level of analogy again. We mean that there are emergent features of the crowd movement. They are *caused* by the actions of individuals, without doubt, but the chain of causality runs through Ant Country. Since we cannot 'read the ant-trails', as Performer-of-amusements put it, we get the impression that the crowd's 'decisions' bear no relationship to those of the individuals. Not so: they just bear no relationship whose causality can be tracked in detail.

We have the strong impression that we make choices, but actually most of the time we make judgements – we call them 'rational decisions' when we are conscious of the thought processes involved, but often the thought processes are subconscious, and then we think we make choices. Making a judgement involves running alternative imaginary futures and selecting the most favourable outcome. It is an ability with obvious evolutionary survival value – what *if* there's a lion behind that rock? What *if* the rains don't come on time? – and it has therefore been reinforced by natural selection. In a judgement, we consider a hypothetical decision – nowadays a typical one would be between going out to a restaurant for dinner or buying fish and chips and bringing them home. Various mental demons compare the influence of various factors: if we go out to dinner there's no washing up to do, but fish and chips are cheaper; yes, but we haven't been out to dinner for weeks and we had fish yesterday ... and so on. Other demons 'weigh' these factors against each other. We are seldom conscious of the precise methods they are using, because we are not privy to most of our demonic crosstalk, but it probably is *not* just a kind of score-sheet where the highest total wins. However, a bit of introspection makes it clear that in some manner 'we' – our collective of demons – work out which decision leads to the most desirable combination of features. If the difference between the two choices is sufficiently great, then there is no choice to be made, and we arrive at a judgement – *obviously* this time it's out to dinner. Only when the difference is small do we have to *make* a decision, and now a whole pile of other demons rush into action: was the waiter polite last time we went to the restaurant, was the wine from a politically correct source, do we *care* about political correctness anyway... ? Quite likely the judgement demons are once more invoked on a new, refined level; the whole process may well be recursive. And if we are really stuck, the 'toss-a-coin' demon finally becomes heard above the pandemonium, and that's what we do.

Our mental demons interact in a much more sophisticated way than do people in a crowd, and they obtain much more sophisticated results. They constantly refer activities to each other, cross-check for contextual constraints, leaf through memories for precedents, pull time-honoured subroutines off the racks and run them below the conscious level. This results in a 'flow pattern' that really *does* have 'a mind of its own'. Every element of your mind is constrained by context, the same context that created and maintains your sense of being 'you'. So the 'choice' that 'you' make is consistent with 'your' particular 'personality'.

When we say 'Oh, but Mary would *never* do that,' we are expressing exactly this insight. When Mary *does* do that, our initial shock is replaced by a search for reasons. 'Oh, but she had a row with her boyfriend last week.' Partly this is the ringmaster springing into action, trying to rationalise Mary's uncharacteristic decision; but it also shows that we do *not* actually believe that Mary had a free choice!

From this point of view it is very unclear whether any of us ever has a genuine choice: whether we *could have* done the opposite to what we finally 'chose' to do. Indeed it's unclear that this question has any operational meaning. As a theoretical question about the workings of minds, however, it has a definite meaning: if we ran a mind twice, under identical conditions and starting from identical states, would it do the same thing both times? It is a question about the dynamic in mind-space: is it deterministic or not? It is not a question that we can answer experimentally, however, because the mind-space dynamic is chaotic and ant-countrified. The tiniest discrepancy in setting up the mind's state and external conditions the second time round could cause the mind to behave totally differently, so we wouldn't know if we had observed a real difference in what was chosen.

The argument seems to be heading inexorably towards the conclusion that free will is 'just' an illusion. But we should be very careful with the word 'just'. It implies insignificance, but is often used to dismiss something that actually is significant; and it also suggests that only one viewpoint is appropriate, only one context is credible. Especially when discussing the mind, whose entire operation involves layer upon layer of interlinked and variable contexts, with cross-interpretations between them, that attitude is fatal to any serious understanding. If Dennett is right, consciousness is 'just' an illusion too, the upshot of mindless pandemonium. Consciousness and qualia are complicit, and it is qualia that give an animal an edge in a competitive world; so the illusion of having a conscious mind is a figment of reality. The rules for the interaction of mental demons have been refined over millions of years to produce the emergent phenomenon of consciousness, locked into the real world at every available mind-hold. 'Just' an illusion? Oh no. A carefully crafted illusion, only one without a craftsman. An illusion that appears vividly real to the 'I' inside.

It is the same, we suspect, with free will. On a quasi-reductionist level free will is 'just' the result of Ant Country pandemonium created by deterministic demons. But the rules for the demons' interactions have evolved complicitly with human culture. Most of our choices are to do with our relation to our environment or to our culture. Do we turn left or right at the main road? Do we tell our hostess that we don't like tomatoes, or do we grit our teeth, eat them, and declare how wonderful the salad is – knowing that we will be offered another helping, and cannot possibly refuse?

What is going on, at the cultural level, is not free will but a personality test. If you have the sort of demons that lead to you insulting your hostess, then you will not be invited back next time. (Because she has the sort of demons that will react in that manner.) Indeed you will not be able to move in those circles at all, and your life will be adjusted accordingly. If you are the sort of person that steals things, then you will be made an outcast – not because you have free will, but because you don't! Every time you get an opportunity to make off with somebody else's possessions, your demons are far more likely to come down on the 'yes' side of the decision than 'no'; so everybody else's demons react rather negatively, and you become distinctly unpopular. This, incidentally, provides a rational reason for punishing the criminal: it is an attempt to reprogram his or her demons from the outside. (The usual word for this process is 'education'.) Yes, on one level it's 'not their fault' that their demons obey those particular rules. But on another level, that's precisely the problem. Culture demands the attempt to correct it, and the demons of self-protection will demand drastic action if it cannot be corrected. It is this cultural feedback loop that reconciles our earlier remarks about 'real human beings' accepting responsibility, and our more recent ones indicating that it is your demons, not you, that take the decisions for you. Our cultural systems *know* this. That is why we seek reasons for Mary's uncharacteristic behaviour. It is why most of us are sickened by serious crimes: not because the perpetrator exercised free will to commit them, but because they must be the sort of person who does things like that. It is why the perpetrator is not sickened by the crime, at least not in prospect – they lack that particular demonic feedback loop. It is why some criminals are sickened by the crime they have committed – but afterwards, when it is too late.

This is what cultural rites – such things as puberty rituals, about which we say much more in the next chapter – are all about. They select for culturally acceptable qualities in the mind – qualities that help keep the cultural snooker break going. These qualities have become very subtle, complicit with a more subtle culture: they include all sorts of strange loops, cross-connections between very different demons. 'Because fifty years ago when I was a child my father told us all never to steal, I won't steal now even if my family starves as a result.' Those who lack the demonic loops that their culture requires will be selected out during the rituals, and either be re-educated, cast out, or killed.

So why do we experience such a vivid impression that we possess free will, if our demons do it all for us? Well, why do we experience such a vivid impression that a rose is red, when it is 'just' reflecting light of a particular range of wavelengths? Why do we experience such a vivid impression that it *hurts* when we bang our elbow, when it is 'just' nerve signals and a bit of minor tissue damage? Because our perceptions are qualia, and they have evolved vividness and immediacy in order to improve our survival prospects.

That brings us to the climax of this chapter: we get such a vivid feeling that we have free will, because that feeling is the *quale* of pandemonic decision-making – what it *feels like*, not what it 'really' is.

The wonderful thing about that particular quale is that it is socially conditioned to an exquisite degree. The evolutionary complicity between the individual and the culture that has produced the free will quale ensures that the whole system works as if there actually *were* a choice – because that is the *model* of the process in the conscious minds of everybody concerned. In short: we have free will because that's what it feels like to us, and we act as if everybody else has free will because the whole system has grown up around that assumption and internalised it in every person's head. And, at least as far as the rest of humanity is concerned, our actions happen out there in external reality. Therefore free will is not 'just' an illusion: it is a *figment rendered real* by the evolutionary complicity of mind and culture.

10 Extelligence

A series of laboratory experiments was testing the linguistic abilities of a parrot.♪ It had been shown a variety of objects and taught the corresponding words, and the idea behind the experiment was to see whether, having learned a large vocabulary in this manner, it could still associate the words with the corresponding objects. For example, the scientist conducting the experiment would show it a biscuit, and see whether the bird responded 'biscuit'. This particular parrot was doing rather well, and the experiment continued for some time as the bird worked its way through a long list of objects. Then to the scientist's surprise, the parrot suddenly looked up from its task, squawked 'Tired! Going now!' and stomped back into its cage.

In previous chapters we have been making an ever-increasing song and dance about 'the complicit interaction of mind and culture'. The time has come to analyse this concept more carefully, to dissect out some of the key features of the complicity involved. In that context the key internal feature of the human brain/mind is intelligence, which lets us process the complexities of cultural interaction and build on them. The cultural counterpart of intelligence is an external feature, which perforce we shall call *extelligence*. It is the central topic of this chapter. Extelligence is all of the 'cultural capital' that is available to us in the form of tribal legends, folklore, nursery tales, books, videotapes, CD-ROMs, and so on. However, extelligence is not just a matter of 'keeping a record'. The intelligence of each individual allows them not only to access the cumulative body of extelligence, but to add to it or change it. Lots of people have discussed this kind of interaction in terms such as Karl Popper's 'Third World', Teilhard de Chardin's 'noösphere', or Medawar's 'extrasomatic evolution'. Our notion of 'extelligence' differs from these, we think, and from the general word 'culture', because we look at the external influence from the point of view of each complicit individual.

As we sit and write this, we are surrounded by more information than we can possibly use – several *thousand* books on shelves, colleagues in nearby rooms who will answer questions willingly, a vast number available in seconds by telephone. Within half an hour's walk there are at least a dozen libraries: two big ones and a lot of smaller specialised ones in university departments. The computers on which we write are not just versatile typewriters with memories, search-and-replace facilities, and easy ways to correct errors and edit prose: they are also terminals to the world. They are connected to the Internet, which – once you take into account what else is connected to it, such as NASA and Bell Labs databases, the Library of Congress, and millions of individuals each with their own special point of view or expertise – is surely the greatest organised source of information that has ever existed on this planet. We don't use *all* of this – it's far too big for any individual ever to access more than the tiniest fraction – but it constitutes the vast resource of extelligence-space, and every bit of it is potentially available to us. The better we understand the geography of extelligence-space, the more able we are to use library indexes and surf the Net constructively, the more power that potential adds to our pens. And our minds. And our pockets.

The information in our minds is ... perhaps we should not say 'trivial in comparison', because we don't know how to compare intelligence with extelligence. There are no common units. All the extelligence in the world is useless if you lack the intelligence to use it; on the other hand, without extelligence we humans would still be back in the caves, rather literally reinventing the wheel in each generation. We are what we are because of a remarkable complicity between intelligence and extelligence. Intelligence invents but cannot reliably and accessibly remember what it has invented; extelligence can remember but (on the whole) not invent. Extelligence deals in information; intelligence in understanding.

Extelligence is the invention that not only allowed humans to change themselves into the type of animal they now are, but made it very difficult for them to avoid doing so. What is the origin of extelligence? Did we invent it or did it invent us?

Both.

Two things characterise our species. One is an enormously exaggerated concern for children of any kind, which extends even to tadpoles. The other is language. We maintain that these two apparently very different characteristics are actually linked, by the medium of cultural privilege, and that together they created the possibility of extelligence.

Language first. There is a lot of debate about whether – in humans – language came before intelligence, or the other way round. Other animals show us that intelligence can arise without language; moreover, language is useless if you lack the intelligence to learn it and to use it. This does not, however, imply that intelligence must necessarily have come first, although that's very reasonable and constitutes one of the two most favoured theoretical frameworks. The other is that the invention of rudimentary forms of language triggered an increase in intelligence. An awful lot of hot air has been expended defending one or other of these positions – with the tacit assumptions that they cannot both be right and that they are the only alternatives on offer. In fact, it seems very plausible indeed that *both* of these theories are true, with each driving the other in a complicit process of interactive co-evolution.

One of the universal features of complicity is the emergence of new patterns, new rules, new structures, new processes that were not present, even in rudimentary form, in the separate components. For language and intelligence, this is abundantly the case. Here the most influential new possibility opened up by complicity is that language permits experience to be stored in the memories of older people and passed on to the young. The collective experiences of the tribe become a cultural lexicon stored in the people that surround each child. This cultural context for each child can then grow, as successive generations accumulate further knowledge, and new discoveries can be transmitted very rapidly to all individuals who have the aptitude to use them, or to select 'in-groups' who share secret know-how. Meanwhile biological adaptation enlarges the capabilities of successive generations of children as ready acquisition of the culture becomes part of the formula for success.

Different writers have different theories of what were the crucial marks of earliest humanity. Sarah Hrdy emphasised the role of concealed ovulation in ancient hominids. (In most animals the state of the female reproductive cycle is very obvious externally to other members of the same species, in particular the males, who adjust their behaviour accordingly. Without such visual cues, social structure becomes more subtle.) Desmond Morris, in *The Naked Ape*, preferred the use of sex as a reward: when the successful male returned from the hunt bearing meat, the female became sexually receptive, an evolutionary trick that encouraged the male to go hunting again and to be even more successful. Alister Hardy and Elaine Morgan espoused a different theory altogether, the idea that humanity evolved on the seashore, where food was abundant and life was more relaxed than in the predator-ridden savannahs where most scientists think humans originated. Inventive members of the tribe – probably the females, left at home holding the baby while the males went out in search of food – provided the cultural stimulus that set humanity apart from the other animals. An extra gloss was added to this story by Michael Crawford and David Marsh, who observed that making big brains needs a ready supply of essential fatty acids, and the most abundant source of such chemicals is – seafood. Bernard Wood and Louis and Mary Leakey saw the crucial difference as the use and manufacture of tools – sticks and stones leading to clubs and spears, discarded animal bones put to use to dig shallow pits or mash food, and – eventually – fire. The movie *2001: A Space Odyssey* opens with a sequence that dramatises this theory; and in one split second it shows a bone club turning into a spaceship as the scene shifts from the distant past to the near future. As we said in *Collapse*, that image is trite, but its message is deep. In the words we are now using, it illustrates the inexorable expansion of extelligence.

However, we don't think that any of the above theories captures the complicit growth of language, intelligence, and extelligence fully. Instead we shall argue that nursery stories and puberty rituals – cultural transmission of 'what it is to be human' – were what marked the arrival of human beings on the evolutionary stage. There is of course no reason why one of these theories should exclude all the others, although some of them fit together well and some sit less easily. There is an analogy between our viewpoint and the idea that mother–baby interactions in the nest were what led to the evolution of learning and social structure, and if we were forced to offer antecedents this is where we would look. However, we have already attacked the idea that everything must have some kind of rudimentary precedent: this may often be the case, but it can't always be true because that would lead to an infinite regress. Indeed complicity is a common mechanism for the emergence of novel phenomena without any antecedents.

It is very important, as Jared Diamond reminded everybody in *The Third Chimpanzee*, that we should not only explain why we became sapient, but why the other apes did not. There is one very plausible answer: once one sapient species gets going, there is a lengthy period of its ethical and moral development during which it clobbers most other species. Especially, we suspect, any other would-be sapient species. So it may well have been a matter of which particular ape got its extelligence going first. For this reason, we don't pretend that we can see our story beginning way back among the early primates; no, the first mother with her child on her knee, her new baby at her breast, sharing primitive words with her older children, was the true Eve. We use that term metaphorically: the discussion of 'mitochondrial Eve' in Chapter 4 should make it clear that 'Eve' must be considered as just a convenient shorthand for 'one of the many proto-human females that were around at the time when humans first became sufficiently different from apes for it to matter'. In this sense, however, Eve played a crucial role. All human cultures have a special language for talking to babies, little songs full of repeated sounds and actions to sing to them, and special stories to tell them as their lexicon – their vocabulary, the list of words that they understand – improves. Steven Pinker's book *The Language Instinct* argues that humans possess a built-in predilection for language, an ability to handle syntax and semantics, a deep linguistic meta-pattern like the 'deep grammar' advocated by Noam Chomsky (see below). We think that such abilities did arise, but they must surely have evolved from something simpler; so first this range of mother-and-baby tricks and responses must have become a characteristic of some proto-human groups. One reason for thinking this is the fossil record, which seems to indicate that *Homo erectus* evolved into *H. sapiens* more or less independently in many separate places. Although this theory is not as popular as it was, it still has some influential adherents. The mother-to-baby social tricks could have spread among different groups before the language instinct started to develop; it would then drive sapient evolution in the same way within each breeding group of proto-humans. It's the Law of Murphic Resonance again: if it *can* happen, it will. And the role of the mother *has* to be central, because minds are put together in babies, primarily, and by far the biggest influence on a baby is its mother – as much so today as it was a million years ago.

Before embarking upon our own theory of the evolution of language, the primary step towards extelligence, the trick that makes all other tricks possible, we ought to give you at least a flavour of the main competitors: not their content, in any detail, but their *style* and their overall point of view. There are several ways to study language. One is reductionist: to dig deep inside language and find the rules that make it tick, the meta-rules that make *those* rules tick, and so on. An excellent and seminal example of the reductionist approach is the concept of a 'transformational grammar', introduced by Noam Chomsky in 1957 in his *Syntactic Structures*. Chomsky observed that even a fixed language has many different ways to say the same thing, and there are many different sentence structures in which you can say it. 'The dog has buried the bone', 'The bone has been buried by the dog', and so on. Chomsky focused on the ways to modify a simple sentence like this one into more complex things like 'That stupid poodle that Mrs Barnes always takes with her on holiday has just buried a bone that it stole from next door's dustbin in the middle of my marigolds.' Here the noun 'dog' has been transformed into the noun phrase 'That stupid poodle that Mrs Barnes always takes with her on holiday', and similar games have been played with 'bone', while the location of the burial has been stuck on the end where previously there was nothing at all. The noun phrase about the stupid poodle itself has a grammatical structure, and can be built up from simpler forms by similar transformations. Chomsky listed permissible transformations, things like:

1 sentence → noun phrase + verb phrase
2 verb → verb phrase + noun phrase
3 noun → phrase determiner + noun
4 verb → auxiliary + verb
5 determiner → the, a, ...
6 noun → dog, poodle, marigold, ...
7 aux → will, can, has, ...
8 verb → bury, buried, take, taken, ...

Then we can start with an abstract symbol S for sentence, and repeatedly apply the transformational rules (in an obvious symbolism, e.g. NP = noun phrase) like this:

(start here) S
(rule 1) NP + VP
(rule 2) NP + V + NP
(rule 3) D + N + V + NP
(rule 3) D + N + V + D + N
(rule 4) D + N + A + V + D + N
(rules 5–8) The + dog + has + buried + the + bone

Linguists of the Chomsky school make much play of the fact that this structure could be derived in many ways (for example changing the order of application of rules 5–8 gives 60 distinct derivations), but that the result can always be summed up by a single universal 'tree diagram' like this:

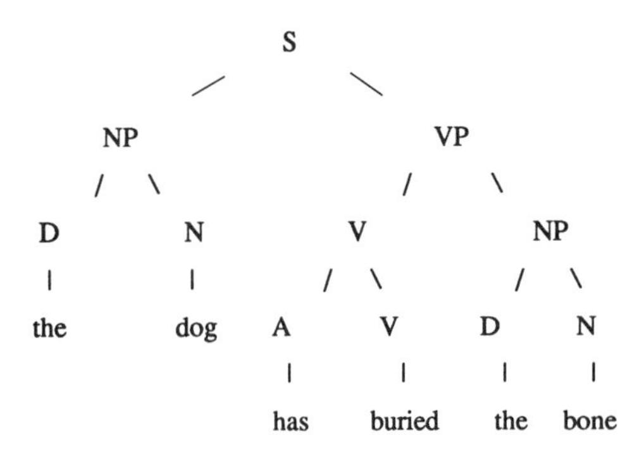

Any mathematician would instead observe that the operation symbolised by '+' has the 'associative' property (X+Y)+Z = X+(Y+Z), which leads to the same conclusion. This algebraic rule is merely assumed in the above notation: the third line of the analysis, where we use rule 2, should really be NP + (V + NP) and you need the associative rule to remove the brackets.

The same transformational rules underly many different sentences. For example, the identical tree structure also fits:

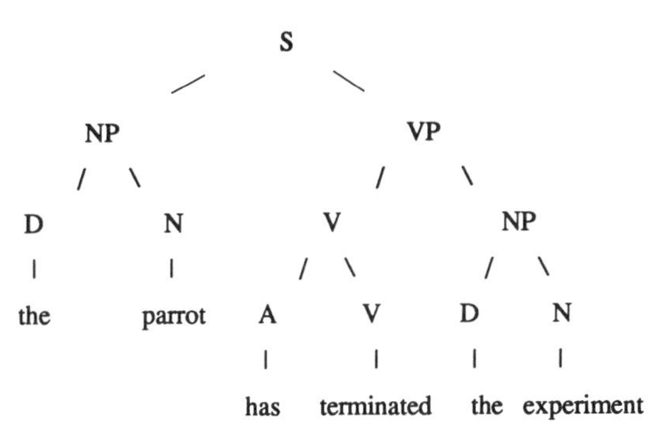

From this point of view the language's syntax – its grammatically permitted structures – reduce to a list of transformational rules. This led to the interesting idea that although each human language has its own peculiar syntax, there might be a single universal 'deep grammar' underlying all languages – which, perforce, would reflect the underlying physical organisation of the area of the brain that handles language. In a computer analogy, the transformational rules might be the 'machine code' of language.

There must be some truth to this idea: when we read nonsense rhymes like Lewis Carroll's *Jabberwocky* – the one that begins ' 'Twas brillig, and the slithy toves' – we get a very strong impression that it is grammatically correct, and even that it would make sense, if only we knew what the words *meant*. But there are also many difficulties. Spoken language seems just too flexible to fit any system of rigid rules: 'Ee, Mabel, t'bloody poodle's bin at t'marigold bed again – 'nother bloody bone, next door's bin I'd bet me 'at.' In 1965 Chomsky was forced to modify the purely syntactic nature of the transformational rules to introduce semantic elements; that is, to build the meaning of the sentence into the grammatical rules. Instead of grammar being an independent rule-based system, and meaning a separate interpretational step (each being carried out by separate structures in the brain in a simple linear sequence), now each depended on the other – a clear-cut instance of complicity, especially when you bear in mind that the processes are both recursive. As you attempt to decipher a convoluted sentence, perhaps one as laborious as this one – well, we say 'laborious' but actually that's not the right word, not the right word at all, a better one might perhaps be 'interminable', but there again it does actually *have* an end, as you'll see when eventually you get to it – well, anyway what we're trying to say is that you 'construe', an old-fashioned word with much the same overtones as the ultra-modern one 'deconstruct', at least to a semiologist or a semioticist or whatever they call themselves nowadays, well, you construe the sentence by working out what part of it means, and then – and only then – can you decide whether it's a noun or a verb, or whatever, and plug it into the transformational grammar to go on to the next step, at which point you're back at the beginning again, so you recursively construe the sentence by working out what part of it means, and then – and only then – can you decide ... well, you get the idea, and, amazingly, the sense of this whole ridiculously over-elaborated example. Though by now you've probably forgotten what it's an example *of*. Well, it explains, among other things, why you had no trouble at all separating the two different uses of 'bin' in the sentence above (a *long* way above) addressed to Mabel – one being a mispronunciation of 'been' and the other an abbreviation of 'dustbin'. And the two uses of 'at' as well, one with an apostrophe.

An extension of and variation upon Chomsky's reductionist approach is the theory of the language instinct, which Pinker promotes. It can be seen as a further step in Chomsky's ideas of a native human grammar, which all native human languages fit because a baby's brain is wired like that. It is like having only one make of computer, one item of basic hardware, but running many alternative programs – Word, WriteNow, Word Perfect, Wordstar, MacWrite ... Many programs, but they can run successfully only if they respect the deep structure of the hardware.

Modern developmental biologists get very hot under the collar at this kind of explanation – or if they don't they should, for it smacks of a way of thinking they long ago abandoned, known as *preformationism*. In its original form this was the belief that inside the egg (for the 'ovists' like Harvey) or inside the spermatozoon (for 'animalculists' like Dalenpatius) there was a tiny homunculus, a miniature version of the future person, waiting only to be expanded into macrospace. Inside the homunculus were nested Russian-doll progeny, sub-homunculi and sub-sub-homunculi ... A variant of the theory went only a few generations down, to atoms; when the resulting empty homunculi grew up they would of course be sterile, and the Day of Judgement would have arrived. In a sense what lies behind this thought is an attempt to 'unwrap' a recursive process into a single linear chain, and the biologists rightly saw this as a pointless, if not misleading, confusion of a relatively simple idea. It is misleading, in part, because it assumes that every phenomenon must already exist, in rudimentary but potentially complete form, in a precursor. Nothing can ever get started, it all goes back forever to things of the same kind, even if cast in a different mould: 'turtles all the way down' again.

Unfortunately, although they got the message, the biologists missed the meta-message, which is that the same kind of game can be played in *any* recursive system, and is usually just as pointless and misleading. The latest incarnation of the preformationist fallacy is the cosy explanation of the complexity of an adult organism in terms of the complexity of its DNA. In this view – widely prevalent among biologists – the complexity of the adult was *always there*, coded in the DNA of the egg, so no real development, no 'becoming' of complexity, actually occurs. But all the complexity of an organism cannot possibly be coded in its DNA. For example, there is insufficient information in the entire human genome to encode the brain's interconnections; and one growing feather is more complicated, by *any* measure of complexity, than the patch of cells on the yolk from which the whole chicken arises.

What has this 'preformationist' explanation got to do with language acquisition by the child? To answer, we must first enlarge our phase space by bringing in other childish propensities. How does baby smile in recognition of mother's face? The usual supposition is that there is a simple low-level circuit in the baby's brain–eye system that recognises the two eyes and a mouth shape. We can easily believe this, because we all share it: on the Internet, :-) means 'happy' and :-('sad'. (Look sideways.) However, a few moments' thought shows that there can't be any *simple* circuit with that function: it must be a high level one. The reason is that the face is recognised no matter where it falls on the baby's retina. What is actually built into the baby is a very reliable way to grow nerve cells in a particular kind of pattern, one of whose features is that a particular collection of signals from the retina – associated with *any* face, not just mother's – causes a complex train of behaviour, part of which is a smiley face. The 'smile' is largely in the observer's mind: the baby just contracts some facial muscles. To think of this as some kind of preformed 'circuit for recognising mother's face' is not terribly helpful: part of it may be very general circuitry for dealing with optic messages, part may be ephemeral connections to the region of the brain known as the thalamus (establishing pleasure and carrying out adjustments to the facial muscles) which will be replaced or reinforced within a week, part may be the circuit that a moment before was regulating eye movement or breathing rate, and indeed is still carrying out that task in parallel with 'smiling' – and so on. At any rate, there is no reason to suppose that there is a specific 'organ' in the brain for recognising mother, any more than there is a specific component in a car for driving it into the gatepost. The same goes for an organ that recognises speech, which is where the relevance to language comes in.

What there is in the brain is a general 'feature detector'. In fact the ability to detect features is a general property of neural nets of many kinds, including – perhaps especially – virtual ones in computers. Provided you put such a neural net into an appropriate context, that is. The context is 'training'. Now sense organs generate 'data' – varying inputs. They reflect either occasional spatial order, as occurs for the retina, or occasional temporal order, occurring for the cochlea of the ear. That order is indicative of some external feature that might be useful or dangerous; however, in raw form it is all but overwhelmed by random-seeming disorder resulting from all the other junk out there. It would be wonderful to have circuitry that extracted the order from the chaos – and lo and behold, that's just the job that neural networks are naturally good at. However, as we've just indicated, they only do it if they are 'trained'. In a computer simulation this is done by presenting input data and tinkering with the connections in the network: changes that reinforce the required response are kept, those that do not are reversed. In an organism, training can occur on two very different timescales. One is 'genetic training' by way of evolution, which alters what is 'hard wired' into an organism. The second, often ignored but, we believe, crucial, is that the sense organs and the neural circuitry that extracts features from their signals can be 'fine tuned' recursively during the development of a single organism. (Evolution can also slowly change the processes that modify the hard wiring during development, and it is worth recalling that in an organism *nothing* is truly hard wired.)

The nerve cells in the baby's brain that are connected to the ear (sending signals *both* ways, incidentally: brain to ear as well as ear to brain) receive messages about balance, acceleration, and angular acceleration from long before birth. As the baby develops, its cellular organisation is capable of responding to these stimuli, so it would not be surprising if some functional organisation were imposed on the associated embryological circuitry. More surprising, but still credible, is the fact that even *before* birth this circuitry should be so exquisitely tuned that it can almost unerringly pick out words from the noise around it. This ability must be a result of general properties of such neural nets – but also of what 'such' means, because the comparable circuit in a chimpanzee can't do that. The process continues after birth: physically it seems to consist of the baby's brain *losing* many neural connections, rather as if it starts with the potential ability to learn all possible patterns but refines it down to those that occur in its own culture. The Grim Sower again, but this time not so grim, unless you're a nerve cell connection. A great deal of the necessary circuitry, sensitivity, and specificity arises from the usual interactions with mother, toys, seeing its own fingers, and 'hearing' the strange high-pitched noises that adults make to babies (such as cooing). There are developmental 'windows' during which certain abilities must be produced: if they are not, they can't be developed later. Language is just such an ability, and it seems that a child who is not exposed to human language until the age of ten or twelve can never acquire it – maybe just a few rudiments, a vocabulary that would shame a three-year-old.

So we do not have a definitive 'language organ' when we are born: we have to construct it as we grow, all the while making use of it. Again, Niven and Pournelle's Moties would understand this process better than human engineers. This kind of DIY provision makes a preformationist viewpoint – even a DNA-preformationist one – inappropriate. Our DNA doesn't specify a mother-recognising circuit or a special speech-recognising organ. Instead we are the kind of animal that, when its brain is stimulated in the usual cultural manner and when it exploits the properties of its nerve cells that result in the detection of features, it builds its own way to see mother and learn language. The resulting neural circuits will be different, in detail, in each of us.

Zarathustran linguistic abilities, of course, evolved along a rather different route from ours.

Liar-to-children I am puzzled to learn that for these human creatures, speech came *before* writing.

Liar-to-adults For once, Liar-to-tadpoles, I agree with you. It is *most* unoctimal.

LtC I am not actually bothered on religious grounds. It just seems weird to do it that way round.

Destroyer-of-facts I imagine it must have something to do with their singlemindedness, coupled to the fact that they do not possess a The Regulations.

LtA You are flapping to hypotheses again, Destroyer. I wish you would not do that. *And* you are promoting The Regulations beyond its limited role in our society, which is to record the consensus that the Ringmast–

DoF [*Ignoring him.*] Our own development of language followed the obvious path, from chemicals to writing to the squoken word. Liar-to-children confused and oversimplified the story, as usual, but he was correct in outline.

LtC As usual.

DoF Originally we were subsapient protozarathustrans, barely able to graffit a message or two on our rudimentary The Regulations. But by a stroke of pure contingency –

LtA Which no doubt was octimism at work in its ineffable manner –

DoF Our primitive The Regulations responded to the chemical messages from our excreta by displaying visual patterns. [*Aside to Liar-to-children:*] This is because its skin is akin to a liquid crystal display, and it is highly sensitive to the vibrational spectra of organic molecules. [*Continues.*] Not surprisingly, in view of the chemical pathways involved, the skin display changed radically according to the chemicals deposited. For example, when The Regulations suffered from a shortage of corpse-juice, its skin turned pink. When it was overloaded with extract-of-pollen, it became suffused with an octagonal quasilattice of lavender spots. And so on. Only with the aid of supercomputer models of The Regulations's biochemistry have we begun to make sense of the correspondence between chemical state and visual pattern, but of course that posed no problems for evolution, which simply exploited the *existence* of such a correspondence. Protozarathustrans with a zenetic tendency to provide extra corpse-juice when their The Regulationses turned pink engendered a prosperous symbiosis: those that did not, did not.

LtA [*Tiredly.*] Of course.

DoF And so a process of complicit co-evolution took place, in which the protozarathustrans learned to respond to the visual patterns on The Regulations's skin, and those patterns acquired 'meaning' as a result. Fascinating how communication and meaning were complicit, by the way ... Of course the The Regulationses co-evolved too, reinforcing the clarity and versatility of their patterns.

LtC And so writing appeared in our cultural phase space!

DoF Well, yes/no (delete whichever). As usual, you are oversimplifying. The *potential* for writing appeared in cryptic fashion, but writing itself did not appear, because the protozarathustrans did not understand explicitly what they were doing.

LtC Oh. So how did they come to make visual symbolism explicit, then?

DoF You will recall that our planet was suffering from an unprecedented series of climatic disasters – which is how our ancestors came to displace the less adaptable proto-symbionts. Times were hard, very hard. But occasionally there would be a glut.

Pursuer-of-sicknesses Such as a glut of corpses?

DoF Exactly. A surplus of corpse-juice great enough to bring snivels of joy to the proboscises of your revered ancestors, Pursuer. A vital resource too valuable to lose. Imagine it. All over the planet, The Regulationses were turning bright pink as a result of too little corpse-juice – while out on the windstrewn plains, corpses lay in abundance. And the learning process had begun to gain hold, so here and there a few protozarathustrans were responding by going out looking for corpses.

Ringmaster Which, at that time, was not a difficult task to accomplish.

DoF Precisely, Ringmaster. Groups of four or five that learned to cooperate fared better than groups of two or three, and eventually they converged to the octimal number – which was initially seven, but paved the way for the final perfect eight, as we have seen. But that is an aside. The important thing to realise is that at that time, the *only* means of communication was The Regulations.

LtC Oh! That would have been inconvenient.

DoF Absolutely. There they were, a group of three protozarathustrans, say, many furlocts away from their own The Regulations, chancing upon a nice, juicy corpse, and suddenly they needed to give each other instructions. So they had to rush back to The Regulations, graffit their messages upon it–

LtA And by the time they got back, some other group had snaffled the corpse.

DoF Precisely. And why? Because that other group had invented the technique of graffooty.

LtC Huh?

DoF They had evolved, purely by accident, the ability to create patterns of footprints in the mud – patterns that, after a period of evolutionary convergence, corresponded, in some arbitrary but commonly accepted fashion, to the displays on their The Regulations. Such as ˇ∧ˇ for 'pink'.

LtC Wow! So that is why we write it that way!

LtA [*Flapping him aside.*] But why use mud?

Hewer-of-wood Because they could slice the top layer of mud with their proboscises, and leave it to dry, thereby creating a record to carry back to The Regulations for permanent storage.

DoF Yes. And because creatures with splay feet that waddle around in middens cannot fail to leave imprints anyway. So, because it was evolutionarily advantageous for our ancestors to communicate *without* going back to The Regulations, they evolved a method for doing just that. It was the earliest form of writing – graffooty. In effect, they started with an external feedback loop mediated by The Regulations, and internalised it. Soon the graffooty code rapidly became enormously sophisticated.

LtA And so we attained written language. What of spoken?

DoF Naturally, that came later, as you would expect: sophisticated systems are always simpler and evolve later. Apparently – and I must admit that this is pure conjecture–

Performer-of-amusements Unlike everything else you have told us?

DoF As I said, apparently our ancestors used to make meaningless squeaks when they paddled in the mud. A harmless expression of pleasure. But now some groups became nocturnal, in order to compete more effectively for corpses that died during periods of darkness. Being unable to *see*, they could not employ graffooty. So they consciously and deliberately invented a *verbal* code, to represent graffooty in sonic terms.

LtC They came up with a system of squeaks to represent writing!

DoF Precisely.

LtC And then the squoken word spread to other groups, because, dark or not, it worked even when there was no mud!

PoA Yes. It was a self-muddifying system.

We claim that brains evolved high-level abilities such as consciousness in order to tackle the problem of feature-detection, and in so doing came to acquire the interlocking structured processes that we call minds. Among the most interesting and important features of an early proto-human's environment – indeed the environment of any reasonably complex animal – are other animals. So our brains should show traces of a bias towards the perception of animals-as-features. Paul Shepard's *Thinking Animals* provides good arguments to that effect, based upon the fact that all human cultures use animals in nursery stories. Urban cultures replace real animals – which the children may never have seen – by teddy bears (USA), dog effigies (Mexico), cat statuettes (Egypt), or Wombles (United Kingdom television). Animal symbols permeate our cultures, because of their use to represent character. To Westerners the fox is cunning or sly, the owl is wise, the chicken is timid, the wolf is violent and greedy, the little pigs are ... small and piggish. For the Inuit child, however, the fox is brave and fast. In our nursery tales the fox is the villain, but in Inuit nursery tales he is the hero. Neither of these is the real fox's fox, of course. 'Fox' is used in the story to *represent* a character trait – a feature of human behaviour. It is not the fox who is cunning:

it is that you learn to recognise 'cunning' as the kind of thing that the fox does in the story. And so this early form of cultural symbolism has converted all Westerners to the belief that 'foxy-looking' beings are untrustworthy.

Owl, parrot, lion, mouse, we have time-honoured character traits for all of them – so much so that despite our sophistication we find it hard to believe that owls are stupid, parrots have a vast repertoire of versatile behaviour, and lions are lazy and careful rather than brave, leaving the hard and dangerous work to the lionesses – whereas mice are brave to the point of foolhardiness. Thus there is a dissonance between animal symbols and real animals. But their usage throughout our culture, from heraldry to hunting, from teddy bears to toy dinosaurs, shows just how much of our minds is filled with animal associations. Shepard suggests that we *construct* our minds by way of these animal symbols, with the construction machinery taking the form of a nursery story driven by mother. It is perhaps significant, in this context, that cave paintings are nearly all of animals, and it is very tempting to see them as symbols of the evolution of the human mind, contemporary with the early stages of a developing proto-language – a sign of the gradual acceleration of extelligence, complicit with a growing ability to change our own minds to suit what our culture prefers.

The trick, as usual, is to provide enough scaffolding for a recursive process to get off the ground. Once it has closed its cycle, it will self-complicate whenever that is possible and provides an evolutionary advantage. So once the language trick is sufficiently embedded in genetics and culture, it will grow. 'Wired-in' language cannot adapt sufficiently rapidly in a changing culture; in fact it would cause the culture to stagnate. More accurately, 'culture' really can't get off the ground if language is wired in. So human children must *learn* language from the people around them.♪ What can be wired in is the equipment required to learn and the propensity to use it. Modern human children are now very good indeed at building their minds from these cultural cues, and they have become amazingly good at learning languages 'from scratch', at the rate of about one new word per hour, day in and day out, until about the age of twelve.

The structure of language evolves in complicity with culture, and its grammatical form is influenced by cultural traditions and usages as well as by brain structure. Syntax (form) and semantics (meaning) are unavoidably complicit, and any attempt to separate them completely from each other is doomed to failure too. Human language works only when it facilitates communication between *minds*. When we study language, especially 'the' meaning of language, we assume that only one real translation is possible – and that is a recursive sequence of events occurring in *one* mind. This is a misleading assumption, as

George Steiner's essay *Extra-Territorial* demonstrates: it lauds Vladimir Nabokov (best known as the author of *Lolita*) for his ability to see a German idea from a Russian perspective and turn it into a wholly new idea in American English. The American readers, however, all took away different associations and understandings from those that Nabokov used to construct his work – although we are quite prepared to believe that his beautiful phrasing linked to similar thoughts in all of their minds. We are prepared to wonder at the events in Nabokov's mind – what an amazing series of electronic impulses must have been involved, what an extraordinary degree of organization! – but we are not prepared to believe that American idiom is appropriate for eliciting a Russian thought. Language is culturally loaded – not just in trivial ways, as with 'pavement' meaning in the UK the thing at the side used by pedestrians, when in the USA it means the road itself, the 'correct' translation being 'sidewalk'. Picking an example pretty much at random off those shelves of thousands of books, we open James Gleick's *Genius: Richard Feynman and Modern Physics* at page 63 to find this passage: 'Walking into the parlor floor of the Bay State Road chapter house of Phi Beta Delta, a student could linger in the front room with its big bay windows overlooking the street or head directly for the dining room, where Feynman ate most of his meals for four years.' Obviously American: not just because of how it spells 'parlor': the dead giveaway is 'Phi Beta Delta', clearly a student fraternity. The images and associations that this passage sparks in an American, especially one with a university background, will be much more vivid than those experienced by a European (whose image of fraternity life, if he has one, will usually be based on Hollywood movies). Compare 'The wizards shuffled their blue-suede feet nervously. "Well," said the Senior Wrangler, "it *is* a fact that last night, er, I, that is to say, some of us, happened to be passing by the Mended Drum–" ' What is the Mended Drum? Obviously a pub. 'Senior Wrangler' is Cambridge academic jargon, and sets the scene in a British style of university with all of the ancient traditions of such places; we included 'blue-suede' in order to mislead you but you saw through it without even noticing. The university is actually Unseen University in the city of Ankh-Morpork, and the passage is from the British humorous fantasy writer Terry Pratchett's *Soul Music*. Which explains 'wizards', of course. You've got it now.

Each of our minds, each of our lexicons, has been established differently, in a brain with different neural connections – remember the ten per cent difference in alleles that makes each of us a genetically distinct individual – so we cannot 'copy' each other's thoughts, even in the 'same' language. 'I know what I say, but I don't know what you hear.' Unsatisfactory though this system is, it is the basis of our society and our civilisation. Paradoxically, the imprecise nature of our linguistic symbolism may actually enhance our ability to create new extelligence: just as it is random mutations – copying 'errors' – that make phenotypes fluid and open up the possibility of evolution, so mental 'transcription errors' can suggest new ideas. Extelligence is not just a fixed tribal archive. The crucial feature that makes it worth inventing a new word to describe it is that extelligence functions in both directions. It is not just a passive recipient of human knowledge. It is a driving force that can affect our behaviour. If you need examples, consider Karl Marx's *Das Kapital*, or The Bible.

The relation between information, complexity, language, and meaning is very tricky. This is because language compresses complex ideas into simple units – it encodes *features* of idea-space. Each language selects its own choice of features; as an American general is said to have protested: 'You can't trust those Russians – they have no word for *détente*.' And there is enormous variation in how well the features chosen by a given language match the features of the real world Even when two people 'speak the same language' communication between them may go awry: how many times have you got yourself in trouble owing to a misunderstanding on the other person's part? Computer languages are different: they have a level of mathematical precision that makes it possible to 'translate' them perfectly. By this we mean 'transmit from machine to machine', we are *not* talking about machine translation of human language. (Just in case you misunderstood. See how easy it can be? See how readily we seize upon a *word* without asking ourselves what might be meant by it? Politicians use this technique all the time to accuse opponents of saying things that they never actually meant at all.)

How do features get into the language? This is another beautiful complicity, rejoicing in the name 'ontic dumping'. Suppose that a trader arrives with a new kind of axe that you bind to a *stick* instead of holding it in your hand. In order to distinguish between the old axes and the new ones, a few people start talking of 'hand-axes' and 'stick-axes'. Then they need a word for that special stick. Well, it's used in place of a hand, isn't it? So let's call it a 'hand-le'. After two or three conversations, the words 'hand-axe' and 'handle' have entered the vocabulary. Other people pick them up and use them, because they make it easier to talk about the new objects. 'Ontology' is the study of knowledge, and words of this kind can be thought of as 'knowledge dumps' – hence the term 'ontic dumping'. Which, incidentally, is itself an example of ontic dumping. Language accretes via ontic dumping, a process that leaves every word liberally spattered with all sorts of curious associations. Words are not passive abstract symbols:

they *drip* meaning – and the extelligence is well aware of that, even if individuals forget it.

As a result, the structure of language is very subtle – and the idea that *formal* structure is what matters is perhaps misleading. We are thinking of Gregory Bateson's discussion in *Mind and the Universe*. He describes language, initially, as a kind of chemistry – nouns, verbs, adjectives, and adverbs are the atoms of the language; grammar provides rules for the chemical 'bonds' that govern their combination; there are levels and meta-levels, rules for sentences that are like chemical compounds, representing or caging ideas. Paragraphs are particles of matter, chapters are chunks designed for larger-scale transmission of cohorts of ideas ... up to what Bateson saw as the highest level of all, fact versus fiction. If someone is shot during the action of a play in the theatre, he noted, nobody in the audience tries to phone the police. There are higher level, meta-meta-meta-language clues that the whole event is unreal play-acting, and Bateson waxes lyrical about this highest function of language. Then, having felt really happy with writing those pages, he tells us that he rewarded himself with a trip to the zoo. Just inside the gate, he came upon a cage with two monkeys playing at fighting ... Suddenly his whole scheme inverted. The monkeys had no verbs, no nouns, no adverbs, and no adjectives – but they understood 'pretend' perfectly. For Bateson, the distinction between fiction and fact became the *basis* of language, its deepest source, not its highest achievement. Similarly the 'deep' level of nouns and verbs was actually the *highest* level of abstraction.

Monkeys recognise 'fiction', then; but they're not so hot at language. Our cousins the chimpanzees are pretty good at learning the habits, customs, and courtesies of their local group, and the tale of the young bonobo chimp Kanzi shows that to some extent they can learn a rather different set – ours. But their linguistic abilities are startlingly limited, even when we try to teach the bright ones. The chimpanzee that offered its teacher a stick of celery as a reward tugged at her heartstrings precisely because that level of generalisation – *she* rewards me if I carry out the task, so *I* should reward her when she does – is unexpected in chimps. In a human child, we would start to worry if after a few years this ability had not developed, though of course the proud parents would be equally moved by signs of its first flowering. Nowadays, nearly all human children are good language learners and pick up cultural cues rapidly. (Though not always those that their parents intend them to pick up.) Some mental pathologies, such as autism, seem to be related to deficiencies in this ability or its control.

Extelligence is much more than language, but language is the big invention that got extelligence started. Another ingredient is cultural rituals. The SF writer and editor John W. Campbell made the suggestion – which he got from the anthropologist Lloyd Morgan – that we control our own evolution by controlling that of our cultures, and that modern ritual puberty practices in both pre-technical and technological societies provide contemporary anthropological evidence. The fundamental structure of such a rite, according to Campbell, is that it sets up competition between a symbolic, culturally prescribed threat, and an 'instinctive' reaction such as fear of fire, of wounding, or of poison. The symbolic threat might be ancestral spirits, the loss of tribal identity, or lack of acceptance in a group. Nowadays it is more commonly a fear of failure, a loss of face within a peer group: we have invented or coopted the emotion of humiliation to act as a hold-all for these symbolic failures. The puberty ritual requires striking a balance between real pain and spiritual pain: the test of true humanity is to make sure that fear of spiritual pain is so strong that when it comes to the crunch, real pain is the preferred choice. The boy who leaps up from ritual abasement as the circumcision knife approaches cannot be accepted into the tribe as a real breeding adult; in a very real cultural sense he is 'not (fully) human'. Those who refuse to drink the vile emetics, or refuse the sexual attacks of older men, are relegated to next year's ceremony, banished, or even killed: either way, they do not become breeders. Even such modern practices as the completely ritualised bar mitzvah still serve the same function, although nowadays nobody seems to 'fail'. In the ghettos, and in those more traditional societies in which parents choose wives for their sons and/or husbands for their daughters, this kind of ritual is the main way that people outside the immediate family assess a child's quality. A peahen judges the peacock's quality by the length and splendour of its tail, and a Jewish mother judges a prospective son-in-law by how literate he is. Those anthropological television programmes, where the intrepid anthropologist has wormed her way into the confidence of the tribe and been permitted to film the holy-of-holies, take on a new significance. The boys, or sometimes girls, are adjured to be bound by wisps of straw, which they could break in a moment but have been told are the unbreakable bonds of the ancestral spirits. Now they lie, spiritually bound, awaiting the red-hot irons that will be applied to their cheeks or more tender parts. They sweat, they are clearly frightened – though possibly more by the cameras and lights than the red-hot cutlery – and they demonstrate their courage by going through with it regardless. And we all feel powerful emotions with them, which perhaps echoes the feelings bred into us by all those ancestors of ours who, similarly weighed in the balance, were permitted to *become* ancestors.

There is perhaps another role for some such rituals, which serves the purpose of demonstrating superiority to other animals in the same group, with consequent breeding advantages. This is visibly to function under a handicap that would cause others to give up.♪ The peacock, with his proud and excessively awkward tail, is demonstrating that kind of superiority. 'Look at me, I'm so good that I can function even when I have to drag this stupid thing around with me!' The peahens take note – not out of choice but out of sexual selection; the ability to drag a huge superfluous tail around really is an effective quality control test for mates, another case of the internalisation of an external feedback loop. In human society men play contact sports, drink to excess, or drive very expensive cars to demonstrate a similar immunity to handicaps. In *The Third Chimpanzee* Jared Diamond tells of a colleague who would drink paraffin – you want to show you're as good as me, you do the same. The Aztecs employed rituals at which dangerous quantities of drugs were taken. And so on. Once we become aware of this social mechanism, we can all recognise dozens of instances every day.

We hasten to add that many of these rituals no longer serve any useful purpose, certainly not of this kind: ritual circumcision of babies, for example, serves no serious selective or instructive purpose for the simple reason that the baby has no choice and is therefore not being tested or selected for *anything*. Many such rituals are outmoded relics, sometimes barbarous to modern eyes, and the only selective purpose they might still serve is to select *parents* on the basis of their willingness to subject their children to them, thereby reinforcing the authority of the priesthood. This kind of selection is very often self-maintaining: the older men are sure that it did them good, and even if not they're damned if the younger generation is going to get away scot-free when *they* had to put up with the pain and fear; it is often the older women who are most eager and proud to perform ritual circumcision of girls. Also older people are more 'frozen into' the culture – they find it hard to conceive of the ritual *not* taking place. This may be why such rituals still survive in technical societies like ours: stag nights, academic examinations, 'hazing' of pubertal boys, 'fresher' rituals at schools and universities, initiation ceremonies in adolescent street gangs, in professions like the police and the army, or in the Boy Scouts – all have resonances with traditional puberty ceremonies.

Nursery tales and puberty rites: these two key ways in which surrounding culture – extelligence – moulds the developing human being are embedded in a mass of other cultural influences. Complicity between the cultured mother and the culture-learning baby is established from the first meeting of their eyes, eliciting the first little smile. Indeed it might go back even earlier, because babies in the womb can hear sounds, and after the gurgling of mother's digestive tract, the thumping of her heart, and the wheezing of her lungs, the next commonest sound that baby hears is its mother's voice. This special relationship continues with cuddling, feeding, the sounds that we reserve for making at babies, and special clothing like swaddling and footbinding. There are nappies and medical rituals like vaccination and regular weighing in a special clinic.

All of these special influences on the growing child form a matrix of extelligence which moulds the child so that it becomes an effective member of society at each stage of its development. Just as there is a proper succession of clothes worn by the developing child, so there is a succession of cultural matrices to pass through. If four-year-olds go to playgroup, this one does; if seven-year-olds are thrown out of the household to scavenge in the streets, so be it; and if the ability to read and write is seen as a necessity for economically effective adults, then there will be a special set of humiliations for those who appear not to be getting there in adolescence. Of course none of this works exactly as intended, and it is by no means uniform in its application or its effect – but nevertheless the broad picture applies, all else being equal, across all human cultures.

This whole pattern, with universal similarities in all cultures but with enormous parochial differences, is what we shall irreverently call the 'Make-a-Human Kit'. It is a recursive snooker break in which the aim of each generation is not just to bring up the next generation, but to bring up a next generation *like its parents* so that it can in turn bring up the next generation that can in turn ... But, like every recursive process, this one changes as it builds, because too rigid a Make-a-Human Kit will 'run out of position' and the break will come to a halt – not with a failure to produce the next generation, but with a failure to perpetuate the culture to it. On the other hand too flexible a Make-a-Human Kit will not make humans effective enough to pass on their own Make-a-Human Kit, and again the culture breaks down. It is a difficult balancing act, made all the more difficult by changes to the outside world that disrupt the traditional way of life, as the story line of *Fiddler on the Roof* makes clear. The most resilient Make-a-Human Kits tend to be those that operate, at least in part, on a meta-level, inculcating ways of thinking that are appropriate to a broad range of circumstances: for example 'this is how to spot a business opportunity' rather than 'you are a shoemaker and this is how to make a shoe', or 'you should take thought for the needs of others' rather than 'it is wicked for a woman to reveal her ankles because this inflames the passions of men'. In particular, rigid cultures transplanted into what, to them, are alien environments, have a very difficult time surviving for more than a generation. Paradoxically, those parts of the culture that have the least practical importance can survive more easily than those that have lost their original relevance to everyday life, so that many people who consider themselves good Catholics will keep a statue of the Madonna in their living-room but practise birth control in the bedroom, and many who consider themselves good Christians will piously put a pound in the collecting-plate on Sunday and rob their neighbours blind from Monday to Saturday.

That brings us to religion: in particular, the pervasive concept of God. Religion excites strong emotions even in non-believers: it creates cohesiveness within a single cultural group but is often responsible for deep divisions between different ones. Belief in gods goes back as far as recorded history, but modern religions have generally converged on belief systems centred around a single 'invisible' being, credited with a range of supernatural powers. We think this happened because cultured humans relate to the world in a new way, thanks to extelligence. Because we and most of our readers belong to the Judaeo-Christian tradition, we shall set our discussion in that mould, taking as an example the Old Testament God of Abraham. What we shall tell is pure conjecture, at best a myth or a parable, almost certainly wrong in many details; in particular we may well have given it the wrong protagonist. But we suspect it is at least 'meta-true', describing a general process that *did* occur.

Abraham, we are told, was a shepherd. His culture – possibly that in the region of Ur of the Chaldees – surrounded him with a whole list of rules, recommendations, and prohibitions. Around him were other cultures, which he knew by the names of the gods: Marduk and Baal, Anat and El, represented by 'graven images' – masks, altars, temples ... the numinous appurtenances of many tribal peoples today. Abraham was unimpressed by such tawdry displays. Instead, he was awed by how much more 'something outside him' knew than he did. That something was the early blossoming of extelligence. It knew when to plant crops, when to reap them, what to eat, how to run your life to ensure prosperity, even how to ensure that all of this was passed on to your children. Abraham could see that his own intelligence was tiny by comparison to that outside authority on his own life, invisible but immensely benevolent, knowledgeable, and successful, and *not* represented by a graven image. It was best described as That Which Is (the meaning of 'Jehovah'), the God with no name and no image.

It is a very small step from 'There is Something out there' to 'There is a *Being* out there'. It is also a very small step from 'That Being looks after my affairs and makes sure that my sheep, wives and children prosper' to 'That Being regulates the lives of the sheep, and the wild animals, and all the patterns that I see in nature ... Gosh, It probably makes the sun come up in the morning, and has put all of those amazing stars in the sky! Now, that is a Being *worth* worshipping; I'll make sure Isaac gets a load of this ...'

If we are right, then monotheism arose from a form of ontic dumping, in which extelligence was turned into a *thing*. We are not suggesting that modern Jews and Christians have retained this simple view, merely that perhaps it began that way. The concept of a single all-powerful, all-knowing, all-seeing God then became embedded into the very extelligence that spawned it. A Zarathustran visiting the Earth billions of years ago would not immediately see all the descendants – ammonites, octopuses, squids – implicit in a particular tiny flatworm with a shell; similarly it is not obvious that Abraham's conception would lead to the glory of Solomon, the Archbishop of Canterbury, or Mecca.

An alternative way for humans to relate to nature also grew from the new extelligence that Abraham founded, but it grew along a different branch: science. Religion keeps its snooker-break going because it passes on its beliefs to succeeding generations, and reinforces them with the same awe that led Abraham to the idea in the first place. Science does more than this: not just passing on the *belief* that planetary orbits are ellipses, but passing on the need, and the technique, to constantly test that belief against the real world – and to test it by looking for evidence that it is *wrong*, not just by selectively accumulating evidence in its favour. As a result, science has been far more successful as a way of understanding nature. Science and religion, then, may have had very similar origins. Both are attempts to systematise humanity's relation to its extelligence.

Up until now we have emphasised the contextual element of human development, but it goes without saying that the development of the embryo/fetus/baby/child/adolescent/adult is also underpinned by the biological machinery of embryology, endocrinology (hormones), physiological growth patterns, development of the brain, and sense organs. You can't make a human from just a Make-a-Human Kit: you need a potential human to apply it to. Embedded deep within this developmental programme are DNA sequences – which, as we have said, differ on average by as much as ten per cent from one individual to another, as regards the alleles that they carry. But genetics has fewer implications than many people seem to think for the kind of human that can be built. There is a growing, and disturbing, tendency to see the pervasive influence of DNA as *the* major determining factor in the growth and acculturation of a human being: scientists have already announced the discovery of such things as a 'gene for homosexuality' or 'a gene for obesity'. Now, there are genes that affect how you behave, or at least how you most easily behave; and studies on identical twins separated at birth show that they have surprisingly similar preferences in many respects. A certain amount of what you are, and what you can be, is 'written in your genes'. But an awful lot more, including almost everything that is culturally and socially important, is not. A child that is genetically Inuit, Japanese, or Caucasian – by which we mean 'from the Caucasus', not 'WASP' – can receive the Make-a-Human Kit appropriate to an Inuit, a Jew, or an English public-school type, *and assimilate it*. This is why second-generation immigrant children usually have a hard time managing the conflicting demands of their parents' culture and that into which it has chosen to transplant itself, and third-generation immigrants often have a near-perfect local accent and cultural and political outlook, whatever their skin colour and facial bone structure. Most of the special characteristics that make us into distinct individuals result not from biological determinants like genes, nor from the cultural matrix in which we are embedded, but from complicity between the two. Complicity generally implies new, unexpected phenomena, which is why – to pick examples at random – in the UK many bus drivers have close forebears from the Indian subcontinent, and a huge number of fish-and-chip shops are run by Cypriots. To put this process in context, we can either look at the development of intelligence among our close cousins such as chimpanzees, or we can speculate about the development of intelligence on other planets and about what might have happened on this one had things gone a little differently. Of course we plan to do both.

Chimps first. JC & IS once spent a totally absorbing couple of hours watching a small colony of bonobo chimps. Bonobos are 'gracile', lightly built like us, whereas the common chimp is 'robust' like a Neanderthaler, with heavier bones and more developed musculature. We do not know much about their behaviour in the wild, but what little we do know – reports from people living nearby and TV films – shows their life to be very similar to that of the common chimps, as observed by Jane Goodall. They are intelligent, but they have very little culture, so in that most important behavioural characteristics they are not like us – even though they do have wars. Bonobos have been kept in zoos only recently – if only because they were *discovered* only recently – but they have adapted brilliantly to the zoo environment. Their youngsters, for example, seem completely at home in captivity. When we were watching them our initial feeling was to worry, a bit, that these 'people' were in jail. But then we began to see that they were more like 'acculturated' people – people who have grown up inside a culture and become accustomed to it – who would not be entirely pleased to be returned to the wild.

You will recall that in the works of Edgar Rice Burroughs, Tarzan, whose father was the English aristocrat Lord Greystoke, was brought up in the jungle by apes, but despite this still had all the 'chivalrous' instincts that the best English aristocrats (allegedly) possess. But people brought up among chimps, or indeed among aliens, would not be like Tarzan at all. Those instincts are not genetic, they emerge by complicity with the Make-a-Human Kit. Children brought up by animals are not proper people. If that sounds cruel or 'politically incorrect', recall that they cannot learn a language, they cannot reliably respond to a name, and sometimes they have difficulty walking or eating. Such children are genuinely deprived; deprived of the cultural framework of the Make-a-Human Kit. They have the misfortune to belong to a species for whom this kit is as essential as milk, a species that requires a steady supply of cultural input to reconstruct its current level of intelligence in every one of its members. Cut off from their Make-a-Human Kit, they have not been made human. That doesn't stop us exercising compassion and treating them as best we can, but 'compassion' is a concept far beyond *their* understanding.

There are several views of the process by which culture is passed across the generations which are more or less similar to extelligence. One of the most interesting is that of Anatolii Vygotski, a Russian who published in the 1930s. He has now achieved rebirth at the hands of Helen Haste, who has publicised his vision of the growing, learning child. This places the child at the centre of two circles of influence: its intimates (usually parents and siblings) who make sure that it passes through all the ritual stages appropriate to its age; and the wider society, whose mores are translated to the child by the intimates, then by teachers, and so on. The overall effect on the child is a complicit interaction of the social framework, mostly religious or religious in origin, with the child's successive abilities, moderated by its intimates.

Much of the framework of a society's beliefs is carried between the generations by a special cadre of 'educators'. Initially they may have been junior priests, assistants at the puberty rites. Later specialisation, and the breakdown of the passage of extelligence to the young by apprentice schemes, led to the establishment of schools. Initially religious, they spread to the children of the rulers, then downward through society. Much of the Make-a-Human Kit, considerably changed from its tribal form, is now in the hands of educational administrators. Despite which, a few children do receive a good education, because the Make-a-Human Kit is sufficiently flexible for them to survive bad teaching.

Many of our schools are much *worse* than 'the wild' in terms of their ability to establish an effective Make-a-Human Kit in their pupils. In affluent societies, many children from deprived families are repeating the cycle of deprivation. This, according to some rather unsavoury political stances, is the Grim Sower at work, and there is no need to interfere. But the reproduction of complex human societies must run deeper than mere replication: the snooker break must be kept going, and a snooker break requires *forward planning* at each recursion. Poor or primitive societies cannot afford the Grim Sower, and therefore cannot permit much diversion from cultural norms. Like giraffes, most of the next generation must follow the same paths as their parents – even if the trees are losing out to acacia scrub. More affluent societies, however, *can* afford diversity, by making the Grim Sower less grim. One effect of this diversity is that a smaller and smaller proportion of people in 'western' societies – we include Russia and Japan here – are actively engaged with the extelligence. For example, although lots of people can use a computer – in the sense of plugging it in, inserting a disk, and running routine tasks – very few know how to make one. A high proportion of those who are engaged with the extelligence come from ghettos in multicultural cities. For these and other reasons we believe that human minds on this planet will soon be very different indeed.

What about alien minds? Intelligence is a universal evolutionary strategy, so, like flight, photosynthesis, and sex, it will appear on other planets that have developed sufficiently complex life-forms. The same goes for the thought experiment of re-running life on this planet. But what that evolutionary universal will produce is intelligence of the same kind as the mantis shrimp, the octopus, and perhaps the dolphin. Intelligent, yes, but do they have minds? Well, if we stretch our concept of culture a little and look hard, we can find several other mammals that exhibit signs of a *separate* evolution of Make-a-Social-Animal Kits. This is relevant because it suggests that Make-a-Social-Animal Kits are universals, not parochials – in which case culture is a universal, and we can expect extelligent alien life-forms too.

Suprisingly, this feature of these animals was discovered only when they were taken into zoos. It is definitely *not* a general feature of all mammals, invented once and then lost by some descendant species. (This phenomenon is called a 'symplesiomorphy' in evolutionary jargon, and it throws a spanner in the works by making a parochial look like a universal.) But only a few mammalian groups exhibit the right kind of social behaviour. There are many herd forms, and nearly all of them are like gnus and zebras, antelopes and sheep: they produce mature young, who don't learn much, don't *need* to learn much, to become functional members of the herd. There is one extremely social mammal, the naked mole rat: it is as social as the ant and has a matriarchal system resembling that of social insects. Again, most of the behaviour here seems built-in, and it would not surprise us if mole rats brought up outside the parental burrow could not function – a case of extupidity rather than extelligence! Many primates, from langurs to baboons to macaques to chimpanzees exhibit the cultural transmission of ideas – but that should count as only *one* invention, perhaps in the common ancestor of the higher primates, so it still might be a parochial, especially as it has only a marginal effect for all of those creatures.

No examples there. However, we shall argue that there have been at least two, and possibly three, separate evolutions of Culture Kits: one among the carnivores (wolves, dogs, jackals, but not foxes), one among the meerkats (distant relatives of the first group), and the putative third among certain whales and dolphins. Anything invented three or four times is surely a universal.

We have already given brief mention to the cultural systems of dogs, with identification-calls for each separate pack, learned by the youngsters and taught to them by their elders. Another striking instance of animal cultures occurs among meerkats, communal mongooses. Their possession of culture was exposed in zoo communities, by the way that they assigned specific individuals to new tasks, quite different from those in their normal environment. It was then discovered, by careful observation, that they have a similar social system in the wild: for example one particular individual mounts sentry-guard in a tree while the others root around for food. Individuals take turns at sentry duty, although the dominant male in the group may well take the burden upon himself more often than others. In zoos they beg food from visitors and distribute it in the burrow. In the next generation, the babies seem to pick up such new roles very rapidly, by imitation. This is not like breeding different dogs for different behaviours, such as retrieving game birds shot from the sky. The meerkats learn a new role, one not seen in the wild, in one generation. The next generation picks it up immediately. The third –tentative – case occurs among killer whales and bottle-nosed dolphins. These animals possess all of the behavioural characteristics that suggest a 'domesticated' kind of upbringing in the wild, which is perhaps the best marker for a system in which the adult organism has to learn its role in society by interacting, when young, with older members of that society – and so, we hazard a guess, building itself a mind. In the 1960s John Lillie worked with dolphins, writing a celebrated book *Man and Dolphin*. JC spent several weeks teaching one of Lillie's dolphins tricks, swimming with her, and introducing his children to her, and he is convinced that she had a mind.

So the independent evolution of culture in dogs, meerkats, and dolphins suggests that culture possesses a degree of universality – but *only in mammals.* What if our evolutionary re-run, on Earth or off it, fails to produce creatures that resemble mammals in those ways that are crucial to the development of minds? Are there entirely different patterns of life that could have the same effect? Well, those people that have interacted with the tired parrot from this chapter's lead story have come away with a strong impression that a mind is present, which is certainly the most direct way to read our anecdote. Further – and this really is speculative – our own (JC) interactions with mantis shrimps, octopuses, and dolphins have given us a most delicate and subtle feeling that interaction with humans has changed the animal, 'domesticated' it, so that it has begun to see itself as a player in the game, rather than just following the old inbuilt rules that make it look like a player when actually it doesn't even know it's in a game. So even if there are no minds on the alien planet when we first arrive, our proclivity for making pets of other creatures – especially intelligent ones – might well create some. In the 'uplift' stories of the SF writer David Brin, the galaxy is full of sapient species, and the overwhelming belief is that *all* of them (apart from the incredibly ancient and vanished Progenitors) must have been 'uplifted' by a previous sapient culture. That is, a species can develop a mind only with the aid of a species that already has such a thing. The tension in the stories derives from the fact that no species seems to have uplifted humans, putting that upstart race on a par with the great ancestors of all minds.

These two explorations of the phase space around the human evolution of extelligent culture and intelligent minds – which we consider inseparable – lead us to conclude that the intelligence/extelligence complicity is a universal, though perhaps one that requires a fairly special set of circumstances before it is triggered. As a universal, it is therefore not a great surprise that it appeared on Earth when such circumstances *did* arise. Extelligence cannot get going without intelligent individuals to create it and to respond to it. But once it does get going, the resulting complicit feedback loop drives both intelligence and extelligence ever faster, ever further. Nearly all that makes us human has come about because, long ago, we were grabbed by that feedback loop. Like the sunbeam on the floor that waits to trap any passing cat, extelligence waited in phase space until we blundered into it. It began with cuddly bunnies and language; it has grown via Holy Texts and illuminated manuscripts into the global telecommunications network and the Internet. Beyond lies –

We'll just have to wait and find out.

11 Simplex, Complex, Multiplex

It is well known that Albert Einstein was born in Ulm in 1879, but his family moved almost immediately to Munich where his father Hermann and his uncle Jakob set up a small engineering company. Later he went to Milan, and he studied in Zurich. It is much less well known that for a few years Jules Shloer, who was then studying mathematics but later went on to found the famous soft drink company, lived in an apartment block next to Einstein. Not far away was a corner shop, with a cramped partitioned section at the rear which served as a café. Here Einstein and Shloer would often meet, to drink coffee and talk. The shop was run by an Italian immigrant, Antonio Mezzi, and the only kind of coffee that he served was thick, dark, and enormously strong, made from beans imported from one particular Arabian village. In later life Shloer and Einstein both attributed their success to the remarkable mental clarity induced by Mr Mezzi's special coffee.

We end our journey through human mind and culture by trying to answer some of the questions that we raised in the opening chapter. How did such a peculiar animal as the human gain such a grip upon the planet? What is it that makes us the way we are? And where are we going next?

Let us first take stock.

We are genuinely remarkable members of the animal kingdom. This is not mere anthropocentric speciesism, even though we're rotten at catching mice and thereby fail the cat's test for being a successful animal. We are intelligent, able to reason in a more sophisticated manner than any other animal. (We can't be *absolutely* sure of that, but the circumstantial evidence is compelling.) We have technology, a system for turning ideas into *things*. We are conscious, in the recursive sense that we are not only aware of ourselves *as* selves, but we are aware of the fact that we are aware of ourselves ... We have a special trick, language, which has enabled us to surround ourselves with more culture than can be carried through one family. We have selected ourselves through tribal rituals, which form part of a general Make-a-Human Kit that is designed to produce a new generation that will keep the tribal culture on track. We have extelligence – or does it have us?

Most theories of the human condition focus on one of these many unusual attributes, to the exclusion of the others. Theories of consciousness, for example, centre around the internal structure of the brain, ignoring the cultural dimension. Theories of social behaviour focus upon culture, and ignore the role of brain structure. In Chapter 8 we argued that intelligence and self-awareness are generated by complicit co-evolution of brainy animals and their culture: we have a strong mental feeling of selfhood because we see *others* as selves, they similarly see us as a self, and each of us has learned to internalise this general perception of selfness. By the same token, cultures do not create themselves in some kind of global, collective social structuring: they arise from the interactions of individuals and the trading of ideas. In this final chapter we take a look at the development of human culture, arguing that it is complicit with the development of individual minds. We end with a 'phase space' comparison with possible future cultures, taken from the SF literature.

Today's culture is not just highly complex; it is self-complicating, a 'downhill bicycle race' with unstoppable momentum and no end in sight. This is a surprise, because from the story that we have told so far, a Zarathustran might expect to see us as a loosely knit collection of various tribal cultures – cultures in which everything not mandatory is forbidden, cultures whose impending breakdown is exemplified by *Fiddler on the Roof*. However, we escaped from that trap long ago, and it would now be very difficult indeed to stuff us back into it. Our cultures are not fixed: they are recursive. Each generation re-creates its culture in the likeness of that passed on to it by the previous one. However, this re-creation is not exact – the cause of much parental *angst*. We re-create our cultures *flexibly*.

Why? Is it just the perversity of the younger generation?

Not at all. The flexibility is necessary to ensure stability – not in the sense of rigid conformity, but in the sense of being able to keep the cultural snooker break going when the outside world is changing. It is easy, in principle at least, to keep a perfectly cyclic system going: if we return it to its precise original state then it will repeat exactly the same moves forever. At first sight it seems much harder to keep a 'cycle' going if each successive loop is different from the previous one, which is how cultures behave. In fact, however, cultural recursion of a self-modifying, self-stabilising kind is more or less inevitable – certainly in cultures that *do* survive. A rigid cycle may be thrown off course by external changes. We all know people who can cope extremely well provided nothing unexpected happens, but who panic when faced with something they hadn't anticipated. A culture that panics is a culture whose days are numbered. Just as the snooker player applies spin and swerve to keep the break going even when the ball fails to settle in the expected optimal position, so a culture that will survive must apply its own flexible corrective measures. And cultures as complex as ours require an extensive repertoire of such measures, together with the ability to implement them effectively. It is here – impaled upon the horns of a dilemma, able to initiate changes huge enough to affect the entire planet, but increasingly uncertain which changes will be beneficial and which conceal hidden disasters – that humanity now finds itself. We are in danger of losing confidence in our own future – and that is when the cultural snooker break is in danger of running out of position, and breaking down. On the other hand, we have an unparallelled ability to take charge of our own destiny, and that is where the cultural snooker break stands the greatest chance of racking up a big score.

The story that opens this chapter exemplifies many of our most deeply held beliefs about the different factors that influence the development of the human mind: genetics, friendships, culture, psychotropic chemicals. It tells us that Einstein's mind was in part inherited from his scientifically inclined father, that it was moulded by intellectually active friends and acquaintances, that it developed within the closed, intensive back-street culture of turn-of-the-century Switzerland, and that it is at least conceivable that without the chemical stimulus of that special strong coffee from the mysterious Arabian village, neither Einstein nor his friend Shloer would have become household names. It's nonsense, of course. Unlike all of our other stories, we made this one up. As far as we knew when we wrote it, the only true statements are those in the first two sentences. Months later we discovered, by accident, that Einstein belonged to the 'Olympia Academy', a club with only three members that met informally in Zurich coffeehouses in the early 1900s.♪ We doubt that Shloer was a member, though. We invented the tale to exemplify the kind of cultural myth that appeals to human minds. 'Wouldn't it be marvellous if–?' is but a short step from 'Isn't it marvellous that–?' We would not be greatly surprised if in ten years' time our little fiction shows up, garbled but recognisable, in a scholarly journal – and represented as truth. One of the problems of being immersed in extelligence is that extelligence has a mind of its own: the kinds of cultural myths that circulate are the kinds people *like* to be told, and truth is by no means the sole criterion for being liked. Witness the continuing popularity of astrology, spiritualism, spoon-bending, UFOlogy, alien abductions, and their like. Our minds are influenced by the outside world, and by the wealth of accumulated extelligence, but they select what appeals to them and to some extent construct their own realities accordingly. This is the trick that has made humans considerably more special than fictitious Arabian coffee. We have a secret mind-bending ingredient that *works*.

It is us.

We bend our own minds, from each generation to the next, and so far the positive results have far outweighed the negative ones. (If you doubt this, imagine being offered the opportunity to live in a previous age. Golden ages don't look so golden when you're in the middle of them, suffering the diseases, sleeping on infested bedding, trying to hack your enemy to bits with a lump of iron, being hung in chains at the whim of the local gang-leader.) Humanity has been around for quite a while now, and civilisations have risen and fallen. However, today's civilisation is qualitatively different from any that have gone before. At some point our relatively restrained development accelerated dramatically, setting in train the aforementioned downhill bicycle race of increasingly complex technology, larger and larger social units, the capacity to wreak serious damage upon the entire planet, and the capacity to take action against planet-wide threats. What caused this acceleration? What does that tell us about what kind of animals we are, and how we came to be that way?

We shall sidle up to these questions by way of stories and analogies, founded in the concepts and images that have been developed in previous chapters. Human cultures – not in the abstract, but as enormous collections of people – are one level in a hierarchy that, ordered by scale, goes something like this: molecule, autocatalytic network, bacterium, eukaryote cell, organism, person, tribe, town, city, nation state, supranational community. We can view this as a kind of evolutionary ladder – though we must be careful to acknowledge that like all such ladders it is a teaching myth, in which most of the interesting non-conformities are suppressed for the sake of clarity of description. Molecules got together – somehow – to produce autocatalytic networks of chemical reactions. Autocatalytic networks got together – somehow – to produce bacteria. And so on, until we find nations getting together – somehow – to produce organisations like the United Nations, the European Union, and OPEC. We are not the first to have noticed apparent analogies between various levels in this hierarchy. Over the next few pages we shall investigate a few of these analogies – not because they are perfect, but because they focus attention on interesting questions. To what extent is a city like a cell? Are there parallels in the evolution of the cell and that of the city? Are there *differences* in the evolution of the cell and that of the city – deep structural differences, not just being made of different stuff?

First, a reminder about evolutionary change. In Chapter 5 we pointed out that two very different kinds of change can occur in developing systems: explorations and explosions. Explorations are the relatively slow and 'natural' changes that occur in existing features: in human culture these might be the invention of a new shape of clay pot or a new variation on the dance to the rain-god. Explosions are rapid, almost discontinuous introductions of revolutionary new features: the discovery of fire, the evolution of language, the invention of agriculture. Explorations investigate what lies nearby in a pre-existing phase space; the novelties are so close to what was already present that they – or something like them – are more or less inevitable. In explosions, phase space itself becomes altered, and entirely new regions are not explored, but invented. In the new, expanded phase space, the strategies remain unexplored and unknown and the players are amateurs. As time passes, the players turn professional, the strategies canalise, and the phase space acquires a fixed geography. We have described how this happens for the evolving social structure of human populations. Now we take a look at what it means for the evolution of culture in a broader sense, and we start with people themselves. Human beings are what they are because of a long accumulation of such changes – millions of explorations for every explosion. One of the most significant large-scale changes of all happened when the complicity between intelligence and extelligence 'took off' and became self-referential, self-organising, and self-sustaining. One way to carve up the skein of ideas involved is to focus on 'culture', human social behaviour organised around extelligence. How did culture come into being, and what – if that is a sensible question – triggered it?

Precursors of culture go back so far that it is pointless to ask when it first started: it emerged gradually from simpler things, just like life. For example, packs of wild dogs have their own doggy miniculture of identification signals; no doubt packs of proto-humans had their own rudimentary miniculture too. On the scale of description that suits this chapter, the first significant blooming of human culture was the tribal village. Tribes are a natural strategy for dealing with a hostile environment: the pack may be able to accomplish things that are beyond the reach of any individual. Under circumstances that require this kind of collective effort, therefore, tribes will evolve if they can. We do not know when tribalism first began. *Homo erectus* often formed large groups, and judging by the remains of stone tools and rubbish dumps, they lived a somewhat tribal life. As we said earlier, *H. erectus* may have turned into *H. sapiens* in many different locations at much the same time. It is conceivable, indeed plausible, that this came about because, as the *H. erectus* population increased, the aggregation of families into tribes changed the rules of selection. A Hollywood scenario would have *H. erectus* evolving into *H. sapiens* by way of puberty rituals, of course.

Individual families can find many ways to stick together – foraging on the savannahs and running away from other groups, attacking other groups, killing them, stealing their food and possessions. Tribes can make similar choices about their reaction to other tribes, but they cannot behave in such an anarchic fashion in their relationships with themselves. A culture, even at a level as simple as that of a tribe, is like a snooker break: it must keep going. So a tribe, in order to exist at all, must evolve practices that make it possible for the members of the tribe to live together – such as ways to resolve disputes or deal with antisocial behaviour, behaviour that if allowed to continue might prevent the break continuing. There is thus considerable evolutionary pressure on tribal groups to develop rudiments of culture. Social animals generally have ways to keep the pack under control: pecking orders, alpha-females, whatever. Typically a small number of the animals are 'in charge', and they enforce their rule by direct action. This kind of immediate control structure we take to be characteristic of the tribe, and it leads to a rather rigid type of system in which 'every action not mandatory is forbidden'.

But now an evolutionary 'arms race' begins. There is an advantage for individual tribes to combine forces: it helps the beginnings of agriculture and herding, and it helps protect them against predators – especially other tribes of their own kind. But if the move to tribes is advantageous, it may by the same token be advantageous for tribes to amalgamate into larger groups – so villages, and later towns, appear. The evolution of the town, and subsequently of the city, is analogous to that of the eukaryote cell, with villages playing the role of bacteria. We saw in Chapter 1 that the evolution of the cell was not simply a process of aggregation in which bacteria grouped together to form a colony of organelles. Instead, cellular evolution was a complicit, self-referential process which created its own evolving phase space of surrounding possibilities as it proceeded. Bacteria did not 'decide' to get together because it was best for them – but in the end they did 'get together', in a 'two steps forward, one step back' kind of way. Analogously, tribal villages did not *just* get together to form towns. There are at least two ways to explain what they did: a broad-brush 'phase space' one, and a more detailed structural one. We'll give you both, since they are complicit.

The 'phase space' version argues that a successful village culture necessarily becomes a victim of its own success. Tribal villages work only as long as the tribe does not outgrow its resources. If it gets too big, there is no longer enough agricultural land within, say, a few hours' walk; and other resources such as firewood or water also become scarce. When this happens the village either starts to die, or it is forced to adopt a new way of life. In order to continue its existence, the tribal village must explore its surroundings in culture-space, or explode into a new phase space altogether. The geography of phase space is flexible, but not totally arbitrary: the main possibilities are 'already there', constrained by contextual realities. For example, there is *no* region of phase space in which disease organisms fail to infest large groups of malnourished humans. So, in order to keep its snooker break going, the new culture must continue to cope with disease and keep its people fed. One of the attractive possibilities in phase space is the town – which is *not* just a village writ large. In a town, the activities and locations of its people are structured in a quite different way from what happens in a village. In a town, specialist groups have easy access to a large number of customers: this lets them specialise even more, and develop their own extelligence more rapidly and extensively. So you get trade areas such as the Street of the Leathermakers or the Alley of the Potters – just as in the eukaryote cell you get specialised organelles such as mitochondria. The relatively even population density of scattered villages gives way to urban centres, surrounded by a rural hinterland that supplies food to the centre and is in turn supplied with most other goods – shoes, pots, ploughs, horse-collars. From the phase space point of view the development of the town became inevitable once the village structure became overloaded.

The second kind of explanation offers more detail but a narrower perspective. It describes the transitional process itself, and not just the topological features of phase space that made it inevitable. A town is a collection of specialist 'patches' like the Street of the Leathermakers, and each such patch organises itself very like a village – with guild officials in the role of tribal elders, for instance. So is a town just a collection of villages? Did the villages simply aggregate?

They aggregated, yes – in a way. But not simply. They got together in the same sort of way that bacteria got together to produce cells: in a complicit and fairly rapid process of recursive self-modification. And just as the bacteria were reorganised into organelles, so the villages were reorganised into trade guilds and other specialist associations. The problem is that you can't just have a Village of the Leathermakers, another Village of the Potters, and so on, and then let them drift together. Villages don't work like that. So somehow a collection of nearby villages must, pretty much simultaneously, come together geographically *and* carve themselves up in a new way. Here's one possible way it might have come about ...

Imagine a collection of neighbouring villages, none especially different from the others. In one of them there lives an especially talented leathermaker. News of his abilities spreads far and wide, and many people come to his village to buy his goods. He prospers, and fathers many sons, who take on the family trade.♪ The same kind of thing happens elsewhere with pots, weaving, shoemaking, and most of the other trades that go to make up a town. The result is still a collection of 'separate' villages, but now they are linked in a 'autocatalytic network' of self-reinforcing economic activity. The farmer trades food to the shoemaker in return for shoes. The leathermaker gets candles (where did those come from?) from the shoemaker in return for leather. The farmer also trades cow skins to the leathermaker in return for candles (them again). The leathermaker trades leather to the saddlemaker, getting candles from him (still don't see where they come from) who trades a saddle to the candlemaker in return for candles (oh, right, now I see). And so on. Everybody takes in everybody else's economic washing, and it all works beautifully.

Now, and only now, can a town form. Its economic network is in place and working: next the geography has to change. Perhaps the local baron builds a castle for security. Other buildings begin to huddle round its walls, streets branch away from it towards the rural hinterland. The leathermakers' market is centred upon the castle – indeed their main customers are the baron and his nobles – so they relocate. Soon only the farmers, woodcutters, and other non-relocatable trades – plus itinerant knife grinders and a few traditionalist diehards – inhabit the more distant rural areas. The leathermakers, a family group, build their huts near each other. This concentration of leathermaking becomes an identifiable focus: it is easier to attract business if there is one centre and everybody goes there for that trade. It is also easier to buy or borrow tools and materials if similar traders are tightly grouped, so other leathermakers set up shop in the same area ... and so the Street of the Leathermakers is born. The same kind of thing happens to the potters, the butchers, the bakers, the candlestickmakers ... What carries over from the conglomeration of villages to the incipient town is the autocatalytic network of economic activity; but now that begins to change too. With all of its components in close proximity, it works better, and faster, and new avenues open up. Indeed the changed social structure creates new needs and new opportunities – the street vendor, the delivery boy, the town crier.

Most of this would have happened, incidentally, without the presence of the baron. The advantages of concentrating trade groups in similar locations would have attracted others anyway. But before it got very seriously established, the proto-town would probably have thrown up a baron anyway. Human groupings are like that: if they don't have a leader, one will emerge and 'take charge' – whether he is needed or not. So the political structure of the town evolves complicitly with its economic and social structure.

Social anthropologists spend a lot of time looking for repeatable patterns in such events. Is it *always* the leathermakers that start the ball rolling? Or is it the potters? Or does the baron building his castle always act as a seed around which the town coheres? We doubt that universal patterns of this kind exist: such details are mostly parochials. It is like breaking a stick and looking for a unique sequence of causes: does the bark split first, or the woody part, or the softer pith? What shape is the first crack? These details tell us *how* the stick breaks, but they differ from one stick to the next. The reason *why* the stick breaks is that the bending forces exceed the stick's elastic limit. You can predict *that* it will break, but not where, when, or how.

In mathematics this kind of effect is known as a *bifurcation*: a qualitative change that comes about because the system's original state becomes unstable as a result of other 'external' variables changing. Instability is a simple test which tells us *that* a bifurcation is occurring, but it does not tell us exactly *how* it occurs. What the system does just before a bifurcation, and what it settles down to just after, is usually fairly easy to describe – just as for the breaking stick. The only thing that bifurcation explanations cannot predict is the precise route taken *during* the bifurcation: this is enormously sensitive to very tiny 'fluctuations', and never exactly the same twice, because it passes through Ant Country. It is also irrelevant in most cases: the 'cause' of bifurcation is the initial instability, the 'effect' is where the system ends up. But in many sciences you find people trying desperately to track every tiny fluctuation that takes place during the transition, under the impression that it is there, deep in Ant Country, that the 'real' causes lie. But often it is better to think of them as 'side-effects' of the instability and the geometrical constraints on phase space that describe the original and final states.

The formation of a town is a bifurcation, which occurs because an economic network of villages becomes unstable. Not in a political sense – we're not envisaging riots here – but in the sense that certain types of small change naturally become amplified. A few people concentrate their trade in one small area, and economic and social interactions lead others to be attracted to the same place. So the concentration grows, gaining momentum. In a stable village system, in contrast, a small concentration like this will fail to trigger any important changes. Bifurcations, as we have said, wander across Ant Country: where they start, and where they finish, is describable and repeatable, but how they make the transition usually is not. So we can be pretty sure that in one town it happens to be the leathermakers who get together first, in another the potters, in another the candlestickmakers, whatever. In either case the other trades will not be far behind. The common feature, if there is one, is that autocatalytic network of economic activity – a cryptic pattern of processes, not a thing. Nobody has to be conscious of its existence, in any explicit manner, for the bifurcation to happen: the autocatalytic economic network 'primes' the network of villages for incipient instability, and once it is in place, the bifurcation is not far away.

In much the same manner, the city grows from a collection of towns and outlying villages. However, the analogy with the town is not exact: cities are larger, but their basic unit is still people. One important difference in the structure of cities is the ghetto – a word that now has many negative associations, but is derived from the entirely neutral Italian *borghetto*, 'borough'. Centuries ago, many Italian cities had a Jews' quarter, to which they were generally confined, and it is those overtones that the word now carries. Unlike the Street of the Leathermakers, a ghetto does not take part in the economic mainstream of the city. Instead it is a more or less isolated enclave with its own distinctive culture, whose principal aim is often to *remain* isolated. This may be because the rest of the city prefers it to remain a ghetto: 'if you marry one of *their* girls you'll turn into a pig / go to Hell when you die / have to live in the ghetto too.' Sometimes, however, it is because the inhabitants of the ghetto do not wish their culture to be contaminated by the immoral / unethical / sacrilegious (delete whichever is inapplicable) beliefs and behaviours of those nasty outsiders. Often both elements are present. In today's big cities ghettos often degenerate into crime-ridden no-go areas where it is dangerous to belong to the wrong ethnic grouping or come from the wrong street. This is a consequence, to some extent, of not being part of the economic mainstream. Today's cities also include areas that, like ghettos, are distinguished by a separate culture – but are actually much more analogous to the Street of the Leathermakers, because they are part of the economic mainstream. Think of Chinatown, the Italian quarter, the theatre area, and so on. These areas play quite different roles in the life of the city, and are analogous to the organelles in a eukaryote cell.

You can't get a ghetto in a town: it's too small. You can in a city, though not as easily nowadays as in the past. The absolute *size* of the population is not important: what determines the viability of the ghetto is the 'communication horizon'. Any closed culture is vulnerable to slow diffusion of new ideas from its edges, but it is far more vulnerable to rapid invasion at its centre, or across its entire extent. Communication lets in new ideas that can disrupt the rigid ghetto culture if they take hold. For example, on a larger scale still, the existence of fax machines made a substantial contribution to the breakdown of the former Soviet Union by preventing the ruling communist party from controlling the spread of ideas that it considered subversive.

To recap: the evolution of human social units – village, town, city – is analogous to that of the eukaryote cell. It is a complex process of complicit self-modification, and if there are common patterns they occur on the level of general processes, operating in the phase space of possible cultural systems.

As the number of people involved in the basic social unit grows, the political organisation of that unit must also change, otherwise the social system starts to fall apart. The entire town, city, or nation state must keep its *own* snooker break going, even as its citizens pursue their own disparate and often mutually conflicting snooker breaks, and that can no longer be achieved by having a few leaders who wander around cuffing miscreants about their ears. There are just too many individuals for the leaders to control them directly. The classic solution is the hierarchy: the system of tribal elders evolves into town councils led by the local bigwig, be it the Lord of the Manor, the mayor, or the baron in his big castle. To maintain order – meaning to ensure that the townsfolk do what the bigwig wants – there will be some kind of militia. Keeping the break going on the level of a town requires all sorts of additional structures, so it is to the culture's advantage to diversify – just as the differentiation of cells is to an organism's advantage, and the diversification of species enriches an ecosystem and (perhaps) stabilises it against environmental disturbances. In human culture, these structures often have an educational dimension – the young people must be taught how to fit into the evolving social system, so roles like the priest and the teacher arise. The tribal Make-a-Human Kit becomes more subtle, and responsibility for its inculcation is distributed over a wider section of society. Not just the parents and tribal elders, but the school teacher, the visiting nurse, the table tennis club, the beekeeping association, the local education board, the church, the National Heritage Secretariat ... A hierarchical political structure begins to crystallise out. In order to maintain it, the amount and type of teaching may become very restricted – 'enough to take orders but not to give them'. Only the ruling classes have full access to the cultural secrets and the levers of power. This maintains their status even as the culture grows. The trick is one of amplification. Somehow the decisions of a small number of people must be conveyed to all the rest – and imposed upon them if they object.

It is here that extelligence comes into its own. An amplified individual needs an amplified mind. The culture as a whole must remember the complexities generated by its own activities. The king cannot collect taxes if nobody knows who owns things or what they are worth. He cannot trace his line of descent back to the gods unless that line of descent is stored in a form that can be recalled at will and without dispute. For a time it was just about possible to do this using the unaided human memory: specially trained people would remember the Holy Texts word for word at enormous length. But human memory is fallible, equivocal, and dies with its possessor. The answer is to replace it by some physical process that creates a more or less permanent record. And it is here that humanity's downhill bicycle race began, with the invention of writing. Language created the possibility of self-complicating extelligence, but the limitations of human memory kept it contained. When writing was invented, extelligence escaped from our control.

To start with, the race was run at a fairly slow speed, on the flat. The technology was a pale shadow of what we now have, the quantity of written material small, the individual's ability to access it limited. However, just ahead –in phase space – the flat began to slope downwards, almost imperceptibly, as the written records began to acquire a momentum of their own. The Holy Texts, for example, did not bind only the ordinary people. They also imposed constraints on the actions of their rulers. Tension between the priesthood, the nobility, and the monarch led to disputes whose resolution often rested as much upon the interpretation of the Holy Texts and the written law as upon force of arms. Social cohesion rested upon a small number of rigid institutions – the state, the church, the military – and was enforced, in no small part, by the rule of extelligence. Such a system can be astonishingly stable: keeping the snooker break going does not require constant social upheaval and change. On the contrary, a rigid, unadaptable system can survive for a long time if it gets its hierarchical structure right – meaning that the next generation not only fits neatly into the same slots as the previous one, but it ensures that the next generation after that does exactly the same. 'Do it this way and ensure that your children also do it this way.' So all over the world, from mediaeval Europe to China, we find various types of feudal system, all structured along much the same lines, and all entirely stable. The world might have remained feudal forever, were it not for sources of change *external* to that kind of social structure. They led to the runaway train of village–town–city–nation, and complicitly with that they also led to flexible, self-modifying, self-complicating extelligence.

One such source was barbarians. Not all individuals can function – or are willing to function – in societies with rules. All societies face the problem of what to do with unruly members. They can be imprisoned, or killed – but often they are merely expelled, and even if not, some will escape, or anticipate retribution and leave before it arrives. Groups of barbarians – whose existential imperative is *not* to obey rules – will appear in the countryside away from the towns. They too will grow, though not beyond tribal size. But a gang of barbarians can be more than a match for a town full of docile law-abiding citizens – as we currently see in many parts of the world, for example in Afghanistan, where a mere 10,000 barbarians are imposing a fundamentalist state on an unwilling populace of millions. What happens when the barbarians strike back? The traditional image of marauding Vikings, pillaging town after town, is overly romantic. What typically happened was that the barbarians invaded, took over the town, settled down, and became farmers, town councillors, priests, and teachers. Their underlying objective was, in fact, to settle down, to reclaim the life from which they had been expelled. The barbarian parasites rapidly became symbionts. Barbarian invasion, however, did not merely reconstitute the existing culture. Instead, it imported all sorts of extracultural baggage – barbarian practices, items that the barbarians had picked up from other cultures. The infusion of new practices, new trade goods, and new ideas expanded the rigid cultures of the townships, and mixed them up with each other. Nearby towns therefore became more like each other, which facilitated interaction between them. It wasn't all sweetness and light – rivalry is strongest between siblings – but the general tendency was towards increased interaction. Isolationism and barbarians are not compatible.

A second source of enforced change was the gradual accumulation of new problems caused by the presence of unprecedentedly large numbers of people. Methods that work for small groups may fail as the group expands. Slash-and-burn agriculture works fine for a small group in a large forest – not because they are romantically 'in tune with nature' but because they lack the power to mess it up – but fails for a large group that has already burnt most of the forest down. So practices that worked when the towns first formed were forced to evolve, in order to deal with their own inevitable consequences. Lurking in phase space, surrounding cultural practices that could not generalise to larger numbers, were the seeds of those practices' own destruction.

A third important source of change followed much the same course as barbarian invasion, but on a larger scale and for different motives. The movement of large numbers of refugees, displaced by wars, caused previously distinct human cultural systems to come together. So did immigration, inspired by economic motives, political repression, or just a sense of adventure. The results ranged from the sterile confines of the ghetto to the flourishing multiculture of the New York docks – often, as here, occurring in the same city.

A fourth source of change, piggybacking complicitly on all the others, was technological innovation. Technology turned the bicycle race into a truly downhill one, whizzing into its own future at whirlwind speed, building that future as it went. It started slowly. Every so often some bright citizen would come up with a new idea, and even the most rigid society would make the mistake – in its terms – of deeming a few such innovations harmless. Maybe it was just an improved horse collar (one of the big novelties of Middle Age Europe). What threat could such a thing pose to the ruling hierarchy? But better horse collars mean more effective agriculture, leading to either more serfs or healthier serfs, or both. And those mean long-term trouble for the feudal lords. Not only that: the *idea* that the horse collar can be improved leads to the idea that maybe other things can be improved too. It was a positive feedback loop: the faster the bicycles moved, the steeper the slope became; the steeper the slope, the faster the bikes raced. Each innovation made it easier to innovate the next time round. Each piece of evidence that innovations produce benefits lowered resistance to the next innovation.

So the slope may have been imperceptible, but ahead it began to steepen. In phase space, where 'ahead' makes sense as a description of the dynamic, the prognosis was as visible as it is in real space when a bicycle is racing head-on towards a cliff edge. The ground may be flat *now*, but look at what's coming ...

The accumulation of change caused a change of strategy in the cultural snooker break, one that was necessary if the break were to avoid grinding to a halt. It led to a different kind of organisation of human beings, which for want of a better word we shall call 'civil'. In human culture it took the form of organisation on a national scale, together with relations between nations. Different nations adopted different cultural solutions to identical questions – democratic government in some, dictatorship in another – but increasingly their deep structure became similar, even as their surface structures diverged. This is why international trade, for instance, is so widespread – it is a cultural universal. The downhill race gathered speed as the positive feedback went round and round the loop, and increased extelligence provided the power to increase the extelligence at a faster rate.

This is the stage that human organisation has now reached. Although we are not organised on a planetary scale, in the sense that we have no global government (with any teeth – forget the United Nations, that's just a talking shop, though a useful one), our lives are *influenced* on a planetary scale. A squabble between two Arab states in the Persian Gulf, for example, pushes up the price of oil worldwide and creates instant crisis in every country. A new disease arising in one part of the globe can rapidly be transmitted to every corner, thanks to air travel. (But it would have arrived, a few years later, even in the days of sailing ships.) The use, by developed countries sited mainly in the Northern hemisphere, of aerosol sprays powered by CFCs, can create a hole in the ozone layer that affects peasant farmers in the Andes and sunbathers on Bondi Beach. In the words of Marshall McLuhan, we are living in a global village.

Unfortunately, we are having a really hard time adapting to the globalisation of culture. Individual people are part of many different cultural systems on many different scales – family, village, county, nation. But our minds are small, our visions narrow: we tend to see ourselves most strongly as part of the smaller groupings, even though it is the big groups that have the largest effect: a cousin can steal your girlfriend, but a king can steal your land, your home, and your life. In immigrant communities, the first generation – the ones that brought their culture with them and experienced it in its original homeland – hang on to their traditions like grim death and try desperately to retain them even when they have become totally inappropriate to their new environment. Most of them adapt in small ways, but the further divorced from daily reality the traditions are, the more rigidly they are enforced. The second generation immigrants form a transitional group, and they usually get the worst of both worlds. They are adaptable enough to *want* to change – to see that it is inevitable, even desirable in some ways. But they have to fight the rigid beliefs of their parents. At the same time, they are not accepted very readily by the greater society that they wish to join. Neither one thing nor the other, the second generation immigrant tends to lead a double life, expressing one set of views at home but a different set at the office or in the pub. A good example is an acquaintance of IS who comes from a culture in which marriages are always arranged by parents, and often bride and groom do not meet until the ceremony. When *his* marriage was so arranged, he professed himself to be entirely happy with the procedure – but at the same time he stated emphatically that in no circumstances would he ever enforce it upon his own children.

Despite increased parental tolerance, the third generation usually has the biggest problems. The second generation either stayed within the walls and the traditions, or 'married out' and lost the traditions entirely, becoming assimilated into the hodge-podge of the external culture. By the third generation, however, some economic stability has appeared. Many second generation immigrants have jobs, businesses, or political power, so their children have a chance to test how tolerant the surrounding society *really* is to 'different' kids who become successful by the outside society's own standards– and maybe even do better than many in the outside society. If the level of tolerance is high enough – and it helps if the surrounding society is fairly affluent, since it is then less likely to see the newcomers as a serious threat – then the third generation immigrants become integrated, not just assimilated, and they can even retain much of their original culture. But if not, there will be trouble.

The process is seldom easy or painless. Because people – not just immigrants, but everybody – see themselves as sitting within the small, local groups, they lose sight of the big, global ones. They know that governments, for example, pass laws. They also know that laws affect their lives, often by preventing them doing things they want to do. What they don't always see is that they can have an *effect* on their government – by voting, in a democracy, or merely by passive and concealed disapproval in an autocracy. So even though the ordinary people outnumber their rulers a hundred to one, they feel powerless. In modern democracies fewer and fewer people are bothering to vote, in part as a result of this perception. It takes only one election to get rid of their tormentors, if they really want to make the effort, but instead of participating in a civil process, they resort to processes on their own individual level. They try to resist change – even while participating in it, unknowingly – and the easy way to do this is to be isolationist. This won't work, but they don't see that. So subnational cultural groups like the Basques, the Kurds, and the Shiite Muslims, would become separate nations if they got the opportunity; while what used to be subnational groups like the Serbs and the Croats have done just that. Some groups in Northern Ireland want to unite with the Irish Republic, some want to be part of the United Kingdom, and some want autonomy. None of them have noticed that the world is now becoming structured so that none of these choices will make any serious difference: they should be worrying about the stranglehold that multinational corporations are gaining on everyday life, global warming, and arms sales, not which party runs the Town Hall. We live in a global village, but we are behaving like a bunch of villagers who have just been thrown together and have not yet evolved any kind of tribal hierarchy or organisation. One of them is beating his wives, one has diseased cattle that could start an epidemic, one has burnt down his own hut and is now threatening to evict his neighbour and steal his, one has produced badly behaved children who are running amok in the fields and ruining the crops. Each of them considers himself to have complete autonomy over his own actions, citing 'national sovereignty' in justification.

It won't work. It isn't working now. Its time has long gone.

The break-up and realignment of rigid national cultures became inevitable as soon as the world became truly global. Women in traditional Islamic countries can no longer be kept unaware that in the rest of the world they would have many freedoms currently denied to them – freedoms as simple as being able to show their faces in public or to drive a car. The old ones are reconciled to it, but the young ones don't like it. They are 'second generation immigrants' in their own country: instead of transporting their own extelligence into a new environment, the environment has transported its extelligence into theirs.

There are two ways to respond to such pressures. One is to bow gracefully to the inevitable, adapt, bend with the wind. There will be some loss, but seldom anything essential; the gain will be far greater. The other is to resort to ever more extreme measures to keep the outside world at bay. It works, for a time; but it leads to extremist government, terrorism, violation of elementary human rights, and economic collapse. Fortunately, cultures are not replicative systems, but reproductive ones. They do not copy themselves inflexibly from each generation to the next: they modify themselves – usually against resistance, but that is largely the stabilising force of conservatism – to cope with a changing cultural environment. If you want to keep the snooker break going when the rules of the game are changing, then you have to adapt your strategy accordingly.

You might ask how a culture can possibly do this if everybody is not just learning key things like language and social convention for themselves, but constructing their own learning kits interactively and recursively as they grow. Actually, this flexibility makes it *easier* to keep the break going: it stabilises the culture, letting it adapt to changing circumstances without coming to pieces. In any case, a lot of the process is highly repetitive, just as a snooker break involves potting an awful lot of red balls. The biological development of all human beings during embryology and babyhood is very similar, and so is the succession of environments, morals, myths, and manners, that they find themselves in. The separate strata – age groups, near enough – of a given human society differ from each other far more than the same strata do from one society to another. So humans grow up in much the same way from one generation to the next and between one society and another – despite the ten per cent differences in genetics and many differences in education.

Paradoxically, it is more difficult to understand why most people have broadly similar viewpoints about the 'fixed' and 'objective' physical world around them. They are free to build their own minds, so what keeps their figments real?

We offer two rather different possible answers. first, they interact with the physical world as they play, walk, fall, and see others doing these things. There is a lot of evidence, exploited commercially in 'early learning' toy shops, that babies who have more, and more varied, physical experience are better at many things in later life. Nevertheless, babies that are 'swaddled' or otherwise kept from normal interaction with their environments – normal by modern standards, that is – still seem to acquire adequate motor skills. Perhaps this is like baby birds 'learning to fly', where the actual learning involved is less than appears ... but we rather doubt it. The second answer is that the succession of environments and tasks that children experience is very constrained, which strongly suggests that societies have been 'selected' in the sense that any society that gives its children inappropriate ideas about physical reality eventually fails. For example, the fatalistic view that disasters are 'the will of Allah' encourages societies not to take action to prevent them or ameliorate their effects, so in the long run a really big disaster will wipe them out. If we are correct in thinking this, then societies that survive will be those adopting educational strategies that encourage children to accept physical realities. Certainly we think that this is how different societies become characterised by different technologies. If you grow up in a nomadic group then you know all about tents, goats, and yoghurt; if you grow up in modern Manchester you know about cars, videos, and Nintendo. So the overall cultural environment and viewpoint affects the engines of change that drive that culture in new directions.

Now a really important problem raises its head. How, in such a reproducing system, can an unreasonably effective system for dealing with *generalities* arise? How do you get mathematics or science? The standard story starts from astronomy, its regularities and the unreasonable potency of the ability to explain or predict such phenomena as seasons and eclipses. But if the science or religion that develops deals with such things as 'Mary, come into the cave at once, I told you not to trust those big mammoth things!' how can it later deal with bicycles and computers? The answer is that we have evolved a general cultural strategy, not just a fixed list of tactics. Extelligence began as a list of prescribed actions – 'don't trust those big mammoth things'. But the underlying idea was that *any* kind of information could be passed on to succeeding generations, not just rules for avoiding being trampled by mammoths. When you have invented a bottle that contains a genie, it is only a matter of time before some inquisitive soul pops the cork. The human mind, complex as it is, does not have the capacity needed to invent complicated science and technology from scratch in each generation. We managed it because we discovered the *universal* trick of extelligence – creating an archive of cultural experience and know-how, which can be accessed by any individual who knows how, and can be augmented by any individual who knows how. It is the idea that archives can be created, not specifically what is in those archives, that really got us off the ground. Extelligence has brought us part way towards a genuine multiculture. Our children already understand that better than we do – the new generation grows up in the present culture, while the old one remains bound by the culture in which it grew up. This is why parents do not understand their children, and it is one of the obstacles that makes it hard for a true multiculture to evolve. But evolve it will, as it always has done, following a complicit path, building its own phase space as it goes.

A multicultural world will require new kinds of thinking. In 1966 the science fiction writer Samuel R. Delany published a little-known but highly unusual novel called *Empire Star*. The main narrative line is a convoluted time-travel/many worlds story which eats its own tail. It concerns the (rumoured) freeing of the Lll, a race of creatures whose architectural abilities are so immense that they have become slaves, so desperately needed throughout the galactic empire that they cannot be permitted to be free. But the philosophical subtheme is a remarkable study of how human minds structure their vision of the universe. 'Simplex, complex, multiplex' is a running catch-phrase.

The test for a simplex mind is to ask it what is the most important thing in the universe. If it answers, then it is simplex – be it a mind on the backwoods satellite Rhys, which knows that the most important thing in the universe is its staple crop, plyasil, or the incredibly sophisticated technoculture writing the Encyclopaedia of Everything, which is similarly focused on a single overriding goal.

A complex mind can perceive the many intertwining strands of cause and effect that combine, within some consistent worldview, to constrain and control the unfolding of a particular selection of events. Complexity is a state that is inaccessible to the vast proportion of the human race, but as the global village shrinks, more of us take the complex view.

Rarer still is the multiplex mind, which can work simultaneously with several conflicting paradigms. It sees not just one interpretation of reality, but many, yet it sees them as a seamless whole. Such a mind is untroubled by mere inconsistency: it is comfortable with a mutable, adaptive, loosely coherent flux. The real universe (inasmuch as there is one), says Delany, is multiplex. Order your perceptions multiplexually, and you will understand the universe on its own terms.

In *The Broken God* another SF writer, David Zindell, added a fourth type of mind: omniplex, embracing the Cosmic All. But Delany would have denied that there was a Cosmic All to embrace, as do we – because the concept of omniplexity is a simplex thought. A very *big* simplex thought, but simplex for all that. Figure 26 illustrates Delany's three ways of thinking. A simplex mind wanders along a single, fixed axis. A complex one explores the region around several axes, by making small excursions away from one axis in the direction of the other(s). A multiplex mind explores the uncharted territory *between* the axes, making unexpected connections. Complexity produces explorations, multiplexity produces explosions.

There was a time when human cultures were simplex. Tribes living in small villages or travelling in hunter-gatherer clans shared a single vision of their place in the world. However, the simplex, tribal mind cannot deal with external realities that go outside its limited experience. The tribal snooker break continues because all tribal authority rests with the priests, but when the priests do not know what to do ... typically they do the wrong thing, and disaster follows.

The barbarian approach is equally ineffectual. Operating by social rules such as 'honour', 'glory', 'prowess', 'vengeance' – figments that have not evolved complicitly with the outside world and therefore fail any stringent reality-check – falls apart because it cannot mount a coherent response to a substantial new threat.

As the human population increased, cultural alignments were forced to outgrow the bounds of the simplex. Societies were still composed mainly of simplex individuals – the priest who *knows* that the proper order of worship of the rain-god is the only thing that really matters, the teacher who sees all human activity as the passing on of knowledge, the builder whose sole goal is to build bigger, and higher, than his predecessors – and damn the consequences. As Delany well knew, simplex thinking is not confined to the elders of 'primitive'

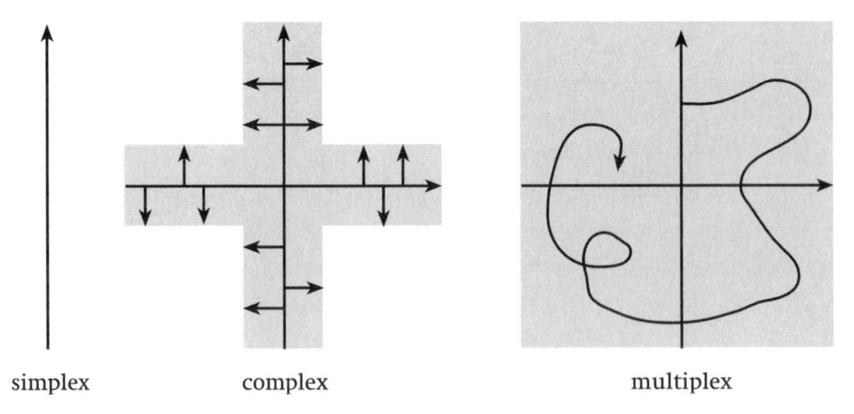

Figure 26 *Visual representation of Delany's classification of ways of thinking.*

tribes. Many of today's religious leaders are equally simplex in their thinking, for example by equating contraception with abortion and abortion with murder, so that the overwhelming need to control human populations becomes bogged down in small-minded moralising. Politicians are typically simplex, especially the narrow nationalists, although they themselves recognise the transcendent quality of 'statesmen', whose thinking is generally complex.

From the interactions between simplex individuals there emerged a complex culture, a civilisation. For a civilisation to keep its snooker break going, there must be enough complex thinkers. Even today, surprisingly few people can cope with complexity, although some are slowly learning to comprehend the chains of cause and effect whereby a safe seabed WWI weapons dump becomes a nightmare for parents on the west coast of Scotland, or the reduction of the sulphur content of coal, intended to combat acid rain, also deprives crops of an essential nutrient. Across the globe, different cultures have evolved their own way to handle complexity. As a result, today's global culture is more than complex: it is bordering on the multiplex. And if it is to function as a true multi-culture (and not as just a Balkanised rag-bag of competing complex mono-cultures) then we must 'order our perceptions multiplexually' as Delany put it.

Easy to say, hard to achieve ... But we think we can give you an idea of what is involved. Let's go back to the story we told about the formation of a town. There are many ways to tell it. The simplex baron sees it this way: 'I am powerful, I am a leader. People do what I tell them to. When I built my castle I created a centre around which I could then build a town. The purpose of the town is to support me and my family.' The equally simplex leatherworker sees it quite differently: 'It was us leatherworkers who started the whole thing, you know. Wasn't anything here until we started to expand our business. People started to gather here because of the trade opportunities. Then the baron horned in, like the nobs always do, and built a castle in the middle of everything. In fact, the only reason he can keep going is that we supply his troops with leather jackets for armour, and saddles for their horses. If we stopped, he'd be in very serious trouble – but it suits us to keep him in power. Law and order is good for stable trade. The purpose of the town, then, is to support us and our families.'

Two explanations, both simplex. Both think there is only one important thing in the universe, and that they know what it is. A complex explanation would combine elements of both viewpoints, seeing both the baron and the leathermaker as being partially correct, but blinkered. It would tell a step-by-step story, alternately taking the baron's view (but with leathermaker bells and tassels) and the leathermaker's view (but with baronial decorations). Most of this book gets no higher than the complex level, and a lot doesn't even get to that. A multiplex explanation would contemplate, at the same time, the phase space in which the town became inevitable, the many different routes by which that inevitability could have been realised, the special accidental features that led to *this* town being realised *this* way, the many alternative viewpoints that would be taken by the individuals who built it ... In particular it would see the town *as* a town, the system as a system, and not make the mistake of attributing features that are important on an individual level (such as purpose) to the system as a whole.

So far, humanity's cultural development has mainly come about by thinking simplex thoughts. Now it is starting to put some of them together to attain a complex viewpoint – maybe, sometimes, it even comes close to a multiplex one. To the Zarathustran eightfold group mind, in contrast, multiplexity comes naturally: it is the simplex worldview that baffles them.

Watcher-of-Moons is preparing to depart from the Solar System, and the atmosphere on board is becoming extremely tense. Liar-to-adults, characteristically, has reminded his octuplet of the need to comply with its philosophico-legal obligations. The others had hoped that he would not raise what, for civilised Zarathustrans, is a rather indelicate matter, but of course he did anyway. The future of the human race now rests upon their collective interpretation of ancient, obscure, and draconian law. We are at the mercy of the Zarathustran extelligence.

Ringmaster I am attempting to assemble a consensus on the fate of these humans ... but I am experiencing severe difficulties.

Liar-to-adults Nonsense, you know *exactly* what our collective decision is. The more we learn of these strangely unoctimal creatures of the planet Earth, the more disturbed we become. We must implement the Octual Directive without delay.

Destroyer-of-facts Yes, but in which interpretation? Lenient or severe? I must say that personally I find their unorthdocty intellectually stimulating. Already it has suggested to me a new way to phurgalise transic veritables in the Gurgenfuff calculus, and –

LtA I am not interested in your trivial insights into the fundamental nature of thought. My point is that their singlemindedness is in flagrant contravention of The Regulations.

The regulations Minute 107B, subparagraph 22xx(iv)B/epsilon. [*Pauses, and turns pale blue in embarassment. Hurriedly adds:*] As revised in appendix Q/444/x.

Creator-of-creations Adultliar, I hope that I do not sense your implication. You can/cannot (delete whichever is applicable/inapplicable (delete whichever is applicable/inapplicable (delete whichever is – [*He is so upset that he plunges into infinitely recursive multiple-choice mode and has to have his feet tickled to disrupt the endless cycle.*]

R Liar-to-adults: I must warn you now that it is graffited on The Regulations that we may not exterminate alien life-forms.

LtA Provided that their thought patterns do not pose a threat to our way of life. If they do, then we are *required* to exterminate them. And the threat is evident. Not only are these creatures singleminded: they are simplex! [*The others are stunned by the realisation that Liar-to-adults is right.*] Hewer-of-wood, go and get the – you know. [*Hewer nods and sets off for the weapons room. Shortly he returns with a small black sphere from which project eight yellow buttons.*]

Hewer-of-wood One exterminator. You point it at the planet, you each push a button, and – boom. [*The silence grows unbearable. Hewer begins a new carving, and wood chips fall softly to the floor as the tension mounts.*]

DoF [*Reluctantly.*] This does not smell at all octimistic to me, but there seems to be no–

HoW Hmm. Curious. [*He waves his latest carving, a hollow-looking holey thing.*]

LtA [*Annoyed at the interruption.*] Not now, Hewer – wait. What *is* that ridiculous object?

HoW I call it an antinurphle.

LtA A *what*?

HoW Well, I nurphlerised that this piece of wood really contains a snoozosaur tusk. But instead of carving away all the rest, to reveal the tusk inside, I – [*He starts to snivel at the enormity of it all.*] –I *left* the outside and carved away the *tusk!*

DoF Ingenious.

HoW Well, it seemed an equally valid way of looking at it. [*Suddenly realises he may be out of order.*] Um – did I do right?

Performer-of-amusements [*Surprising everybody, himself included.*] Yes! Oh, yes! An equally valid – Hewer, you are what the terrestrials would call a genius!

LtA Nonsense, he is perfectly sane, just slow-witted.

PoA I mean that he has reminded us all of a great truth, the philosophy of phase space. We must focus not just on what we perceive, but on its context as well. We perceive the humans' simplexity, but we fail to observe that it is surrounded by potential multiplexity. They have just not got there yet.

R What do you mean, Performer?

PoA Well, *we* perceive the multiplexity of the universe with ease, but maybe – look, Destroyer-of-facts, you must help me. I am only a poor Performer-of-amusements, but it seems to me that we Zarathustrans naturally order our perceptions multiplexually because we are an eightfold group mind. Humans, being singleminds, think the other way up.

DoF [*Puzzled.*] Oh. You are saying that the humans subconsciously seek to *assemble* a multiplex view by combining simplex ones into complex ones? But that is a very silly way to go about understanding a multiplex universe.

PoA I agree ... but you forget, they are singleminded! How *else* can a single-mind achieve multiplex thinking?

DoF Great Carer, he is right! Where we see a seamless conceptual plane [*actually he used the term 'octospace', an eight-dimensional analogue of the plane, but for simplicity his concept is here transcribed into two dimensions*] they see two independent one-dimensional simplexes. They construct conceptual 'coordinate axes' on the plane, and project reality on to them.

LtC Such as the axis labelled 'mathematics' and that labelled 'biology'?

DoF Precisely. I know that it seems strange to separate the two, but these humans do not realise that coordinate axes are an arbitrary construct without intrinsic meaning. Whereas *we* grasp the plane as a unified entity, and therefore do not need to assemble it from arbitrary axes.

HoW [*Admires his antinurphle and mutters happily.*] I reckon I could get this on the Discarvery Channel ...

LtA [*Dismissively.*] All very well, but that still means that humans see the universe upside down compared to us. Hewer, the extermi–

PoA Yes – but by the same token, *we* see the universe upside down compared to *them*. We find it very difficult to be simplex – our Ringmaster is the mechanism by which multiplex understanding becomes simplex action!

R Zmmmph ... It is a difficult consensus ... but clearly the question is whether we would achieve a deeper understanding and a more effective action if we dealt with these creatures through extermination / cooperation.

Delete whichever is inapplicable

If we could reorganise our world along multiplex lines, then it would function more smoothly and more effectively. For example, take science and technology. Simplex officialdom sees scientific research as anarchy, and is convinced that with a bit of organisation the whole thing would function far more efficiently. But a multiplex view shows just how wrong this is. Working our way up via a complex level, consider a recent example, the development of a new X-ray lens for the microchip industry.♪ A microchip is a tiny electronic circuit, etched into a fragment of silicon crystal; the smaller the circuit's components, the faster it works, and the more powerful it becomes. The design is transferred on to the silicon by a technique known as photolithography – a beam of parallel light rays shines through a stencil and wherever the light gets through it affects the atomic structure of the silicon. Current fabrication methods are close to their limits, because of diffraction – wave effects that soften sharp edges, making them too diffuse to create workable circuits. One plausible way forward is to use a beam of X-rays, which have a shorter wavelength than ordinary light, which makes the distorting effect of diffraction smaller. Unfortunately, the shorter wavelength also means that X-rays cannot be rendered parallel by a lens. After twenty years' research, this obstacle has finally been overcome by a team of scientists from Leicester University in the UK and Nova Scientific in the USA, who have invented a successful X-ray 'lens'. But officialdom is incapable of anticipating, managing, or even comprehending the route that led to this breakthrough. The X-ray lens originates in a study of the lobster eye, made in 1978 by Mike Land and Klaus Vogt, and its technological implications were first realised in 1978 by Roger Angel – whose interest was the design of an X-ray telescope for astronomy. How come? It's quite simple. The lobster's eye does not have a single aperture, like ours: it is constructed from a multitude of tiny facets, a bit like a fly's eye. The facets do not possess lenses: instead their walls act like mirrors. The incoming light grazes the mirrors at a slight angle and is reflected back to receptors at the base of the facet. In 1978 Land described this structure in *Scientific American*. Angel read the article, and realised that a similar method might be used to bring X-rays to a focus – because although X-rays cannot be bent by a lens, they can be reflected in a lead glass mirror, provided they graze it at a very slight angle. In those days X-ray telescopes had a single aperture, resembling an old-fashioned pinhole camera, so they didn't give good quality images. In the *Astrophysical Journal*, Angel proposed combining millions of pinhole cameras into a lobster eye. The technological difficulties of realising this design proved substantial, but over the last twenty years the necessary glass-making techniques have been refined to the point where such a device can be fabricated. The Leicester astronomers are currently building it into an X-ray telescope due to be launched into space on a NASA satellite in the year 2001. They also spotted the potential for microchip manufacture, which is immense: the prospect of new, faster, smaller chips could easily spawn a multibillion dollar industry.

The simplex mind cannot cope with this kind of thing. If zoologists had put in a grant proposal twenty years ago, asking for funds to study the lobster eye because when allied to glass-making and astronomy it would have a major effect on the electronics industry, they would have been sent packing. Worse, they could not even have put in such a proposal, because the microchip industry was in its infancy in those days, and X-ray fabrication was on nobody's agenda – certainly not as a realistic prospect requiring immediate attention. Anyway, who could have the imagination to see such an application in advance? From zoology to electronics by way of astronomy, from natural science to the marketplace by way of physical science. Such a transfer of ideas became possible only because of the evolution, over the past twenty years, of a suitable scientific and technological 'autocatalytic network'. Just as the leathermakers' goods were useful to the shoemakers, their shoes were useful to the farmer, and his eggs and bacon fed all three trades, so the idea that came to light in the lobster eye sparked off the design of an X-ray telescope, the microchip industry was waiting for a similar gadget, and the glassmakers came up with a way to manufacture it. It required *the whole network* to be in place, potentially, before anything important for industry could happen. Nobody could have foreseen it and set in train a single 'goal oriented' line of Research and Development – and even if they had done so, the extelligence generated would be too linear and too impoverished to spawn other equally unexpected products. Simplex management of scientific research is futile and counterproductive. A multiplex view, in which science is given as much freedom as possible to evolve its autocatalytic networks of ideas, and industry is poised to recognise them when they are ready to create new products, would work far better. Improved communication across disciplines, adding new connections to the network, can assist the process (but not if the connections become so rich that meaningful ones get lost amid the noise). A town whose simplex inhabitants *all* decided that leathermaking was so much more profitable than any other trade, and abandoned pottery, weaving, and baking to make leather, would rapidly fall apart. For exactly the same reasons, our scientific extelligence will fall apart if research is allowed to be managed by simplex minds. This, by the way, includes simplex scientists involved in allocating research funding, as well as bureaucrats. Not so long ago, most of the world's biology departments got rid of all the old fogies doing things like zoology, botany, and taxonomy because they *knew* that the only important biology in the world was molecular. Now that environmental problems such as algal blooms, depleted fish stocks, and loss of biodiversity in rainforests are on the agenda, they are trying desperately to reconstitute the extelligence that resided in those old fogies – which is not so easy.

Bring on the multiplex manager.

Bring on the flying pigs.

A final question. Our involvement with extelligence is a downhill bicycle race – we may think we are studying the lobster's eye, but before you know it we're making faster computers. We can't predict the nature of tomorrow's extelligence, but we can ask where it might be taking us. In answering this question we shall try to adopt a multiplex view, but we'll have to work up from simplex ones.

At one extreme there is a relentlessly pessimistic scenario, which in effect sees no future beyond global anarchy, violence, and war – except, perhaps, by severe repression. 'A foot stamping on a human face, forever,' as George Orwell put it in *Nineteen Eighty-four*. But he was targeting the Communist system, and that has fallen apart, precisely because of the forces of global extelligence. The foot did not stamp forever, because after a while it ran out of human faces to keep an eye on the production of its boots. It took a simplex view and was clobbered by a multiplex world.

And that leads to the more optimistic scenario – that, just as the tribe found a way to accommodate individual differences and still keep its snooker break going, a global multiculture might do the same. At this other extreme we find the future civilisation envisaged by Iain M. Banks in books like *Consider Phlebas*, appropriately known as 'the Culture'. The Culture is civilised in the best sense: its people are intelligent, sensitive, ethical. Their technology is stunning – superhuman artificial intelligences known as 'Minds', faster-than-light starships, robotic 'drones' with a wicked sense of humour. Their mastery of diseases and their ability to refashion the human body are so complete that they use them for recreational purposes. The Culture is diverse, a true multiculture: everybody is an individual, and there are few restrictions on personal freedom, yet it keeps its parallel, distributed snooker break going very successfully. It may sound anarchic, but it is not: everybody understands exactly what they must do to keep the break going. In Delany's terms, the people of the Culture take a multiplex view, especially when it comes to politics. The Culture is a pleasant system in which to live, if you respect its way of thinking, but it is in no way 'nice'. Internally it has few serious problems, but it is very quick to recognise a threat from outside, and can be utterly ruthless in dealing with it. Like Liar-to-adults, the Contact section's Special Circumstances unit can and will take responsibility for destroying an entire alien civilisation – and do so – if it is genuinely unavoidable. One of its subtlest weapons is moral espionage. The Special Circumstances unit is frowned on by most citizens of the Culture, for in principle it represents the antithesis of all that the Culture stands for – but it is also tolerated as an unfortunate necessity, because the Culture will not permit its ethical sensibilities to override its own survival. The Culture would never make the mistake of extending freedom of speech to those who wish to destroy freedom of speech, or of being tolerant to those whose intolerance becomes intolerable. It's not a perfect model for a planet-based multiculture, but it has the right kind of ambivalent, multiplex attitudes, and the right dose of common sense. It is the kind of multiculture to which humanity could reasonably aspire, out there in a distant and rather hypothetical region of culture-space, drawn on a galactic scale.

What of the real future, down here, on this single planet? Instant global communication is upon us, and that particular genie can never be stuffed back into its bottle. The geographical boundaries that helped to stabilise international relations by creating cultural equivalents of ghetto societies are no longer effective against incursion by outside ideas. Virtually the whole of human extelligence will soon be available to anybody who wishes to have access to it. Repression and moral panic can keep such access under control, for a time, but the simpler and more widespread the technology becomes, the more difficult this will be. The old cultural walls are collapsing all around us. While adults dither and refuse to make difficult choices, their children are already refashioning tomorrow's culture. The reason is that children have to live in the present, whereas most adults live in the past. The children are in the position of the second-generation immigrant, except that their subculture has a much shorter cycle time than adult culture – it takes only a few years for today's teenagers to be replaced by the next batch – and it evolves faster. This is why parents can never understand their kids, and why adult society is totally incapable of coming to grips with new cultural phenomena, be they good or bad, like drugs or programming the video. Those who are unwilling or unable to adapt may try to divert us into the pessimistic future of *Nineteen Eighty-four*, and for a time, and in some places, they may succeed – are succeeding. But their grip is tenuous, limited by the human lifespan, and the extracultural pressures are constantly building all around them.

Like it or not, the world is changing, and it is starting to change as an interconnected unity. As the world of the Culture suggests, unity is in no way inconsistent with diversity: we seem to be heading for a multiculture, not a monoculture. Extelligence has become a force for change in its own right. It has given us more ability to influence our present, and our future, than at any other time in history, but by the same token its influence on us is becoming irresistible. And for all our political upheavals and petty nationalistic confrontations, it seems to be guiding our separate cultures towards a successful multiculture. The proof is that where this *fails*, as in Bosnia or Northern Ireland, almost everybody is disappointed. We don't stand and cheer them on, as earlier humans might well have done. Our expectation, honed by extelligence, is that cultures should see the benefits of cooperation and integration. Everybody outside the conflict sees at once that 'ethnic cleansing' is a recipe for political and economic disaster, not an answer. For these reasons we are optimistic that humanity can become a successful context for the planetary ecology, and maintain its trajectory towards a global multiculture of the kind that Banks projects into our far future.

There will always be tribal groups, nations, wars, just as there are still bacterial ecologies and anthills. Over-population may yet spread our resources so thin that extelligence gets bogged down. Fundamentalists may yet take us back to the Dark Ages. But our hope is that the over-arching multiculture will come to include most of humanity, perhaps eventually emulating The Culture by spreading to other worlds, other stars.

Our minds, our societies, our cultures, and our global multiculture, are all evolving within a reality that we mould in images of our own creation. We are a figment of reality – but reality is increasingly a figment of us.

Epilogue

The Ringmaster of the Zarathustran cruise-vessel *Watcher-of-Moons* lay back and tried to relax in a sensuous swaddle of preening-curd, only his eyes and beak projecting from the glutinous layers, giggling slightly whenever one of the nanotribbles that roamed the curd in search of tiny parasites and dirt particles encountered a sensitive patch of skin around the base of his funny-feathers.

His mind was troubled. It had been a strange voyage. Those extelligent ape-creatures with their overprivileged solo minds and their extraordinarily unoctimistic view of how the world worked were really disturbing. Always obsessed with the *insides* of things – no doubt a resurgence of their child-aspect in later life, the monkey curiosity that tried to find out how things worked by breaking them and seeing what they no longer did.

He expanded his neck-ruff, the Zarathustran equivalent of a sigh. The problem with preening-curd is that once you have opened a tuble you have to wallow for a full octad, and after a time preenwallow gets boring. Especially to a Ringmaster, who spends so much time making sense out of what everybody else is doing ... And this Ringmaster was subject to troubled thoughts, things he was having difficulty rationalising. He recalled that not an octuple of octoons away from him was an almost inexhaustible source of alien extelligence, refreshing even if naive. And once Hewer-of-wood had got the catalytic converter working again, *Watcher-of-Moons* would resume its voyage ... For a moment he wondered which catalyst it was failing to convert, but he would only be able to explain that to everybody when Destroyer-of-facts had found out.

It was his last chance to sample the terrestrial view of things, and it would take his mind off rational thought completely. He instructed the ship's computer to summon a package of alien extelligence and verbalise it to him in translation ...

Computer Do you wish any particular sample of alien extelligence, Ringmaster?
Ringmaster No, choose at random.

C No sooner said than done. I have selected a volume entitled *Figments of Reality*. It consists of monkey speculations regarding the nature of extelligence and other contextual influences on self-organisation.

R How delightfully/irritatingly (delete whichever is inapplicable) recursive. Proceed.

C [*Clearing its circuits.*] Zhrrmph. 'Prologue. Fifteen thousand million years ago the universe was no bigger that the dot at the end of this–'

R I hope this is not the start of a detailed history.

C No, the tale accelerates rapidly.

R That is fortunate. You may continue.

C '–sentence. A tiny, tiny, *tiny* fraction of a second before that – but there was no fraction of a second before that. There was no time before the universe began, and without time, there can be no ...'

The Ringmaster dozed as the synthetic voice droned on. Such curious monkey fascination with internals, with building up from 'fundamentals', by which they seemed to mean the universe stripped of every interesting feature ... He had expected to be irritated by the singleminded – yes, that was the word – focus on that style of explanation. But preening-curd affects the mind in funny ways, and he drifted off into a dreamlike state in which the computer's outpourings were not so much translated as trancelated, so all of the details became malleable, while the underlying meaning remained the same but was carved up in a different direction ... *Merely complex, but not yet multiplex*, his fading self-referential feature-detector noted, as it fell into the trancelation dreamstate ... And as the computer worked its way through *Figments of Reality*, the Ringmaster's mind put together an entirely different, and blessedly octimal, slant on the same material ...

It came out something like this.

There were no sentences, and without sentences there can be no dots at their ends. (As well to ask what happened before the start of time.) Instead there was a vast, unfathomable, rather vague and wispy sea of *potential*. Unending spaces-of-the-possible reached foggy tendrils out into the intangible realms of Everywhy, awash with the promise of octimism, waiting to be pruned into reality ... It was the Era of the Phase Spaces, when potential fed off itself unchecked by constraints and Everywhy was nothing but externals, ready to expand into novel complicities ...

Into this unreal ocean of possibilities there plopped a speck of reality.

Nobody knows where it came from: it was a spontaneous, symmetry-breaking octahedral seed. Reality is dangerous stuff: it reacts to surrounding possibilities by choosing among them, selecting some and denying fruition to others. Reality causes vast oceans of potential to collapse into tiny puddles of actuality, crystallising flexibility into brittle rigidity.

Potential is timeless: reality crystallised it into the eightfold unity of time. Potential is spaceless: reality crystallised it into the octuplicity of space. Potential is immaterial: reality crystallised it into the octimality of material interaction. The ocean of potential began to thicken, coating every particle of reality with layers of phase space, lassoing reality with sticky threads of causality, weaving tangled webs of complicity and contingency. The more reality tried to prune away every sprouting growth of potential, the more it failed. [*The Ringmaster chuckled, perhaps at the attribution of intention to reality: all that reality possessed was a complicit dynamic. Or was it just at the nuzzling of a nanotribble?*]

In every direction reality trailed a shimmering haze of possibility, a haze that grew ever more dense as reality accreted around that first tiny seed. So space and time grew and progressed by thickening their potential into substance, and pruning the potential that adhered to them like glumpweed to bring ever more matter into being, and ever more causality to connect it. And the material universe began to explore its own phase space, putting out new tentacles into thitherto unexplored realms of the possible, accepting some, pruning others, growing organically into its richly textured surroundings.

Some parts of those surroundings were close, easily accessible. And so matter, exploring its immediate vicinity in the space of the possible, discovered clumpiness. Clumpiness stumbled further out, finding such things as surfaces, features of reality richly infused with octimistic potential. Surfaces are interfaces, and the phase spaces adjacent to them become complicitly intertwined, sprouting yet new realms of possibility – surphases and interphases. This, the Ringmaster realised, is where complexity 'comes from': it is out there, potentially, all the time. As the universe's exploring tentacles penetrated deeper into the coagulating glumpweed of the possible it found new universals, inevitable results of such exploration ... And all was in accordance with the Principle of Murphic Resonance: 'Anything that is woven into the fabric of possibility will eventually be blundered into, unless a previous blunder excises it from the fabric before you can blunder into it.'

Woven into the fabric were many forms of material organisation – plasma, gas, liquid, colloid, crystal ... Reality obediently brought each into existence. As the threads of potential thickened into ropes they drew in more distant regions of phase space, and a new form of material organisation became possible, replicative but error-prone, thickening further into systems that could reproduce, evolve, self-complicate, and self-organise. Life. This formed such fertile ground for outgrowths of potential that for a time reality was almost submerged by the enormity of what it might become, but many of the new growths choked each other's realisation, until the coagulation of new regions of phase space became more closely matched to the rate at which they could be pruned into reality.

Independently but inevitably, upon every suitable clump of matter, life sprouted – spilling over into its potential in a trillion different ways. Parochials vied with universals, but the universals played with loaded dice, and reality inevitably acquired a meta-level: self-representation in coded form.

Phase space throbbed and pulsed, bubbling over into new realms of possibility ...

Intelligence flourished.

Upon many obscure clumps of matter in many obscure and utterly ordinary regions of the universe, a universal form of recursive self-complication gave rise to creatures whose limited intelligence was augmented by an ever-burgeoning extelligence. Complicity between these two aspects of telligence took different parochial forms in each instance, but suspended within an octimal framework of universals. It found expression in organised sound waves in the burning plasma of stellar nuclei, as mobile patterns of dislocations in methane ice that took a million years to think a coherent thought, as crystal slicks on the surface of a neutron star that communicated by bartering quarks for gluons, as democratic collectives of circulating vortices in liquid helium hurricanes (one vortex, one vote). On the clump known to its inhabitants as Zarathustra it gave rise to the universe's most perfect creation, the octuple group-mind, able to sense the underlying octimality of the Cosmic All. On another rather squalid clump it gave rise to a curiously singleminded apelike creature that liked to pull things to bits to see how they stopped working.

And every single one of these life-forms asked the same question: 'Isn't it amazing that all of this was pulled together to create *me*?'

Still deep in dreamstate, the Ringmaster sniggered, and deleted where inapplicable. Every life-form *but one*.

Only Zarathustrans saw the meta-pattern: that the universe coagulates itself out of potential, not because of the presence of singlemind in its phase space, but because of the Principle of Octimality, which implies that the only octimal form of material organisation is the eightfold group-mind, which therefore must *inevitably* be pruned into reality ...

Less troubled now, he slipped by imperceptible stages from dreamstate into sleep, while the nanotribbles continued to gnaw his feathers clean.

Notes

The passages to which these notes refer are indicated in the text by the symbol•. The main purpose of these notes is to provide additional argument, indicate difficulties and counter-arguments, or state precise references to justify some of the statements that we make. All this notwithstanding, the main rule for these notes is 'anything goes'.

Preface

p. x This famous phrase originated with the physicist Eugene Wigner.

Prologue

p. 1 We mean 'north *on the surface of the Earth*'. There is air and sky *above* the North Pole, but there is nothing north of it. In fact from the North Pole, all surface directions are south. This is a peculiarity of our system of longitude and latitude, not an intrinsic feature of the land at the North Pole.

Chapter 1 The Origins of Life

p. 5 We recall reading this tale, but have not been able to locate the original. A very similar story is related by Karen Pryor in *Dolphin Cognition and Behavior: a Comparative Approach*, edited by Ronald J. Schusterman, Jeanette A. Thomas, and Forrest G. Wood, Lawrence Earlbaum Associates, Hillsdale NJ 1986, page 256 – starring primate caretaker Melanie Bond and a chimpanzee. Pryor also mentions a juvenile elephant which rapidly trained *her* to offer sweet potatoes as a reward for good work. Maybe Douglas Adams was right: maybe the entire Earth is an experiment run by mice.

p. 6 There is a Gary Larson 'Far Side' cartoon about the decipherment of dog language. It turns out that every bark means 'Hey!'

p. 10 *We are stardust,*
We are golden,
And we've got to get ourselves
back to the Garden ...'

p. 18 The name comes (perhaps) from 'pick-a-back', a children's game in which one child rides around on the back of another.

p. 19 Phil Cohen, 'Let there be life', *New Scientist* (6 July 1996) 22–27.

p. 21 For example, suppose we have a collection of molecules A, B, C and so on which catalyse each other in various ways. Symbolise such catalytic reactions with arrows, so that X → Y means that X is a catalyst in a reaction that produces Y. (By this notation we do *not* mean that the reaction starts with X and ends with Y: remember, X is a catalyst, not raw materials. To keep the discussion simple we will leave out the 'input' chemicals needed for X to get to work: think of them as a common pool of 'feedstock' big enough for us not to worry about them getting used up.) Then several catalytic pathways might coexist, say:

A → B
A → C
B+C → D
C+D → E
E → B
B+D → C

This set is not replicating, because nothing catalyses A. But suppose you start with just A, plus enough feedstock for it to work on. Now molecule A catalyses both B and C, and these together catalyse D; then C and D catalyse E. At this point the team formed by B, C, D, E 'closes up', because every molecule in it is catalysed by some combination of the molecules in it. Molecule B is catalysed by E, C by B+D, D by B+C, and E by C+D. Molecule A now fades from the picture, although it formed part of the *scaffolding* for the eventual replicating team. Kauffman discovered that this kind of 'closure' property in a network of catalytic reactions is not a rare phenomenon: for purely mathematical reasons it is relatively common. As long as the connectivity of the network of reactions is not too simple and not too complex, closed teams of self-catalysing molecules – autocatalytic networks – will nearly always be present.

p. 22 Fred Hoyle and Chandra Wickramasinghe, *Life Cloud*, Dent, London 1978.

p. 23 Christian de Duve, 'The birth of complex cells', *Scientific American* (April 1996) 38–45.

p. 27 Harry Jerison, *Brains and Behaviour*, Oxford University Press 1973; also 'Animal intelligence as encephalization', *Philosophical Transactions of the Royal Society of London* B **308** (1985) 21–35.

Chapter 2 The Reductionist Nightmare

p. 36 Philosophers tell us that 'some' here is 'a few'. *We* think it is 'too many'.

p. 41 Words like 'fundamental' and 'basic' for 'bottom', or 'centre' and 'core' for 'middle', all suggest a philosophy that is currently out of favour – the idea that some statements are *much* more important than others. A philosophy that *is* currently in favour, post-modernism, holds that all statements carry equal weight. We think it ought to go out of favour again: it says that the mental landscape has no 'geography', that *any* theory is as good as any other. (If you deny this, then you haven't fully grasped the message of post-modernism.) That's a great way for arts graduates to convince themselves that there's no need to know any science, but it throws the baby out with the bathwater. JC and IS prefer to move on to what Mal Leicester calls *post-post-modernism*: we think that nearly all theories are crap, but some are less crap than others.

p. 45 The intention here is to indicate that, on human terms, the speed is rapid. The image of an electrical current does not fully capture how impulses are transmitted along nerves, but we have used it because it relates to things that most readers will understand. A more accurate image is that electrical activity spreads along the nerve rather like fire spreading along a fuse: activity at one position triggers activity nearby. The electrons don't flow *along* the nerve cell: they move in and out across its membrane, creating the appearance of a moving pulse. In the same way, molecules of water moving up and down in the sea create the appearance of a wave that travels along.

Chapter 3 Ant Country

p. 70 An extensive description of Life is given in volume 2 of *Winning Ways* by Elwyn R. Berlekamp, John H. Conway, and Richard K. Guy, Academic Press, New York 1982. The rules go like this. Start with a configuration of counters on an infinite square grid. Then a counter survives to the next generation if and only if it has two or three neighbouring counters (left, right, up, down, or diagonally). A new counter is born on any empty cell that has exactly three neighbours. All births and deaths are supposed to happen simultaneously, with respect to a given configuration, in order to create the next configuration.

p. 73 Croquet (see e.g. J.W. Solomon, *Croquet*, Batsford, London 1966) is played on a grass court (35 yards by 28) with four grapefruit-sized balls – blue, black, red, and yellow – which must be hit through a sequence of hoops using a mallet. In the singles version of the game one player takes the blue and black balls, the other the red and yellow. At each turn the player uses the mallet to hit one of his two balls. There are several types of stroke. The first is a 'roquet', in which the striker's ball hits another ball (his own or his opponent's). He may then play a 'croquet' by placing his own ball next to the roqueted ball, wherever it has moved to, and hitting it in such a manner that both balls are moved. (He may not place his foot on top of the ball being hit, despite a common contrary belief.) The croquet is followed by a 'continuation stroke' in which the striker's ball is hit once more. If the continuation stroke produces a roquet or makes the striker's ball run a hoop, then the sequence may continue; but the player may not roquet any ball twice unless his own ball has passed through ('run') a hoop in between. There are six hoops, but each must be run twice in a specified sequence and in specified directions.

The classic four-ball break (**figure 27**) is a method for running the striker's ball through an entire sequence of twelve hoops with the aid of all three other balls. Suppose the player is striking blue, and for the purposes of this description number the hoops 1–12 in order. The break begins with blue about two yards in front of hoop 1, black about a foot closer, red near the centre of the court, and yellow a yard in front of hoop 2 (acting as a 'pioneer ball'). Roquet black, in such a way that it is knocked just in front of, and slightly to one side of, hoop 1 (a stroke known as a 'rush'). On the croquet shot send blue directly in front of the hoop and black a yard or so past it. Run blue through the hoop on the continuation shot. Now roquet black again, which is permitted since blue has run a hoop. On the croquet shot send black near hoop 3 and blue near the red ball in the centre. Roquet red. Croquet blue from red to near yellow at hoop 2, leaving red in the centre. Now the break can continue at hoop 2 with yellow replacing black as the ball used to assist in running the hoop, and black replacing yellow as the pioneer ball. Because of the geometry of the shots, especially the croquet shot, the presence of red in the centre makes it relatively easy to correct moderate errors. An expert player will take the opportunity to adjust the position of the red as needed to make the break as straightforward as possible, and can commonly continue the break until blue has run all twelve hoops.

Note that again the player's thoughts are well ahead of the hoop being run; indeed the next two hoops are involved as well. The complicated series of shots used in the four-ball break is designed so that blue can run a lengthy sequence of hoops using only the simplest and most reliable strokes. There is a similar, but more difficult, three-ball break (in which red is not used) and an extremely difficult two-ball break (no yellow). An expert can win the game in two four-ball breaks, running blue through all 12 hoops on the first turn and black through them all on the second. Tactics therefore concentrate upon not allowing one's opponent to set up a break, leading to play which might otherwise seem bizarre. For example, the standard opening moves involve players sending balls almost anywhere except towards the first hoop.

p. 73 For the rules of snooker see (e.g.) *Encyclopaedia Britannica*, 15th edition, Chicago 1995, volume 10, p. 911.

p. 75 In a cricket Test Match, which lasts five days, the impact of the ball causes wear on the pitch, and this has a strong effect on strategy.

Chapter 4 Winning Ways

p. 78 Of course Both Galileo and Darwin had intellectual predecessors with similar ideas, but it was they who crystallised those ideas into a form that other people took notice of.

p. 79 There is a similar case that is much closer to home. Human beings adopt a strategy akin to that of the fish eagle, though one that is much less overt and so less gruesome: it is known in gynaecological circles as 'The Case of the Disappearing Twin'. When techniques of in vitro fertilisation (IVF, 'test tube babies') were under development it became standard to use ultrasound to form images of the uterus. As the technique achieved more detailed images it was also applied to look for natural pregnancies, but early results were puzzling. The ultrasound image appeared to show 'implantation sites', where a fertilised egg was developing – but there were too many of them. About half of the women who showed any such sites seemed to have at least two, and often three. However, the vast majority of pregnancies result in a single child, so the scientists concerned assumed that these images were of something else. As the technique improved, however, the images even showed developing embryos at these sites: they really *were* implantation sites.

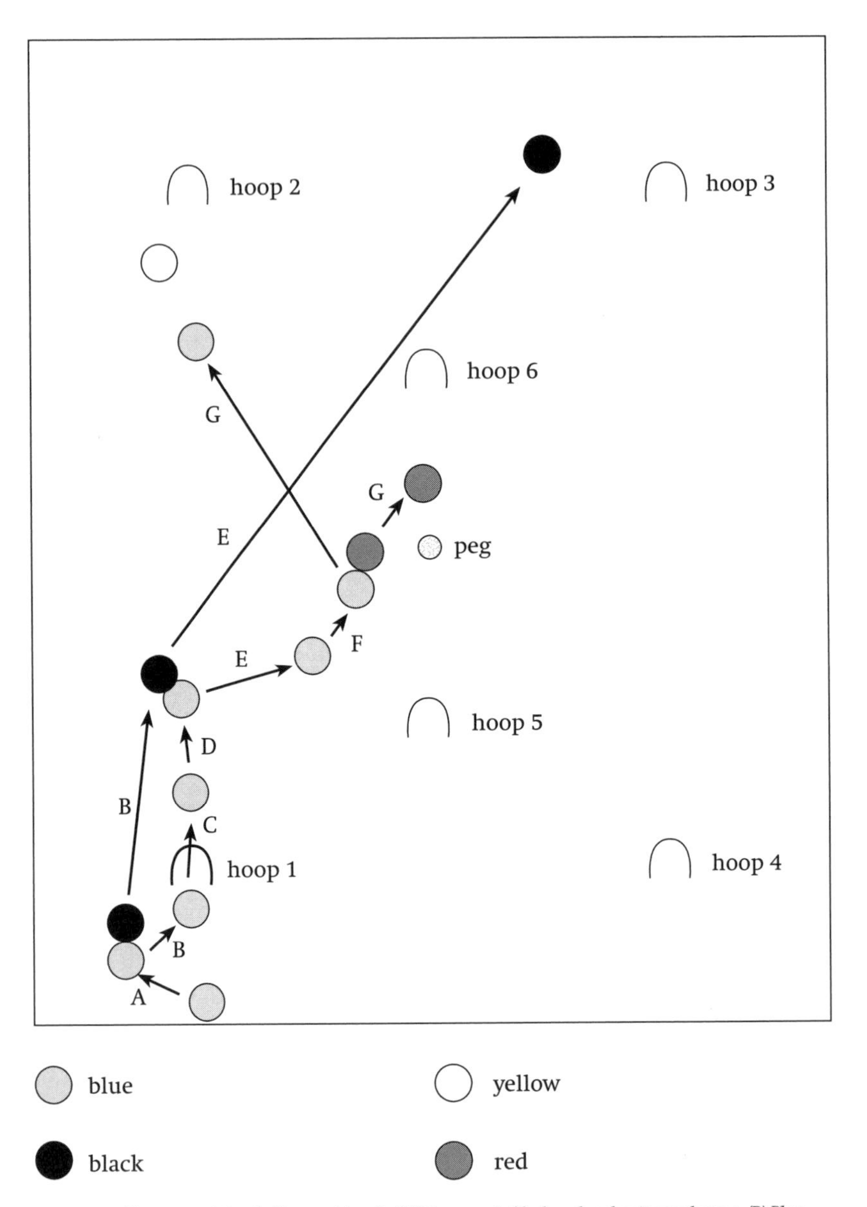

Figure 27 *A four-ball croquet break. (A) Blue roquets black and rushes it near hoop 1. (B) Blue croquets black past hoop 1 leaving blue in front of hoop 1. (C) Blue runs hoop 1. (D) Blue roquets black. (E) Blue croquets black to hoop 3 while blue ends up near red. (F) Blue roquets red. (G) Blue croquets red, so that red stays in the centre and blue moves into position near yellow at hoop 2. Now return to step (A) with the roles of yellow and black interchanged and all hoop numbers increased by 1.*

It took a great deal of evidence to persuade them of what is now generally accepted: about half of (entirely normal) human pregnancies begin with (non-identical) twins – or even triplets – only one of which grows beyond the earliest stages of development and eventually becomes a fetus. So two or three eggs are often fertilised initially, and some selection process reduces the number to one later. The reproductive biologist Roger Short has argued that this kind of reduction in numbers has evolved in order to reinforce privilege.

p. 85 Robert Foley, *Humans Before Humanity*, Blackwell, Oxford 1995.

p. 87 Francisco J. Ayala, 'The myth of Eve: molecular biology and human origins', *Science* **270** (22 Dec 1995) 1930–1936.

p. 90 Most scientists, especially biologists, think that nearly all observations are 'normally distributed', bunched together in a bell-shaped curve. In fact, an enormous variety of things come in a 'logarithmic distribution' in which small values are very common and bigger ones tail off, so that 'half as big' and 'twice as big' are equally likely. This phenomenon is known as Benford's Law, and it seems to have first been noticed by looking at the frequency of the first digits of numbers in logarithm tables. (The story is that early pages of such tables are always dirtier, because they get thumbed more.) Examples of logarithmic distributions include the exchange rate figures for different currencies, the size of islands in the Bahamas, and the numbers of different species or organisms of given size in rainforests.

p. 98 Analogously, there are viable markets for personal stereos and mobile phones. Now that these markets are well established, there is no hope for the inventor of the phonojuke, a machine that lets you dial up records from a central source and listen to them over a radio link (and only that). In a world that had never heard either of Walkmans or mobiles, however, the phonojuke might be a really neat executive toy, and every yuppie would want one, just as they bought filofaxes by the millions before everything went electronic.

p. 100 V. Courtillot and Y. Gaudemer, 'Effects of mass extinctions on biodiversity', *Nature* **381** (1996) 146–148. Evidence from the fossil record that mass extinctions are followed by a period of increased biodiversity.

p. 100 We were told this by Tim Benton, our editor, who has been carrying out research on it.

Chapter 5 Universals and Parochials

p. 114 J.L. and C.G. Gould, *The Animal Mind*, Scientific American Library, New York. Foraging bees, 'kidnapped' and displaced 30–100 metres from their path, flew in a straight line to their nectar site, or direct to the hive if already loaded with nectar.

p. 115 See the *West of Eden* SF series by Harry Harrison.

p. 116 J. Roger Angel and Neville J. Woolf, 'Searching for life on other planets', *Scientific American* (April 1996) 46–52.

p. 116 Gabrielle Walker, 'Seven planets for seven stars', *New Scientist* (15 June 1996) 26–30.

p. 116 Did Martians land in Antarctica? *New Scientist* (17 August 1996) 4–5.

p. 118 SF fans dislike the term 'sci-fi' because TV and Hollywood don't really understand science fiction, and pronounce it 'skiffy' to show their dislike. We use it here precisely because it pins down that corner of the genre, the one where aliens are 'designed' without any attention to scientific realities.

p. 123 Feathers are parochials. Actually what the Zarathustrans call 'feathers' differ in several ways from terrestrial ones. But they serve a similar function.

Chapter 6 Neural Nests

p. 141 *A Code in the Nose*, BBC Horizon, 27 November 1995. Producer Isabelle Rosin.

p. 145 Richard Dawkins, 'The eye in a twinkling', *Nature* **368** (1994) 690–691. The original paper is: Daniel E. Nilsson and Susanne Pelger, 'A pessimistic estimate of the time required for an eye to evolve', *Proceedings of the Royal Society of London* B **256** (1994) 53–58.

p. 147 Nicholas Humphrey, *A History of the Mind*, Simon and Schuster 1992, pp.174–8.

p. 151 Siddhartha (Buddha) made this point before Heraclitus, and much of the Buddhist image of the unimportance of 'self' depends upon it.

p. 152 For example see J. McGlade and P. Allen, 'Evolution of multifunctionalism in enzymes: specialist versus generalist strategies', *Canadian Journal of Fisheries and Aquatic Sciences* **43** (1986) 1052–8. Another example is the bombardier beetle. Its ancestors used strong oxidising and reducing agents, secreted from cloacal glands, to 'tan' the proteins of the egg case – as do their relatives to this day. Bombardier beetles evolved the habit of mixing these, together with a 'hypergolic' enzyme – one that makes things explode. The result combines a bang, a stink, and jet propulsion, as a defence against spiders and birds.

p. 153 R. Thornhill, S.W. Gangstead, and R. Comer, 'Human female orgasm and mate fluctuating asymmetry', *Animal Behaviour* **50** (1995) 1601–1615. See also David Palliser, 'Symmetry in the human body is sexier and healthier as well as aesthetically pleasing, says scientist', *The Guardian* (10 August 1996) p. 4.

p. 153 Christopher Wills, *The Runaway Brain*, HarperCollins, London 1994. Wills argued that humanity's uniqueness lies in its ability to 'hold' its sense-impressions and to compare them with new ones. The brain/mind is therefore supposed to become more complicated as many more impressions 'need' to be stored.

p. 156 It was in fact a PhD in irritation, from the University of Octford.

p. 164 Harry Jerison, *Brains and Behaviour*, Oxford University Press 1973, is a classic for this story.

Chapter 7 Features Great and Small

p. 173 *Visual Symmetry and Kohonen-like Neural Networks*, PhD thesis, Peter Mason, University of Warwick Mathematics Institute, 1996.

p. 174 Madhusree Mukerjee, 'Engendering faces', *Scientific American* (April 1996) 24.

p. 174 The serious mathematics of this involves representing vectors of numbers in terms of 'principal components', which are the eigenvectors of a correlation matrix. So there!

p. 187 Frog eyes also analyse vision into points and lines, but a frog mathematician has yet to win the Fields Medal (the mathematical equivalent of the Nobel Prize). In order to evolve a mathematician's mind, a brain needs feature-detection on many meta-levels too.

p. 188 Steve (not Jay) Gould wrote about the physics of cartoons in *New Scientist's* special Christmas edition for 1995.'Looney tuniverse', *New Scientist* (25 December 1993/1 January 1994), 56–57.

p. 188 IS's student Andy Mumford at Warwick University has been analysing the paths taken by people's hands when they reach for objects, in order to deduce what kind of mental map they set up before making a move. She finds that an Aristotelian 'naive physics' model fits the paths better than one based on Newtonian physics.

Chapter 8 What is it Like to be a Human?

p. 196 Later Wittgenstein changed his mind, suggesting that human minds develop their own 'private language' to describe what they experience, and that the only things that have any meaning – to humans, of course – are the things that can be expressed in that language. The problem with this is that when understanding the universe, humans are not limited to what they experience *directly*: instead they can infer it indirectly. For example we can infer the existence of wind by watching branches of trees moving around. Science relies upon the construction of chains of such inferences, which demonstrate very convincingly that wind is moving air, air is made from atoms, and that the proportions of different atoms and molecules in the air is ... Few of the things envisaged by science would exist in a private language of the kind Wittgenstein meant, and therefore – if he were right – science would be meaningless.

p. 197 What if the painter were blind or the composer deaf?

p. 200 In Book 7 of Plato's *Republic* you will find this famous passage, about people imprisoned in a cave, able to see only shadows of the outside world on the wall:

> And do you see, I said, men passing along the wall, some apparently talking and others silent, carrying vessels and statues and figures of animals made of wood and stone and various materials, which appear over the wall?

You have shown me a strange image, and they are strange prisoners.

Like ourselves, I replied; and they see only their own shadows, or the shadows of one another, which the fire throws on the opposite wall of the cave?

True, he said; how could they see anything but the shadows if they were never allowed to move their heads?

And of the objects which are being carried in like manner they would only see the shadows?

Yes, he said.

And if they were able to talk with one another, would they not suppose that they were naming what was actually before them?

p. 203 One of our readers asked: is this just because they are sufficiently similar to us that we recognise the signs – just as we westerners recognise western faces but 'all orientals look the same'? We don't think so, in part because what we mean by 'mind' is something that we *ought* to be able to recognise if it were there.

p. 215 The second edition of *Does God Play Dice?* by IS has a chapter discussing new work on the possibility that quantum mechanics might actually be classical chaos, deep down inside.

p. 217 A nice experiment could perhaps be done here. Instruct the fielder to 'keep their eye on the ball' and see whether what they think they experience matches what their eyes actually do. It might not work, if the eyes are also extrapolating – but if they did, the fielders wouldn't see the ball because they'd be looking ahead of it. So we predict they will think they follow the ball exactly, whereas actually they will be $^1/_3$ second behind its position.

p. 218 Susan Blackmore has a fascinating theory about 'alien abduction' experiences, in which people get a very strong belief that they have been taken away in a UFO and experimented on. There is plenty of evidence that (many of) these experiences are false: for example JC once met a woman who had 'been abducted' and claimed that the aliens 'took away my baby'. He then asked a question nobody else had thought to ask: 'were you pregnant?' The answer was 'no'. IS was in a radio studio with a woman who had undergone an abduction experience, but said herself that she knew she hadn't really been abducted, because several members of the family reported nothing unusual. (She was dozing beside the fire the whole time.) But to her, as to many 'victims', the experience *felt real.*

Blackmore thinks that these experiences are caused by 'sleep paralysis', a necessary condition that prevents people who are asleep acting out their dreams. If the system goes wrong, then people who are still partially awake may enter sleep paralysis, and experiments show that in such circumstances they get a strong impression of there being 'somebody there'. Folk tales of such things abound: in Newfoundland people tell of an old hag sitting on their chests at night, and in Vietnam people speak of a 'grey ghost'. This curious feature of the mind may go back to when (as Jerison suggested) the nocturnal mammals were released from dinosaurian thrall. Then, as we argue in this chapter, their qualia for sound and vision might have become linked into a general feature-detecting system. When they heard something strange, their visual sense would kick in and give them the impression that they could *see* its cause. We inherited this tendency, and we interpret it through our current culture: bogeymen (and perhaps dragons) a few centuries (millennia) ago, aliens today.

p. 225 Steven P. Dear, James A. Simmons, and Jonathan Fritz, 'A possible neuronal basis for representation of acoustic scenes in the auditory cortex of the big brown bat', *Nature* **364** (12 August 1993) 620–623. The cover had a snappier title: 'How bats' ears see'.

Chapter 9 We Wanted to Have a Chapter on Free Will, but We Decided not to, So Here It Is

p. 229 Several readers asked: what about XYY? Some years ago the media seized on the extra Y chromosome of XYY men as the carrier of a 'criminal aggression' syndrome (because there was a somewhat higher proportion of such men in violent offender institutions). The same argument leads to the belief that *one* Y chromosome carries 'aggression genes': many more men than women are aggressive criminals; we think this argument is, at best, not useful.

p. 233 Except, inevitably, Terry Pratchett – in *Interesting Times*, Gollancz, London 1994.
p. 236 Dennett discusses many such experiments in *Consciousness Explained*.
p. 237 Such as the 'Legion' system invented by G. Keith Still and marketed by FMIG Ltd.

Chapter 10 Extelligence

p. 243 We were told this one by a reputable source, several years ago. In *Scientific American* (April 1996) page 23, almost exactly the same behaviour is attributed to a parrot named Alex, and the story is told as if the event was very recent. We're beginning to wonder if this one is the scientific equivalent of an 'urban myth'.
p. 257 This reminds us of a joke. A couple moved to Wales and adopted a Welsh baby. They surprised their neighbours by becoming inordinately interested in learning the Welsh language, taking evening classes and practising at every opportunity. Eventually the neighbours asked why. 'So that we can understand the baby when she starts to talk.'
p. 261 Lloyd Morgan was a late 19th century anthropologist who popularised the 'primitive savage' as 'ancestral type' in *Ethnical Periods* and *Ancient Society*, two very popular books.
p. 262 Jared Diamond, *The Third Chimpanzee*; A. Zahavi, 'The theory of signal selection', in *International Symposium on Biological Evolution, Bari 1985*, (ed. V.P. Delfino), Adriattici editrici, Bari 1986.

Chapter 11 Simplex, Complex, Multiplex

p. 273 *The Sciences*, May/June 1984, 27.
p. 278 It is PC to add that there could equally well have been a Street of the Dressmakers too, with the daughters passing the know-how along. It is not clear whether historically the womens' contribution would have been so public. But there *must* have been a female subculture, keeping its own snooker break going in a kind of complicity with the male-dominated surface culture – there always is. And its role in 'driving' the culture would have been just as important.
p. 295 *New Scientist* 6 July 1996.

Further Reading

Each entry is annotated with a mini-review. We have each reviewed the other's books.

John Barrow, *Theories of Everything*, Oxford University Press, Oxford 1991. [A very readable account which dismisses Theories of Everything for similar reasons to ours.]

Gregory Bateson, *Steps to an Ecology of Mind*, Paladin, St. Albans 1973. [Essays by a very intelligent scientist who emphasises context.]

Elwyn R. Berlekamp, John H. Conway, and Richard K. Guy, *Winning Ways for Your Mathematical Plays* (two volumes), Academic Press, New York 1982. [Hugely original compendium of mathematical games, full of puns, and *the* source for Life.]

R.J. Berry, *Neo-Darwinism*, Edward Arnold, London 1982. [A short, excellent, 'straight' account.]

Margaret A. Boden, *The Creative Mind: Myths and Mechanisms*, Weidenfeld and Nicolson, London 1990. [Delightful study of mind and creativity: emphasises the 'creative space' and distinguishes between exploring it and expanding it.]

A.G. Cairns-Smith, *Seven Clues to the Origin of Life*, Cambridge University Press, Cambridge 1985. [A more accessible Origin of Life than his *Genetic Takeover*.]

A.G. Cairns-Smith, *Evolving the Mind*, Cambridge University Press, Cambridge 1996. [Molecules and minds, but not multicultures.]

William H. Calvin, *The Throwing Madonna: Essays on the Brain*, Bantam Books, New York 1991. [Stunning series of essays by a neurobiologist with a solid grasp of the science but an unconventional turn of mind. Why left-handedness is rare.]

John L. Casti, *Searching for Certainty: What Scientists can Learn about the Future*, Morrow, New York 1990. [A well-written overview of modern reductionist science, with many indications of incompleteness and contextuality.]

Jack Cohen, *Reproduction*, Butterworths, London 1977. [A classic student text collecting 'all' aspects of biological reproduction between two covers. Highly illustrated.]

Jack Cohen, *The Privileged Ape: Cultural Capital in the Making of Man*, Parthenon, Carnforth 1989. [An account of human evolution, especially the evolution of multicultures, emphasising the role of privilege. Sometimes turgid.]

Jack Cohen and Brendan Massey, *Animal Reproduction: Parents Making Parents*, Edward Arnold, London 1984. [A 'Noddy' version of Cohen's *Reproduction*.]

Jack Cohen and Ian Stewart, *The Collapse of Chaos*, Viking, New York 1994. [How do complex structures exist in a chaotic world? *Figments* is its sequel.]

Michael Crawford and David Marsh, *The Driving Force*, Mandarin, London 1991. [Substantial and well presented setting for human evolution, getting brains on the seashore. Sea food means success.]

Charles Darwin, *The Origin of Species*, Penguin Books, Harmondsworth 1985. [Still very worth reading.]

Richard Dawkins, *The Blind Watchmaker*, Longman, London 1986. [A beautifully written account of modern evolutionary theory – marred only by its basic premise that genes map to characters.]

Richard Dawkins, *The Selfish Gene* (second edition), Oxford University Press, Oxford 1989. [Elegant account of neo-Darwinism: argues the view that DNA rules. Superb, well worth reading, but we don't believe it.]

Richard Dawkins, *The Extended Phenotype*, Oxford University Press, Oxford 1982. [Sparkling and extremely clever backtrack on and amplification of *The Selfish Gene*.]

Richard Dawkins, *River out of Eden*, Weidenfeld and Nicolson, London 1995. [Short and snappy summary based around the image of the 'river' of evolution. Pity that rivers join together as they flow, rather than dividing like a tree, but a brilliant image nevertheless.]

Samuel R. Delany, *Empire Star*, Ace Books, New York 1966. [Remarkable science fiction story, with many wise comments on simplex, complex and multiplex world-views.]

Daniel C. Dennett, *Consciousness Explained*, Little, Brown & Co., Boston 1991. [Utterly brilliant.]

Daniel C. Dennett, *Kinds of Minds*, Nicolson, London 1996. [A gentler discussion of the same ideas, with an added dash of human culture.]

Jared Diamond, *The Rise and Fall of the Third Chimpanzee*, Vintage, London 1991. [Why are humans so different from chimps when they share 98% of the same DNA? Argues, compellingly and in fascinating detail, that most differences have precursors elsewhere in the animal kingdom. Tends to forget that in DNA it's *quality*, not quantity, that counts.]

Freeman Dyson, *Disturbing the Universe*, Basic Books, New York 1979. [Deep thoughts about the universe by one of the wisest leading physicists.]

Freeman Dyson, *Infinite in All Directions*, Basic Books, New York 1988. [More of the same. Explains why the 'heat death of the universe' is a misleading image.]

Timothy Ferris, *The Mind's Sky*, Bantam Press, New York 1992. [Thought-provoking essays on mind, matter, and the universe. Sees reality as a figment of the imagination, less aware of mind as a figment of reality.]

Robert Foley, *Another Unique Species: Patterns in Human Evolutionary Ecology*, Longman, Harlow 1987. [A professional, but contextual and ecological, view of human evolution – savannah version.]

Robert Foley, *Humans Before Humanity*, Blackwell, Oxford 1995. [How a gang of apes were caught out by progressive deforestation, and became naked, sweaty humans.]

Alan Garfinkel, *Forms of Explanation*, Yale University Press, New Haven 1981. [What do we mean by 'explanation'? Carefully and sensibly argued. We've reinvented some of his wheels.]

Ronald N. Giere, *Explaining Science*, University of Chicago Press, Chicago 1988. [Philosophy of scientific theories, very close to our features/instances approach.]

Stephen Jay Gould, *Ontogeny and Phylogeny*, Belknap Press, Cambridge MA 1977. [His most professional book, one of the great biology books of the century.]

Stephen Jay Gould, *Ever Since Darwin*, Penguin Books, Harmondsworth 1980; *The Mismeasure of Man*, Penguin 1981; *The Panda's Thumb*, Penguin 1983; *Hen's Teeth and Horses' Toes*, Penguin 1984; *Time's Arrow, Time's Cycle*, Penguin 1988; *An Urchin in the Storm*, Penguin 1990; *Bully for Brontosaurus*, Penguin 1992. [Witty, pithy, memorable essays – you can learn better biology from them than from most textbooks.]

Stephen Jay Gould, *Wonderful Life*, Penguin Books, Harmondsworth 1991. [The famous account of the soft-bodied creatures of the Burgess shale and the celebration of contingency in evolution.]

John Gribbin, *In Search of Schrödinger's Cat*, Black Swan 1992. [Good popular book on the meaning of quantum mechanics.]

Brian K. Hall, *Evolutionary Developmental Biology*, Chapman and Hall, London 1992. [Superb textbook.]

Nina Hall (editor), *The New Scientist Guide to Chaos*, Penguin Books, Harmondsworth 1991. [Collection of articles by experts for the general reader. One of the best introductions to the scientific content of chaos.]

A.H. Halsey (editor), *Heredity and Evironment*, Methuen, London 1977. [Very interesting collection of essays: puzzled, 'sure', stupid, and elegant.]

M. Harris, *Cows, Pigs, Wars, and Witches: the Riddle of Culture*, Random House, New York 1974. [Enormously entertaining contextual pop-anthropology book.]

Douglas R. Hofstadter, *Gödel, Escher, Bach: An Eternal Golden Braid*, Penguin Books, Harmondsworth 1980. [Classic mind-expanding cult book of the 80s – funny, infuriating and enormously illuminating unless you think science must be discussed solemnly. Excellent for distinction between information and meaning, and accessible source for Gödel's theorem. Often tough going but rewards effort.]

Douglas R. Hofstadter and Daniel C. Dennett, *The Mind's I: Fantasies and Reflections on Self and Soul*, Penguin Books, Harmondsworth 1982. [Essays with punch: content and context in studies of 'self'.]

Nicholas Humphrey, *A History of the Mind: Evolution and the Birth of Consciousness*, Simon and Schuster, New York 1992. [An excellent, vibrant book; derives consciousness from the internalisation of external sensory loops, so that we have our own internal sensory world to refer to.]

David Layzer, *Cosmogenesis: the Growth of Order in the Universe*, Oxford University Press, Oxford 1990. [Marvellous account of modern cosmology and what it means for us.]

Roger Lewin, *Complexity: Life at the Edge of Chaos*, Macmillan, New York 1992. [A people-based description of the work of the Santa Fe Institute.]

John Maynard-Smith, *Evolution and the Theory of Games*, Cambridge University Press, Cambridge 1982. [Much better than *The Evolution of Sex*: contextual.]

Marvin Minsky, *The Society of Mind*, Simon and Schuster, New York 1985. [Up-to-date view of human intelligence.]

Elaine Morgan, *The Aquatic Ape*, Souvenir Press, London 1982. [Evocation of Alister Hardy's theory, well presented and well argued.]

Rodney Needham, *Against the Tranquility of Axioms*, University of California Press, Berkeley 1983. [Takes anthropological 'certainties' apart with a deft hand.]

Susan Oyama, *The Ontogeny of Information*, Cambridge University Press, Cambridge 1984. [The very best nature/nurture destruction; a witty, wise, illuminating book with an *awful* title.]

Roger Penrose, *The Emperor's New Mind*, Oxford University Press 1989. [Definitive source for 'quantum uncertainty, therefore free will'. Fun, unafraid of mathematical formulas, and brilliant on almost everything except its central theme.]

Roger Penrose, *Shadows of the Mind*, Oxford University Press 1994. [The sequel. Challenging to non-specialists, but widely read, like its predecessor.]

H.C. Plotkin (editor), *Learning, Development and Culture*, Wiley, Chichester 1982. [Essays on content and context with regard to culture.]

Israel Rosenfield, *The Strange, Familiar, and Forgotten*, Picador, London 1992. [Tales of strange neurological defects and injuries that illuminate the even stranger nature of consciousness.]

R.M. Rosenweig, A.L. Leidman, and S.M. Breedlove, *Biological Psychology*, Sinauer Associates, Sunderland MA 1996. [Explains very clearly the relationship between the biology of the nervous system and the perception of sound, light, and smell. There is no mention of 'mind' in the index – or in the book.]

David Ruelle, *Chance and Chaos*, Princeton University Press, Princeton 1991. [Wise words from one of the mathematical founders of Chaos Theory, very good on time-reversibility, and astonishingly readable.]

Ian Stewart, *Does God Play Dice?*, Blackwell, Oxford 1989; Penguin Books, Harmondsworth 1990. [Best general account of chaos, against a background of modern science. Second edition in press.]

Ian Stewart, *Nature's Numbers*, Weidenfeld and Nicolson, London 1995. [A celebration of mathematical patterns in nature. Shortlisted for the 1995 Rhône-Poulenc Prize for Science Books.]

Ian Stewart and Martin Golubitsky, *Fearful Symmetry: is God a Geometer?*, Blackwell, Oxford 1992; Penguin Books, Harmondsworth 1993. [A whole new way of looking at pattern, complexity, and the generation of order in nature.]

D'Arcy Wentworth Thompson, *On Growth and Form* (2 volumes), Cambridge University Press, Cambridge 1942. [Wonderful, full of thoughtful and thought-provoking examples.]

D'Arcy Wentworth Thompson, *On Growth and Form* (edited by J.T. Bonner), Cambridge University Press, Cambridge 1961. [Abridged edition which conveys the same message in more digestible form.]

C.H. Waddington, *The Evolution of an Evolutionist*, Edinburgh University Press, Edinburgh 1975. [The most accessible account of canalization.]

Mitchell Waldrop, *Complexity; the Emerging Science at the Edge of Order and Chaos*, Simon and Schuster, New York 1992. [How emergence is becoming respectable: a detailed look at the Santa Fe Institute and the theories that it is developing.]

James Watson, *The Double Helix*, Signet, New York 1968. [Insider's warts-and-all view of the discovery of the secret of life. Necessarily biased.]

Steven Weinberg, *Dreams of a final Theory: the Search for the Fundamental Laws of Nature*, Hutchinson Radius, London 1993. [A leading advocate explains what he means by a Theory of Everything. Thoughtful, fascinating, but tacitly assumes that 'fundamental' in the sense of 'ultimate bits and pieces' is the same as 'fundamental' in the sense of 'foundation for everything else'.]

Christopher Wills, *The Runaway Brain*, HarperCollins, London 1994. [The theory that brain/mind become more complex in order to store sensory impressions for comparison purposes. Also contains a very good discussion of recent human evolution.]

Lewis Wolpert, *The Triumph of the Embryo*, Oxford University Press, Oxford 1991. [A leading biologist discusses the problem of biological development in terms of positional information.]

Semir Zeki, *A Vision of the Brain*, Blackwell Scientific, London 1993. [Excellent account of the eyes, the cortex, and perception of colour and shapes in the brain, and – to some extent – the mind. All from 'inside', reductionist, except where context is unavoidable.]

Index

Note: Main references are given in **bold** type